# Springer Tracts in Civil Engineering

**Springer Tracts in Civil Engineering** (STCE) publishes the latest developments in Civil Engineering - quickly, informally and in top quality. The series scope includes monographs, professional books, graduate textbooks and edited volumes, as well as outstanding PhD theses. Its goal is to cover all the main branches of civil engineering, both theoretical and applied, including:

- Construction and Structural Mechanics
- Building Materials
- Concrete, Steel and Timber Structures
- Geotechnical Engineering
- Earthquake Engineering
- Coastal Engineering; Ocean and Offshore Engineering
- Hydraulics, Hydrology and Water Resources Engineering
- Environmental Engineering and Sustainability
- Structural Health and Monitoring
- Surveying and Geographical Information Systems
- Heating, Ventilation and Air Conditioning (HVAC)
- Transportation and Traffic
- Risk Analysis
- Safety and Security

**Indexed by Scopus**

To submit a proposal or request further information, please contact:
Pierpaolo Riva at Pierpaolo.Riva@springer.com (Europe and Americas) Wayne Hu at wayne.hu@springer.com (China)

More information about this series at https://link.springer.com/bookseries/15088

Androniki Miltiadou-Fezans ·
Theodosios P. Tassios

# Mix-Design and Application of Hydraulic Grouts for Masonry Strengthening

Springer

Androniki Miltiadou-Fezans
School of Architecture
National Technical University of Athens
Athens, Greece

Theodosios P. Tassios
School of Civil Engineering
National Technical University of Athens
Athens, Greece

ISSN 2366-259X ISSN 2366-2603 (electronic)
Springer Tracts in Civil Engineering
ISBN 978-3-030-85964-0 ISBN 978-3-030-85965-7 (eBook)
https://doi.org/10.1007/978-3-030-85965-7



This Springer imprint is published by the registered company Springer Nature Switzerland AG
The registered company address is: Gewerbestrasse 11, 6330 Cham, Switzerland

# Preface

1. It is commonly admitted that grouting of existing masonry building elements has a merit to completely respect the initial form of the element to be repaired or strengthened, while improving its resistance and integrity. In the specific case of cultural heritage structures, the respect of the "monumentic" value of the form is of particular interest. That is why, grouting is accepted also in fissured monolithic architectural members, as well as in detached and cracked mosaics or frescoes or surface decoration jointing mortars and plasters; hence the broader significance of this method of structural intervention of existing masonry structures.

This book has a clearly practical scope: to assist the person in charge to design and apply such grout mixes. It is however important to note that masonry grouting is a rather delicate method of structural intervention. Strength improvement is achieved only if grouts of appropriate properties reach the fine discontinuities (voids and microcracks) of the main mass of masonry; simple filling of large voids and cracks is not sufficient. That is why in several cases, masonry grouting was reported to be ineffective, due to grout mixes:

- unable to *penetrate* into finer discontinuities, or
- *not fluid* enough, or
- excessively bled or segregated (i.e., deprived of *stability of the grout*).

That is why, grout-mix-design cannot be carried out by means of rough empirical rules: The number of intervening parameters and their nature do not allow such a solution. Consequently, in order to serve its practical scope, this book should describe in detail the aforementioned performances of a correct grout, in relationship with the internal structure of the masonry. The extent of such descriptions, however, and the necessary scientific handling of the related analysis should not be considered as an unnecessary sophistication; on the contrary, it is presented in a way to help a better understanding of the related phenomena, thus contributing to the best application of this strengthening method.

By its scope, the book is addressed to a rather large audience, such as:

- Structural engineers
- Material engineers
- Chemists
- Architects
- Construction contractors
- Students in the respective disciplines
- Archaeologists specializing in restoration of monuments may also find an interest in reading a good part of this book.

2. The book attempts to contribute toward to a holistic methodology for the mix-design and application of hydraulic grouts,

   - taking into account their injectability (as well as their strength and durability),
   - considering the conditions of the masonry to be repaired or strengthened, and
   - having in mind the targeted properties of masonry after its grouting.

   To this end, the comprehensive concept of injectability is first analyzed as the resultant of three grout performances:

   - Penetrability, the capacity of the grout to pass through the effective finest width of discontinuities (voids, microcracks, interfaces, joints),
   - Fluidity, the capacity of the grout to overcome lateral frictions and to flow up to lengths larger than the distance of consecutive grout tubes,
   - Stability, the capacity of the grout to avoid harmful bleeding and segregation along its flow or when in its final place.

   Regarding the conditions of the existing masonry, a practical expression of the permeability of its mass is introduced, by means of the concept of nominal width ($W_{nom}$) of its fine discontinuities affecting its compressive strength.
   The required strength of the grout is roughly estimated by means of calibrated empirical formulae, based on the targeted after grouting strength of the masonry.
   Finally, durability issues are discussed, depending on both environmental conditions and chemical properties of the constitutive materials of the masonry.
   Based on these concepts, the book describes decision-making methods, resulting in the quantitative composition of the appropriate grout. Due to the multi-parametric nature of the problem, it is obvious to refer also to several preliminary tests, in order to face the inevitable uncertainties accompanying the entire procedure. However, the rationality of the followed methodology insures the rapid convergence of the whole procedure.
   The book is concluded with a long chapter, offering practical recommendations for the execution of grouting, together with indispensable quality assurance procedures.

3. A short presentation of the contents of this volume may assist the reader in understanding various aspects of the conceptual design and execution of grouting, as well as in better organizing the reading of the book.

Chapter 1 is an *Introduction* assisting the reader in understanding basic issues of the grouting technology. First, the categories (simple, binary, ternary) of grouts are presented in combination with examples of their use. Subsequently, the main performances required from a hydraulic grout are enumerated; thus, the corresponding design parameters of a grout mix become clear. A design procedure is then described, with reference to the relevant parts of the book, where the reader will find detailed assistance. Finally, it is reminded that an experimental verification of the performances of the "trial mix" has to be carried out in laboratory, and in all cases in situ, during pilot grout preparation and application.

Chapter 2 deals with the first component of injectability, i.e., the *penetrability* of the grains of the grout through the effective finest width $W_{nom}$ of masonry discontinuities (voids, microcracks, interfaces, joints). As it is known, grouting is intended to fill voids, fissures, and open joints of the masonry as a system, producing a "dendrite" (a three-dimensional skeleton), directly contributing to the strength of the masonry as a whole. However, to do so, the grout should be able to pass through the "narrowest" possible width of such discontinuities, in order to reach the maximum possible internal volume of masonry and open joints, avoiding most of possible blockages. In the specific case of three-leaf masonries, the most decisive result of the grouting is expected to be the strengthening of the bond along the interfaces between the external layers and the infill; the rather small voids, as well as pre-existing fissures along these interfaces, have to be penetrated.

In this chapter, the penetrability of hydraulic grouts is discussed, and relation ships between (i) two characteristic diameters of the grains of the solid phase of the grout and (ii) the nominal minimum width of fissures and voids of the structure to be injected are proposed. Furthermore, the beneficial role of replacing part of the cement or hydraulic lime with ultrafine materials in order to improve penetrability is presented, and related criteria are proposed. Extensive experimental verifications of the proposed quantitative models are finally offered.

Chapter 3 refers to *fluidity*, i.e., the second component of injectability of the grout: The grout should be able to easily flow along the sinuous paths of internal interconnected discontinuities, up to a distance larger than the distance between consecutive grout tubes. To this end, appropriate water-to-solid ratio (W/S), superplasticizer content (SP), and mixing technique should be selected. A new practical (but scientifically significant) fluidity measurement is proposed (the fluidity factor test—FFT); thus, a "fluidity factor" is defined. It is proved that the follow-up of this factor as a function of the water-to-solid ratio may reveal fundamental characteristics of the grout composition under design.

The influence of the mixing method and superplasticizer on grout's fluidity is also experimentally studied. The chapter concludes with a case study to highlight the practical use of the proposed test.

Chapter 4 deals with the final component of grout's injectability, i.e., the *stability* of the suspension against excessive bleeding and segregation. The grout should keep its homogeneity up to its setting, as much as possible. When the bleeding remains lower than a certain limit, the grout can be used; otherwise, the grout may be unable to flow in and adhere to the internal surfaces of the discontinuities. Similarly, because of inadequate internal cohesion, larger solid particles of the grout may quickly settle, resulting in segregation phenomena, i.e., creation of a non-homogeneous layered structure of the mix. Excessive bleeding and segregation may produce blockage of the flow and sudden dangerous increase of the pressure.

The most prevailing parameters shaping stability characteristics are water content and percentage of ultrafine materials. After a brief literature survey, an oversimplified predictive model of bleeding is firstly proposed and then its validity is confirmed using the results of an experimental study. The role of superplasticizers is also discussed. In both cases, with and without superplasticizer, semi-empirical formulae are proposed, that are useful for the design of a grout composition. Finally, the chapter presents experimental results demonstrating the role of water and superplasticizer content in the appearance of segregation; some empirical formulae are also proposed for the estimation of the critical water content initiating segregation.

Chapter 5 describes the categories of possible internal discontinuities of masonry; it is because of such discontinuities (pores, local interface detachments, local slidings, cracks) that masonry strength may be reduced. The filling of these discontinuities by means of an appropriate grout may increase masonry strength, provided that the grout was able to penetrate the body of masonry, to reach most of these discontinuities and flow along each of them. In order to decide the necessary "penetrability" capacity of the grout, a rough evaluation of a *critical value* "$W_{nom}$" of the opening of these discontinuities of masonry is needed.

This chapter examines several possibilities of quantification of such a representative opening value for several categories of masonry. Finally, an easy to apply practical approach of "opening classes" is proposed, and relevant examples are given. Thus, for each specific case, the selection of grout solids' grading is facilitated, in order to satisfy penetrability requirements established in Chap. 2.

Chapter 6 deals with grout-mix-design issues related to the targeted *strength* of the masonry to be grouted. Only two parameters enter the discussion: targeted $f_{wc,s}$-value and corresponding required $f_{gr,c}$-value. The chapter explores how a range of required $f_{gr,c}$-values suffices to be related to a targeted $f_{wc,s}$-value. This loose correlation is due to the fact that grout-to-stone bond properties are shaping the final structural behaviour of grouted masonry. Thus, tensile rather than compressive strength of the grout is the relevant parameter. Besides, dehydration of grout entering the masonry takes place; consequently, some additional rules regarding mix-composition are respectively derived.

Finally, several empirical relationships are offered predicting masonry compressive strength before and after grouting. Obviously, among the grout compositions resulting in $f_{gr,c}$-values within the required strength range, those mixes will be retained respecting the other required performances.

The chapter ends with a long Appendix presenting detailed strength results (both tensile and compressive) for several grout compositions, described in the literature.

Chapter 7 is related to *durability* issues. Physical effects are considered first, referring to the consequences of water introduced in masonry by the grouting (freezing and dissolution of soluble phases). Subsequently, chemical effects are considered, such as sulfate reactions, alkali–silica reactions, and possible chlorides' attack and leaching. The chapter ends with a brief presentation of the literature results of durability tests and with a guide for the selection of binders vs. durability.

Chapter 8 refers to some inevitable *contradictory* requirements of a grout, especially related to ultrafines' content (penetrability against fluidity), superplasticizer's content (fluidity against stability), type and content of binders to achieve sufficient bonding with masonry materials, without however imparting to the masonry disproportionally high stiffness (strength against ductility) and without jeopardizing durability (strength against durability) or level of grouting pressure to avoid local rupture of very low masonry strength.

Optimization of grout performances is needed in most cases. To this end, a simple first procedure is proposed following a step-by-step selection of mix-design parameters (Sect. 8.2). First, the type of binder is selected, based on strength and durability demands. An appropriate percentage of fines is then selected, based mainly on penetrability criteria, in combination with strength and durability aspects as well. Using experimental results already presented in Chaps. 2–6, diagrams combining penetrability, fluidity, stability, and strength characteristics vs. W/S ratio are traced. Thus, a "usable" area of water content appears; its compatibility with the required grout strength range will be checked. Normally, several satisfactory solutions are thus found. Otherwise, the necessary correcting measures are discussed in the chapter for each specific case, mainly by means of increased percentages of fines and addition of superplasticizers. This way, the rationality of the proposed methodology is understood. Thus, the designer is in better position to handle the final design of the grout, by means of a better knowledge of the interplay of the intervening numerous parameters.

In Chap. 9, the scientific approach followed in this book finds its justification: The rational and detailed examination of all properties of a grout offers now the possibility to follow a practical step-by-step procedure of mix-design, permitting to handle numerous parameters in a logical sequence. Thus, Chap. 9 contains more practical guidelines for the mix-design of grouts used in masonry strengthening. The guided use of *Tables* and *empirical formulae* presented in Chaps. 2–7 greatly facilitates the selection of (i) the type of binders and the final grading of the solids, (ii) the minimum acceptable fluidity factor, depending on the finest discontinuity width class ($W_{nom}$), (iii) the zero-bleeding and the critical-bleeding (W/S) expressions (with or without superplasticizers), as a function of the calculated average specific surface $S_A$ of the solid phase, *and* (iv) the limit value W/S against segregation. Subsequently, a practical procedure for the selection of the final (W/S) ratio of the mix is described, respecting all the aforementioned limits. Corresponding remedy measures are presented in case such a complete respect is not feasible. Moreover, for masonries of minor historical importance and with representative $W_{nom} \geq 0.25$ mm, a Table is offered, containing approximate compositions of grouts for four different required grout strength ranges. Experimental verification of the required grout performances will be in any case necessary.

Chapter 10 presents a set of recommendations for the *execution* of grouting, regarding grout tubes installation, description of equipment needed, in situ preparation of the grout, and in situ control of injectability characteristics, as well as in-time evolution of grout's strength. Moreover, methods of in situ checking of injection procedure are presented, together with a description of the data that should be recorded during grouting operations and their evaluation.

The chapter concludes with the assessment of the efficiency of grouting; overall quality management is finally described, together with a detailed presentation of laboratory and in situ nondestructive control tests.

A final observation may be needed here.

To a not yet experienced reader, the described procedure for the design of a correct grout may seem too long. But it should not be forgotten that the final product aimed at by the design, depends on 6 parameters: penetrability, fluidity, bleeding, segregation, strength, and durability—some of them being contradictory to each other, and thus needing optimization. Mathematically *thinking*, it is expected that the solution of a system of 6 equations with 6 variables is not a short procedure; besides, after appropriate modifications, the system should be solved several times, in order for an overall optimum solution to be achieved. With this analogy, the proposed relatively long procedure for grout design may be better justified. Without such a disciplined method, the design may become chaotic.

Obviously, experienced designers will continue employing their own design method; but it is believed that, even experienced designers, may better recognize the nature of some phenomena, thanks to the preceding analysis.

In conclusion, this book may be considered both as a *rational synthesis* and as a *practical guide*.

Athens, Greece

Androniki Miltiadou-Fezans
Theodosios P. Tassios

# Acknowledgements

The authors express their gratitude to Anna Kalagri, M.Sc., Chemical Engineer and Conservator of Art, for her continuous assistance with the experiments and the graphics of this book, as well as for her assistance regarding durability matters and, above all, for her critical reading of the entire book.

Thanks are also due to Sophia Anagnostopoulou, M.Sc., Chemical Engineer, Martha Savvidou, M.Sc., Chemical Engineer, and Panagiota Psymogerakou, M.Sc., Civil Engineer, for their help with the experiments and the graphics of this book. The assistance of Michalis Delagrammatikas, M.Sc., Chemical Engineer, regarding durability-related matters is also acknowledged.

Similarly, thanks are due to Euthymia Delinikola, Architect Restorer, and to Fotios Fytilis, Architect, for their kind contribution in the preparation of some of the graphics.

A considerable amount of experimental data included in this book is due to the research works of (i) the Laboratory of Restoration Materials and Techniques of the Directorate for the Restoration of Byzantine and Post-Byzantine Monuments (1995–2004) and (ii) the Laboratory of Building Materials of the Directorate for Technical Research on Restoration (2004–2007) of the Hellenic Ministry of Culture and Sports, carried out for the structural restoration of various monuments in collaboration with several Services of the Hellenic Ministry of Culture and Sports, such as: the Acropolis Restoration Service, the Directorate for the Restoration of Byzantine and Post-Byzantine Monuments, the Directorate for the Conservation of Ancient and Modern Monuments, and the 1st Ephorate of Byzantine Antiquities. We gratefully acknowledge the kind contribution of their scientific and technical Staff.

Particular thanks are due to the Acropolis Restoration Service and the Committee for the Conservation of the Acropolis Monuments, the Ephorate of Antiquities of West Attica, and the Ephorate of Antiquities of Chios of the Hellenic Ministry of Culture and Sports for their kind permission to publish photographs of structural and decorative elements strengthened with grouting, belonging to some very well-known monuments, namely the Parthenon Opisthodomos of the Acropolis of Athens, the Byzantine Monastery of Daphni in Attica, and the Byzantine Church of Panagia Krina in Chios Island. It has to be noted that the rights on the depicted monuments belong to the Hellenic Ministry of Cultute and Sports (Law 3028/2002)/ Hellenic Organization of Cultural Resourses Development. The aforementioned monuments are under the responsibility of the Ephorate of Antiquities of the City of Athens (Acropolis), the Ephorate of Antiquities of West Attica (Daphni Monastery) and the Ephorate of Antiquities of Chios Island (Panagia Krina), respectively.

The authors acknowledge with gratitude the scholars who kindly agreed to reproduce in this book part of their published or unpublished material: Prof. Giorgio Macchi and Dr. Simone Ghelfi; Nikolaos Delinikolas, Architect Restorer; Prof. Elizabeth Vintzileou and Dr. Vassiliki Pallieraki; Dr. Xavier Derobert and Dr. Philippe Côte ; Prof. Ioanna Papayianni; Prof. Maria Rosa Valluzzi; Prof. Antonella Elide Saisi; Prof. Charalambos Mouzakis; Prof. Mojmir Uranjek and Prof. Violeta Bokan-Bosiljkof; Dr. Filippe Van Rickstal; Alkimos Papathanasiou, M.Sc., Structural Engineer; and Aris Velissarios Raptis, Architect.

Similarly, thanks are due (i) to Dr. Anna Melograni, Coordinatore della Redazione of Bolletino d' Arte, Ministero per i Beni e le Attività Culturali of Italy, for the permission to reproduce some graphics regarding the grouting intervention in the Pisa Tower, as well as to Antonella d' Aulerio and the Università Iuav di Venezia, Archivio Progetti, Fondo Studio Tecnico prof. ing. Giorgio Macchi, dott. ing. Stefano Macchi, for providing us with high-definition scans of some of the Pisa Tower graphics, (ii) to Prof. I. Papayianni, Associate Prof. M. Stefanidou, and Dr. V. Pachta (eds) of the Proceedings of HMC2016, for their permission to include in this book our paper published in the Proc(s) of HMC2016, as well as (iii) to Prof. J. Jasienko (ed.) of the Proceedings of the 8th Inter. Conf. SAHC 2012 for his permission to reproduce in this book a graphic of a paper included in the Prod(s) of 8th SAHC 2012.

Special thanks are finally due to the colleague Gerard Fezans for his continuous support and encouragement, as well as for his talented representation of many diagrams and graphics of the book.

# Contents

# Notations

| | |
|---|---|
| A | Area of the cross section of the nozzle |
| b | Normalized bleeding value or bleeding capacity |
| B | Height of bleed water |
| C | Cement |
| c | Grout-to-stone cohesion (grout-to-stone shear bond strength under zero normal stress) |
| $C_v$ | Cement volume |
| C.I. | Cohesiveness index |
| d | Diameter of grains |
| $d_v$ | Diameter of monogranular ultrafine material |
| $d_{ch}$ | The equivalent diameter of the "channel" (the void) |
| $d_{85}$ | Diameter of the grains of a granular material (of the solid phase of a grout), corresponding to 85% passing |
| $d_{99}$ | Diameter of the grains of a granular material (of the solid phase of a grout), corresponding to 99% passing |
| $d_{100}$ | "Maximum" diameter of the grains of a granular material (of the solid phase of a grout) |
| $D_{15}$ | Diameter of the grains of a soil or other granular medium to be injected, corresponding to 15% passing |
| DSF | Densified silica fume |
| $f_{bc}$ | Compressive strength of blocks |
| $f_{gr,c}$ | Compressive strength of grout |
| $f_{gr,t}$ | Tensile strength of grout |
| $f_{gr,s}$ | Shear resistance of the grout |
| $f_{gr,b,t}$ | Grout-to-stone tensile bond strength |
| $f_{gr,b,s}$ | Grout-to-stone shear bond strength |
| $f_{mc}$ | Compressive strength of mortar |
| $f_{wc,requ.}$ | The required compressive strength of the masonry after strengthening |
| $f_{wc,0}$ | Estimated compressive strength of masonry before intervention |
| $f_{wc,s}$ | Estimated compressive strength of masonry, after strengthening |
| $f_{wc,e}$ | Compressive strength of the external masonry leaf |

| | |
|---|---|
| $f_{wc,\ i}$ | Compressive strength of the infill |
| $f_{wc,\ i,\ s}$ | Compressive strength of the infill after grouting |
| $f_0$ | A reduction of masonry strength due to the inhomogeneity of construction, depending on non-orthogonality of blocks |
| f | Fines (ultrafines) |
| $f_m$ | Mass percentage of fines (ultrafines) |
| $f_{mp}$ | Minimal mass percentage of ultrafines to replace a part of the basic binder, in order to ensure penetrability |
| $f_v$ | Volume percentage of ultrafines |
| $f_{vp}$ | Minimal volume percentage of ultrafines to replace a part of the basic binder, in order to ensure penetrability |
| FFT | Fluidity factor test |
| $F_l$ | Fluidity factor |
| max $F_l$ | Maximum possible fluidity factor value (for a given composition of solids) under stable conditions |
| min $F_l$ | Minimum value of fluidity factor for a grout (with appropriate grain penetrability) to be injectable through a sand column with nominal lower value of widths of voids equal to "$W_{nom}$" |
| $G_{gr}$ and $G_w$ | Weight of the injected grout and initial weight of the single leaf wall, respectively |
| $h_0$ | Height of the initial specimen of the grout at starting time $t_0$ |
| $h_w$ | Height of bleed water at prescribed intervals |
| H | Normalized height of bleeding |
| HT | High turbulence mixing |
| IE | Instability estimator |
| $IE_{crit}$ | Lowest acceptable level of instability |
| k | Permeability coefficient |
| $k_i$-variables | Expressing the planeity of blocks' sides ($k_1$); the filling of joints by mortar ($k_2$); and the visible cracks and on mortar and block/mortar detachments ($k_3$) |
| $k_0$ | Ratio of the volume of mortar to the volume of masonry (one-leaf masonry) |
| $\ell$ | Total length of the grouted "channel" |
| L | Lime |
| $n$ | Numerical factor larger than unity, to compare $W_{nom}$ and $d_{85}$ (Chap. 2) |
| n | "Equivalent number" of solid cubes, with an average width of "d" |
| $n$ | Ratio of the volume of grout embodied to the masonry, normalized to the total volume of the mortar (Chap. 6) |
| NHL | Natural hydraulic lime |
| P | Pozzolan |
| $P_{max}$ | Maximum grouting pressure |
| Q | Volume of the grout (to pass through the nozzle of a Marsh cone) |
| $r_{32}$ | Volume percentage of initial cement "retained" on 32 μm |
| $R_{32}$ | Volume percentage of blend material retained on 32 μm |

| | |
|---|---|
| $S_A$ | Calculated average specific surface of a blended material |
| $S_b$ | Specific surface Blaine |
| SP | Superplasticizer percentage |
| SF | Silica fume |
| SE | Santorini earth |
| $S_v$ | Solid volume |
| S | Solid mass |
| $s_u$ | Critical slip leading to maximum shear bond resistance |
| $t_f$ | Flow time through the nozzle of a Marsh cone |
| $t_{500ml,\ d=4.75}$ | Flow time of 500 ml out of 1000 inserted in the Marsh cone with a 4.75 mm nozzle diameter |
| $t_0$ | Thickness of intergranular adsorbed water |
| $T_{36}$ | The time the grout takes to reach the top of the sand column (of a height of 36 cm) |
| US | Ultrasound dispersion mixing |
| UF | Ultrafines |
| $V$ | The velocity of flow |
| $V_0$ | Variable expressing the "permeability" of the building mortar and the infill material of masonry |
| $V_c$ | Volume of cement |
| $V_{ext}$ | Volume of the external leaves of masonry |
| $V_m$ | The volume of mortar |
| $V_w$ | Volume of bleed water |
| $V_i$ | The volume of the initial infill material within the entire masonry |
| $V_{i,\ gr}$ | Intial volume of the grout at the begining of bleeding test |
| $V_w$ | The volume of the entire masonry wall |
| W | Water mass |
| $W_i$ | Water to solids ratio |
| $W_v$ | Water volume |
| W/C | Water-to-cement ratio by mass |
| W/S | Water-to-solid ratio by mass |
| $W_v/S_v$ | Water-to-solid ratio by volume |
| $W_{nom}$ | Nominal lower value of the aperture of fissures or orifices to be injected |
| $W_i$*-values* | Values of the aperture of fissures or orifices to be injected |
| $\left(\frac{W}{S}\right)_{0,\ bl}$ | The minimum water-to-solid ratio able to initiate bleeding |
| $\left(\frac{W}{S}\right)_{0,\ bl,\ SP}$ | The minimum water-to-solid ratio able to initiate bleeding, in the case the grout contains superplasticizer |
| $\left(\frac{W}{S}\right)_{0,\ f}$ | The minimum water-to-solid ratio able to initiate flow |
| $\left(\frac{W}{S}\right)_{bl,\ cr}$ | Water-to-solid ratio leading to 5% bleeding |
| $\left(\frac{W}{S}\right)_{bl,\ cr,\ SP}$ | Water-to-solid ratio leading to 5% bleeding, in the case the grout contains superplasticizer |
| $\left(\frac{W}{S}\right)_{sand.-col.}$ | Water-to-solid ratio for which the grout starts to be able to be injectable in the sand column |

| | |
|---|---|
| $\left(\frac{W}{S}\right)_{segr.}$ | Water-to-solid ratio for which the grout starts to exhibit segregation |
| $\left(\frac{W}{S}\right)_{u,\ f}$ | An ultimate value of water-to-solid ratio, which practically results in the maximum possible fluidity factor (for a particular composition of solids) without instability effects |
| z | An Instability estimator (IE)-related quasi-constant |
| $\alpha$ | Rate factor |
| $\beta$ | Coefficient reflecting the substrate type (Chap. 6) |
| $\alpha,\ \beta,\ \gamma,\ \delta$ | Rheology-related constants, depending on the roughness of the walls and the form of the cross section of the channel, as well as on rheological properties of the grout (Chap. 3) |
| $\delta$ | Constant |
| $\delta = t_e/t_i$ | Ratio of the respective thickness of external leaf and infill (in case that the external leaves have the same thickness) |
| $\delta_{e1} = t_{e1}/t_i$ | Ratio of the respective thickness of external leaf “1” and infill |
| $\delta_{e2} = t_{e2}/t_i$ | Ratio of the respective thickness of external leaf “2” and infill |
| $\Delta H$ | Bleeding capacity (settlement per unit original height) |
| $\Delta f_0$ | Reduction of the masonry inhomogeneity factor (counteracting the severity of the initial $f_0$-value included in the expression of $f_{wc,\ 0}$) |
| $\Delta f_m$ | Additional quantity of fines |
| $\Delta w$ | Additional quantity of water |
| $\Delta p$ | Pressure loss along a length “$\ell$” |
| $\zeta$ | Constant |
| $\eta$ | Plastic viscosity |
| $\lambda$ | Correction factor |
| $\lambda$ | Mortar-to-stone bond factor |
| $\lambda_1,\ \lambda_2$ | Numerical factors |
| $\lambda_3$ | A constant, reflecting the lowest acceptable level of instability $IE_{crit}$ |
| $\lambda_e,\ \lambda_i$ | Correction factors in order to take into account the interaction between the external leaves and the infill |
| $\lambda_\iota$-variables | In function to the type of the wall (one-leaf, two-leaf, or three-leaf) |
| $\mu$ | Water equivalence constant of superplasticizer |
| $\mu_{max}$ | Maximum friction coefficient |
| $\mu_{res}$ | Residual friction coefficient |
| $\xi$ | A factor expressing the adverse effect of thick mortar joints |
| $\rho_{ms}$ | Average density of the solids |
| $\rho_w$ | Water density |
| $\rho_{mf}$ | Average density of fines |
| $\rho_b$ | Density of the basic binder |
| $\sigma$ | Normal stress acting on a sheared interface |
| $\tau_0$ | Yield stress of a fluid |
| $\omega$ | Equivalent W/S, accounting for SP |

# Chapter 1
# Introduction

**Abstract** This Chapter is an *Introduction* assisting the reader in understanding basic issues of the grouting Technology. First, the categories (simple, binary, ternary) of grouts are presented in combination with the cases of their use. Subsequently, the main performances required from a hydraulic grout are enumerated; thus, the corresponding design parameters of a grout-mix become clear. A design procedure is then described, with references to the relevant parts of the book, where the reader will find detailed assistance. Finally, it is reminded that the experimental verification of the performances of the "trial mix" has to be carried out in laboratory as well as (occasionally) in situ.

## 1.1 The Significance of the Subject

Injections with hydraulic grouts constitute one of the most important techniques applied for the repair and strengthening of masonry structures or fissured architectural members, when interconnected voids are present. In the case of three leaf masonry, hydraulic grouts are very often applied to improve the mechanical properties of the interior weak leaf, as well as to re-instate the collaboration between external and internal leaves. Furthermore, this type of grouting is well accepted even in the case of historic structures, since it has the advantage of improving the continuity, cohesion and strength of the damaged structures without altering their morphology and load-bearing system (Rocard and Bouineau 1982; Miltiadou 1985; Penelis et al. 1989). Grouting is as well used for consolidating fissured and detached mosaics, frescoes and other surface decorative elements (Ferragni et al. 1982, 1984; Biçer-Şimşir et al. 2010; Biçer-Şimşir and Rainer 2013).

Laboratory research and in situ applications undertaken the last four decades have proven that the use of hydraulic grouts can be efficient under the condition that the grouts and the injections have been adequately designed and implemented, on the basis of specific performance requirements (Paillère and Guinez 1984; Bouineau 1985; Van Gemert 1988; Penelis et al. 1989; Paillère et al. 1989; Miltiadou 1990; Tomazevic and Apih 1993; Binda et al. 1993a, b, 1997, 2003; Atkinson and

 
A. Miltiadou-Fezans and T. P. Tassios, *Mix-Design and Application of Hydraulic Grouts for Masonry Strengthening*, Springer Tracts in Civil Engineering,
https://doi.org/10.1007/978-3-030-85965-7_1

Schuller 1993; Vintzileou and Tassios 1995; Valluzzi 2000; Van Rickstal 2000; Toumbakari 2002; Miltiadou-Fezans et al. 2005, 2007, 2008; Vintzileou and Miltiadou-Fezans 2008; Kalagri et al. 2010; Mazzon 2010; Vintzileou 2011; Bras 2011; Silva 2012; Miltiadou-Fezans and Tassios 2012, 2013a, b; Uranjek et al. 2014; Papayianni and Pachta 2015; Baltazar et al. 2015; Jorne et al. 2015; Vintzileou et al. 2015; Luso and Lourenco 2016, 2017, 2019; Mouzakis et al. 2017). These requirements involve injectability, strength and durability aspects, and they are set on the basis of an overall consideration of the structure to be repaired, before and after intervention. Among the above requirements, the injectability capacity of the grout constitutes a key parameter for a successful intervention, since injecting the interior of masonry is a challenging matter for many reasons: non homogeneity, presence of cracks, voids and discontinuities of different width, high water absorption by the bricks, mortars and stones, etc.

As aforementioned, the adequate grout composition has to be designed following a rational methodology (Miltiadou 1990; Miltiadou-Fezans 1998; Miltiadou-Fezans and Tassios 2016). The scope of this book is to establish such a holistic methodology, on the basis of already developed design criteria and empirical formulae for the prediction of some grout properties (Miltiadou-Fezans and Tassios 2012, 2013a, b). Nevertheless, an optimization of grout composition is needed, which is very helpful in such a poly-parametric decision making. The interdependence of various grout performances and the primordial role of some parameters are also highlighted, together with practical guidelines for in situ grouting application and quality control.

## 1.2 Categories of Grouts for Masonries

There is a large variety of grouts to be used for repair or strengthening of existing masonries, depending on the type of masonry, the nature of discontinuities to be filled and the scope of the structural intervention (Miltiadou-Fezans and Tassios 2013a).

1. Strong one-leaf masonries damaged by middle intensity earthquakes or by differential settlement or other causes, may exhibit several thin cracks. If the reinstatement of the initial high strength and stiffness of such walls is decided (e.g., in bridge piers), *cement based binary* grouts may be the solution, provided that chemical incompatibility and efflorescence problems are not at issue. This is also the indicated solution in case of columns or other architectural members or walls of Greco-Roman antiquities with fissured blocks of high strength and low porosity (Miltiadou-Fezans et al. 2005). In these cases, strength and bonding reinstatement is the main scope of grouting, under the obvious condition however, that larger grains of the solid phase will be able to penetrate the aforementioned thin fissures.
2. Similarly, large and open voids in masonry may be merely filled by means of any binary or ternary grout based on cement or hydraulic lime.

3. In both the aforementioned cases, low strength and high porosity stone blocks or possible chemical incompatibilities may exclude or drastically reduce the use of cement in grouts. This is also the case with efflorescence sensitive faces of masonry. The most suitable solution in these cases is *hydraulic lime-based* grouts or *ternary grouts with low cement percentage*, appropriately designed as to respond to durability requirements and to be able to penetrate through thin or large discontinuities (Penelis et al. 1989; Miltiadou-Fezans 1998; Binda et al. 1994; Valluzzi et al. 2004; Toumbakari et al. 1999a; Binda et al. 2003; Miltiadou-Fezans et al. 2007; Kalagri et al. 2010; Bras and Henriques 2012; Badogiannis et al. 2012; Papayianni and Pachta 2015; Baltazar et al. 2015; Jorne 2016; Luso and Lourenco 2016).
4. The very important case of three-leaf masonries necessitates special consideration. Two categories of "discontinuities" are encountered in this case. First, the relatively large voids inside the infill: If these voids are open enough, they may be filled with almost any category of grout. If, however many of these voids cannot be directly reached (unless the grout obligatorily follows complicated paths through other discontinuities of small openings), the basic criterion for the selection of the grout is its penetrability. A second category of discontinuities in three-leaf masonries refers to the interfaces between external leaves and the infill: Relative research (Binda et al. 1993c, 1994, 2006; Egerman et al. 1993; Toumbakari et al. 2005) has shown that traction and shear bond along these interfaces constitute the basic mechanism of integrity and resistance of three-leaf walls. Consequently, the main problem in strengthening such historical masonries is to design grouts able to penetrate the "contact- areas" between external and internal leaves. Due to the inevitable wall-effect, several large voids are also present along these interfaces; nevertheless, several other "intimate" contact-areas are also present there. Moreover, an old three-leaf masonry may also contain dangerous small width "dilatancy"-cracks, along planes parallel to its faces, and located inside the infill or along the external leaf-to-infill interfaces (Miltiadou and Abdunur 1990; Miltiadou et al. 1993; Binda et al. 1993a; Egermann 1993a, b; Vintzileou and Tassios 1995; Valluzzi 2000; da Porto et al. 2003; Toumbakari et al. 2005; Oliveira et al. 2007; Pina-Henriques et al. 2005; Vintzileou and Miltiadou-Fezans 2008; Mazzon 2010; Vintzileou 2011; Silva 2012; Vintzileou et al. 2015; Mouzakis et al. 2017). Regarding the case of two-leaf masonry, discontinuities similar to three-leaf masonries may be encountered (Binda and Saisi 2001; Binda et al. 2005; Corradi et al. 2008; Uranjek et al. 2014; Yüzer et al. 2015; Luso and Lourenco 2019). Thus, in order to ensure the necessary bond, the grout should also penetrate through these intimate contact areas.

Since, on the other hand, sufficient bond strength may be offered by grouts of relatively small compressive strength (Miltiadou et al. 1990; Miltiadou-Fezans 1998; Toumbakari et al. 1999b, 2007; Adami and Vintzileou 2008, 2010; Luso and Lourenco 2017), appropriate binary and ternary grouts with low cement content or hydraulic lime-based grouts are suitable to this end.

In any event, the critical design requirement in the case of three-leaf masonries is the injectability of the grout through relatively small width discontinuities.

## 1.3 Basic Data Influencing the Design of a Masonry Grout

Before designing a masonry grout, the entire structural problem should first be restated.

- Description of the masonry to be strengthened, and design targets:
  - Constituent materials (nature, chemical and mineralogical composition, soluble phases, porosity, strengths).
  - Building technique (one, two or three-leaf masonry, constitution of courses) and geometry.
  - Roughly estimated percentage and effective minimum width of voids included in masonry. A direct measurement of these quantities is hardly possible; a set of practical rules could be used instead, such as those given in Chap. 5.
  - Estimated compressive strength of the masonry, as a whole ($f_{wc,o}$); to this end, at this stage, only rough estimations can be made via empirical formulae relating block-strength, mortar-strength and average width of joints (i.a. Tassios 1988; Tassios and Chronopoulos 1986; Tassios 2004, see Chap. 6).
- Desirable compressive strength "$f_{wc,s}$" of the masonry as a whole (after strengthening), as dictated by the structural behaviour envisaged (see Chap. 6).
- Description of the microenvironment; physical-chemical conditions possibly jeopardizing the durability of the structure after strengthening, should be described (moisture and/or freezing cycles, air pollution etc.), in order to envisage the appropriate mix-design.

In real life problems, however, some of these data may be missing; nevertheless, their importance should be appreciated—be it in a qualitative way.

## 1.4 Main Performances of a Hydraulic Grout

In order to design a hydraulic grout composition adequate for a specific case study, the various performances of a suitable grout should be ensured.

- The necessary compressive and tensile strength of the grout should first be estimated on the basis of the data of Sect. 1.3. To this end, previous experience under similar conditions is needed; empirical expressions could be used relating $f_{wc,s}$, $f_{wc,o}$ and $f_{gr}$, accounting however *injectable* grouts only (e.g. Vintzileou and Tassios 1995; Tassios 2004; Valluzzi et al. 2004; Vintzileou and Miltiadou-Fezans 2008; Vintzileou 2011; Silva et al. 2014, see Chap. 6). Moreover, it has to be noted that disproportionately high compressive strength grouts may not be desirable. First because it is mainly the grout-to-masonry materials bonding that matters, and second because of the possible brittleness that may be induced to the grouted masonry.

- At this stage the grout durability performances have to be stated, on the basis of the data of Sect. 1.3; this will influence the initial selection of the type of the basic binder (cement or hydraulic lime), especially if the grout has to be sulphate resistant. Some experimental results on this matter can be found in Miltiadou (1990), Paillère et al. (1993), Karaveziroglou et al. (1998), Toumbakari et al. (1999c), Miltiadou-Fezans et al. (2005, 2021), Kalagri et al. (2010), Papayianni et al. (2010), Biçer-Şimşir et al. (2010), Biçer-Şimşir and Rainer (2013), Biçer-Şimşir (2016) and in Chap. 7.
- The injectability of the grout should now be considered. In what follows, the basic property of "INJECTABILITY" of a grout will be defined as the resultant of the following grout properties:
  - Satisfactory *penetrability* characteristics, i.e., appropriate effective maximum grain size of the solids of the grout, versus the effective lowest width of the voids of the wall, as well as appropriate grain size distribution of the solid phase of the grout.
  - Sufficient *fluidity*, i.e., the easiness of flow through the fissures and voids of the masonry, with the minimum possible pressure-losses, throughout the whole intervention.
  - Satisfactory *stability* of the suspension, i.e., appropriately low bleeding, as well as avoidance of harmful segregation of solid grains, during the whole intervention.

## 1.5 Design Procedure

Subsequently, the following matters should be taken into account in designing a grout.

- In order to ensure sufficient *penetrability*, the necessary grain size characteristics of the solid phase of the grout have to be selected on the basis of the "effective minimum width" of the voids of the wall. To this end, a rough estimation of the "effective maximum" grain size of the solid ingredients of the grout should first be made. The procedure explained in Chap. 2 is suggested to be followed, in order to make a first selection of suitable grain size distribution of solid materials of the grout (Miltiadou-Fezans and Tassios 2013a).
- Subsequently, a suitable *fluidity* should be decided as a function of the same "effective minimum width" of the voids of the wall; water-to-solid ratio (W/S), superplasticizer content (SP) and mixing technique should be selected, as described in Chap. 3 (Miltiadou-Fezans and Tassios 2012).

- *Stability* of the suspension against segregation or excessive bleeding should also be ensured since, otherwise, blockage may soon appear and the quality of the intervention could be severely affected: lack of filling of the voids after water's evaporation, heterogeneous adherence and strength characteristics, lower durability and resistance to environmental actions, since the solid grout will have a less dense structure. The most predominant among the parameters shaping fluidity and stability characteristics, are water content and percentage of ultrafine materials. Chapter 4 deals with these matters. Specific semi-empirical formulae useful for the design of a grout composition regarding the relationships between the most predominant parameters shaping stability characteristics (i.e. water content and percentage of ultrafine materials), are also proposed (Miltiadou-Fezans and Tassios 2013b).
- As aforementioned, the "*effective minimum width of voids*" included in masonry is a basic design parameter, influencing the penetrability and the fluidity characteristics. As the direct measurement of this quantity is hardly possible, a set of practical rules that could be used instead, are given in Chap. 5.
- The estimation of the *required strength* of the grout taking into account the desired final strength of the masonry under consideration and the selection of appropriate grout compositions able to develop such strengths, (mentioned already in 1.4), is also a delicate operation. Strength related data of grouts are presented in Chap. 6 to assist the designer in this endeavour.
- *Durability* matters, expressed mainly as chemical compatibility with materials existing in the masonry or its surface decoration elements, play as well un important role in selecting the appropriate grout composition. In Chap. 7 a short guidance is given on this performance requirement.
- The parameters influencing the aforementioned interrelated injectability, strength and durability properties, are often *contradictory*. Chapter 8 deals with this matter, in order to seek an *optimal solution* each time.
- Based on the preliminary grout characteristics assessed in the preceding Sect. 1.4, a trial composition may now be attempted, using the guidelines offered in Chap. 9. Thus, a first selection of the mix parameters W/S, SP%, and composition of the solid phase of the grout can be made. Experimental verification of injectability and strength properties may be necessary, as proposed in Sects. 9.7 and 9.9, respectively.
- At this stage, the durability of this trial composition should also be checked (Sect. 1.4), in terms of:
  - Chemical compatibility between grout and the materials of the existing masonry.
  - Resistance of the grout against the anticipated physical chemical actions.

  Appropriate composition modifications may be needed at this stage.
- A "*trial*"*mix-design* of the grout is now possible; in some cases, further experimental verification of its performances may be necessary, i.e., injectability (penetrability, fluidity and stability), strength and durability tests. Under some conditions, however, a complete experimental investigation may not be needed. In any

event, a "final" in-situ trial grouting preparation and application will be necessary, as presented in Chap. 10.

## 1.6 Practical Guidance for In-Situ Grouting Application Methodology

The appropriate design of a grout composition is only a part of a successful completion of the grouting intervention in situ; particular care has to be taken of the adequate execution of the entire intervention on a daily basis.

Based on the experience acquired from in situ applications, practical guidelines for the execution of grouting are presented in Chap. 10, including specific instructions for the preparation of masonry, worksite controls of injectability characteristics of grout, injection procedure, data to be collected during the execution of works, as well as their evaluation. Finally, post intervention assessment of grouting is discussed, and practical guidance is also offered.

## References

Adami CE, Vintzileou E (2008) Interventions to historic masonries: investigation of the bond mechanism between stones or bricks and grouts. Mater Struct 41(2):255–267

Adami CE, Vintzileou E (2010) The potential of medium and low mechanical properties hydraulic grouts for the retrofit of historic masonries. In: Proceedings of 8th international masonry conference, Dresden, Technische Universität Dresden and International Masonry Society

Atkinson R, Schuller M (1993) Evaluation of injectable cementitious grouts for repair and retrofit of masonry. In: Melander JM, Lauersdorf LR (eds) Masonry: design and construction, problems and repair. ASTM, Philadelphia, PA, USA

Badogiannis E, Miltiadou A, Kalagri A, Ipsilanti E, Dedeloudis C, Vintzileou E (2012) Mix design and performance evaluation of grouts with superfine natural pozzolan. In: Jasieńko J (ed) Structural analysis of historical constructions. DWE Wrocław Poland, pp 773–780

Baltazar L, Henriques FMA, Cidade MT (2015) Contribution to the design of hydraulic lime grouts for masonry consolidation. J Civ Eng Manag 21:698–709

Biçer-Şimşir B (2016) Preliminary study on the sulfate resistance of lime-based hydraulic injections grouts for the conservation of architectural surfaces. In: Papayianni I, Stefanidou M, Pachta V (eds) Proceedings of the 4th historic mortars conference, Santorini, 10–12 Oct 2016, pp 548–555

Biçer-Şimşir B, Rainer L (2013) Evaluation of lime-based hydraulic injection grouts for the conservation of architectural surfaces. The Getty Conservation Institute, Los Angeles, p 105

Biçer-Şimşir B, Griffin I, Palazzo Bertholon B, Rainer L (2010) Lime-based injection grouts for the conservation of architectural surfaces. Stud Conserv. 101179/sic.2010.55.Supplement-1.3

Binda L, Saisi A (2001) State of the art of research on historic structures in Italy. Department of Structural Engineering, Politecnico of Milano, Milano

Binda L, Baronio G, Tiraboschi C (1993a) Repair of brick-masonries by injection of grouts: experimental research. J Struct Eng 20(1):29–44

Binda L, Modena C, Baronio G (1993b) Strengthening of masonries by injection technique. In: Proceedings of 6° NaMC, vol I, Philadelphia, pp 1–14

Binda, L, Fontana A, Mirabella Roberti G (1993c) Modelling the mechanical behaviour of multiple-leaf stone walls. In: International symposium on computer methods in structural masonry, Swansea, G.B.

Binda, L, Fontana A, Mirabella Roberti G (1994) Mechanical Behaviour and stress distribution in multiple-leaf stone walls. In: Proceedings of 10th international brick/block/masonry conference, Calgary, pp 1–9

Binda L, Modena C, Baronio G, Abbaneo S (1997) Repair and investigation techniques for stone masonry walls. Constr Build Mater 11(3):133–142

Binda L, Baronio G, Tiraboschi C, Tedeschi C (2003) Experimental research for the choice of adequate materials for the reconstruction of the Cathedral of Noto. Constr Build Mater 17:629–639

Binda L, Cardani G, Saisi A (2005) A classification of structures and masonries for the adequate choice of repair. In: Proceedings of international RILEM workshop on repair mortars of historic masonry, Delft, The Nederland, 26th–28th Jan 2005, pp 20–34

Binda L, Pina Henriques J, Anzani A, Fontana A, Lourenco PB (2006) A contribution for the understanding of load-transfer mechanisms in multi-leaf masonry walls: testing and modelling. Eng Struct 28(8):1132–1148

Bouineau A (1985) Renforcement des maçonneries par injection des coulis. In: Proceedings of Vth international congress on deterioration and conservation of stone, Lausanne, pp 939–947

Bras A (2011) Grout optimization for masonry consolidation. PhD thesis, Universidade Nova de Lisboa, Lisbon, Portugal

Bras A, Henriques FMA (2012) Natural hydraulic lime-based grouts—the selection of grout injection parameters for masonry consolidation. Constr Build Mater 26:135–144

Corradi M, Tedeschi C, Binda L, Borri A (2008) Experimental evaluation of shear and compression strength of masonry wall before and after reinforcement: deep repointing. Constr Build Mater 22:463–472

Da Porto F, Valluzzi MR, Modena C (2003) Investigation for the knowledge of multi-leaf stone masonry walls. In: First international congress on construction history, vol II, Madrid, Spain, 20–24 Jan, pp 713–722

Egermann R (1993a) Investigation on the load bearing behavior of multiple leaf masonry. In: Proceedings of IABSE symposium structural preservation of the architectural heritage, Rome, Italy, pp 305–312

Egermann R (1993b) Stone masonry buildings: research and applications at the University of Karlsruhe. In: Murature Sicurezza Recupero, Proceedings of the conference of ITEA, Trento, Italy, pp 70–95

Egerman R, Frick B, Neuwald C.M (1993) Analytical and experimental approach to the load bearing of multiple leaf masonry. In: Brebbia RJB, Frewer CA (eds) Structural repair and maintenance of historical buildings III. Computational Mechanics Publications, Southampton Boston

Ferragni D, Malliet J, Di Martino S, Forti M (1982) Essais de laboratoire sur des coulis à base de ciment. In: ICCROM (ed) Proceedings of the international symposium on mortiers, ciments et coulis utilisés dans la conservation des bâtiments historiques, ICCROM, Rome, 3–6.11.1981, pp 185–203

Ferragni D, Forti M, Malliet J, Mora P, Teutonico JM, Torraca G (1984) Injection grouting of mural paintings and mosaics. In: Proceedings of the international symposium on adhesives and consolidants, Paris, 2–8 Sept, pp 110–116

Jorne F (2016) Injectability of hydraulic lime grouts for old masonry consolidation. PhD thesis, Universidade Nova de Lisboa, Lisbon, Portugal, May 2011, p 207

Jorne F, Henriques FMA, Baltazar LG (2015) Injection capacity of hydraulic lime grouts in different porous media. Mater Struct 48:2211–2233. https://doi.org/10.1617/s11527-014-0304-9

Kalagri A, Miltiadou-Fezans A, Vintzileou E (2010) Design and evaluation of hydraulic lime grouts for the strengthening of stone masonry historic structures. Mater Struct 43:1135–1146. https://doi.org/10.1617/s11527-009-9572-1

Karaveziroglou M, Papayianni I, Penelis G (1998) Mortars and grouts in restoration of Roman and Byzantine monuments. In: Biscontin G, Moropoulou A, Erdik M, Delgado Rodrigues J (1998)

Compatible materials for the protection of European cultural heritage, PACT 55 1998, Technical Chamber of Greece, pp 219–245
Luso E, Lourenco PB (2016) Experimental characterization of commercial lime-based grouts for stone masonry consolidation. Constr Build Mater 102:216–225
Luso E, Lourenco PB (2017) Bond strength characterization of commercially available grouts for masonry. Constr Build Mater 144:317–326
Luso E, Lourenco P (2019) Mechanical behaviour of two leaf masonry wall-strengthening using different grouts. J Mater Civ Eng 31(7):04019096
Mazzon N (2010) Influence of grout injection on the dynamic behaviour of stone masonry buildings. PhD thesis, University of Padova
Miltiadou AE (1985) Grouting as a method for the repair of masonry monuments. Master of Arts in Conservation Studies, Institute of Advanced Architectural Studies, University of York, UK, p 325
Miltiadou AE (1990) Étude des coulis hydrauliques pour la réparation et le renforcement des structures et des monuments historiques en maçonnerie. Thèse de Doctorat de l'Ecole Nationale des Ponts et Chaussées. Pub. by LCPC in Collection Etudes et recherches des Laboratoires des Ponts et Chaussées, série Ouvrages d'art, OA8 ISSN 1161-028X, LCPC, Décembre 1991, Paris, France, p 278
Miltiadou AE, Abdunur C (1990) Rôle mécanique des coulis d'injection sur le comportement d'un mur en maçonnerie – Une approche numérique. In: Proceedings of the international conference on structural conservation of stone masonry, 31st Oct–3rd Nov 1989, Athens, Greece, Pub. ICCROM, pp 147–154
Miltiadou AE, Paillère A-M, Serrano JJ, Denis A, Musicas N (1990) Formulation de coulis hydrauliques pour l'injection des fissures et cavités des structures en maçonnerie dégradées. In: Proceedings of the international conference on structural conservation of stone masonry, 31st Oct–3rd Nov 1989, Athens, Greece, Pub. ICCROM, pp 299–312
Miltiadou AE, Durville J-L, Martineau F, Massieu E, Serrano J-J (1993) Etude mécanique de mélanges cailloux-mortier-influence de l'injection de coulis. Bull liaison Lab Ponts Chaussées-183-janv.-févr., Réf. 3677, pp 75–84
Miltiadou-Fezans A (1998) Criteria for the design of hydraulic grouts injectable into fine cracks and evaluation of their efficiency. In: Biscontin G, Moropoulou A, Erdik M, Delgado Rodrigues J (eds) Compatible materials for the protection of European cultural heritage, PACT 55 1998, Technical Chamber of Greece, pp 149–163
Miltiadou-Fezans A, Tassios TP (2012) Fluidity of hydraulic grouts for masonry strengthening. RILEM Mater Struct. https://doi.org/10.1617/s11527-012-9872-8
Miltiadou-Fezans A, Tassios TP (2013a) Penetrability of hydraulic grouts. RILEM Mater Struct. https://doi.org/10.1617/s11527-012-0005-1
Miltiadou-Fezans A, Tassios TP (2013b) Stability of hydraulic grouts for masonry strengthening. RILEM Mater Struct. https://doi.org/10.1617/s11527-012-0003-3
Miltiadou-Fezans A, Tassios TP (2016) Holistic methodology for the mix design of hydraulic grouts in strengthening historic masonry structures. In: Papayianni I, Stefanidou M, Pachta V (eds) Proceedings of 4th historic mortars conference (HMC2016), 10th–12th Oct 2016, Santorini, Greece, pp 580–587
Miltiadou-Fezans A, Papakonstantinou E, Zambas K, Panou A, Frantzikinaki K (2005) Design and application of hydraulic grouts of high injectability for structural restoration of the column drums of the Parthenon Opisthodomos. In: Brebbia CA, Torpiano A (eds) Structural studies, repairs and maintenance of heritage architecture IX, WIT Press, Southampton, pp 461–472
Miltiadou-Fezans A, Kalagri A, Delinikolas N (2007) Design of hydraulic grout and application methodology for stone masonry structures bearing mosaics and mural paintings: the case of the Katholikon of Dafni Monastery. In: Arun G (ed) Proceedings of international symposium on studies on historical heritage, Antalya, Turkey, 17–21 Sept 2007, pp 649–656
Miltiadou-Fezans A, Kalagri A, Kakkinou S, Ziagou A, Delinikolas N, Zarogianni E, Chorafa E (2008) Methodology for in situ application of hydraulic grouts on historic masonry structures.

The case of the Katholikon of Dafni Monastery. In: D'Ayala D and Fodde E (eds), Proc. of 6th international conference on Structural Analysis of Historic Construction, Preserving Safety and Significance, 2-4 July, Bath, CRC Press/Balkema, Taylor and Francis Group, London UK, Vol. II, pp 1025–1033

Miltiadou-Fezans A, Delagrammatikas M, Kalagri A, Vassiliou P (2021) Evaluation of performance of matured hydraulic grouts: strength development, microstructural characteristics and durability issues. In: Roca P, Pelà L, Molins C (Eds) Proceedings of the 12th International Conference on Structural Analysis of Historical Constructions SAHC21, on line event 29th Sept–1st Oct 2021, International Centre for Numerical Methods in Engineering (CIMNE), Barcelona, Spain, pp 2480–2491

Mouzakis C, Adami CE, Karapitta L, Vintzileou E (2017) Seismic behaviour of timber-laced stone masonry before and after interventions: shaking table testes on two-storey masonry model. Bull Earthq Eng. https://doi.org/10.1007/s10518-017-0220-9

Oliveira DV, Lourenco PB, Garbin E, Valluzzi MR, Modena C (2007) Experimental investigation of the structural behaviour and strengthening of three-leaf stone masonry wall. In: Lourenco PB, Roca P, Modena C, Agrawal S (eds) Proceedings of the 5th international conference on Structural analysis of historical constructions SAHC2006, 6th–8th Nov 2006, New Delhi, India, Macmillan India Ltd, vol 2, pp 817–826

Paillère A-M, Guinez R (1984) Recherche d'une formulation de coulis à base de liants hydrauliques pour l'injection dans les fines fissures et les cavités. Bull liaison Lab Ponts Chaussées, Paris (130):51–57

Paillère A-M, Buil M, Miltiadou A, Guinez R, Serrano JJ (1989) Use of silica fume and superplasticizers in cement grouts for injection of fine cracks. In: Proceedings of the third international conference on use of fly ash, silica fume, slag and natural pozzolans in concrete, vol 2, Trondheim, Norway, SP-ACI, pp 1131–1157

Paillère A-M, Serrano JJ, Miltiadou A (1993) Formulation des coulis hydrauliques pour l'injection de fines fissures et cavités dans les structures dégradées en béton et maçonnerie. Bull Liaison Lab Ponts Chaussées, 186-juil.-août, Ref 3676, pp 61–78

Papayianni I, Pachta V (2015) Experimental study on the performance of lime-based grouts used in consolidating historic masonries. Mater Struct 48:2111–2121. https://doi.org/10.1617/s11527-014-0296-5

Papayianni I, Stafanidou M, Pachta V (2010) Grouts for injection of historical masonries: influence of the binding system and other additions on the properties of the matrix. In: Proceedings of 2nd historic mortars conference & RILEM TC 203-RHM Repair mortars for historic masonry. Final workshop, 22–24 Sept 2010, Prague, RILEM, pp 1123–1134

Penelis G, Karaveziroglou M, Papayianni I (1989) Grouts for repairing and strengthening old masonry structures. In: Brebbia CA (ed) Structural repair and maintenance of historical buildings. Computational Mechanics Publications, Southampton, UK, pp 179–188

Pina-Henriques J, Lourenco PB, Binda L, Anzani A (2005) Testing and modelling of multiple-leaf masonry walls under shear and compression. In: Modena C, Lourenco PB, Roca P (eds) Proceedings of the 4th international seminar on structural analysis of historical constructions, 10–13 Nov Padova, Italy, Taylor and Francis Group, London, vol 2, pp 299–310

Rocard J, Bouineau A (1982) Renforcement des maçonneries par Injections des coulis dans la region Nord-est de la France. In: Proceedings of international symposium on mortars, cements and grouts used in the conservation of historic buildings, 3–6.11.1981, ICCROM, Rome, pp 165–184

Silva B (2012) Diagnosis and strengthening of historical masonry structures: numerical and experimental analyses. Ph.D. thesis, University of Brescia, Apr 2012, p 407

Silva B, Pigouni AE, Valluzzi MR, da Porto F, Modena C (2014) Calibration of analytical formulations predicting compressive strength in consolidated three-leaf masonry walls. Constr Build Mater 64:28–38 (August 2024). https://doi.org/10.1016/j.conbuildmat.2014.04.044

Tassios TP (1988) Meccanica delle murature, Ed. Liguori

Tassios TP (2004) Rehabilitation of three-leaf masonry. In: Evoluzione nella sperimentazione per le costruzioni, Seminario Internazionale. Centro Internationale di Aggiornamento Sperimentale – Scientifico (CIAS)
Tassios TP, Chronopoulos M (1986) Aseismic dimensioning of interventions on low-strength masonry buildings. In: Proceedings of Middle East and Mediterranean regional conference on earthen and low-strength masonry in seismic areas, Middle East University, Aug 31st–Sept 6th, Ankara
Tomazevic M, Apih V (1993) The strengthening of stone-masonry walls by injecting the masonry friendly grouts. In: European earthquake engineering, vol 7, no 2. Patron Bologna, pp 10–20
Toumbakari EE (2002) Lime-pozzolan-cement grouts and their structural effects on composite masonry walls. Ph.D thesis, Department of Civil Engineering, KU Leuven, p 364
Toumbakari EE, Van Gemert D, Tenoutasse N (1999a) Microstructural evolution and mechanical properties of pozzolanic injection grouts. In: Pietersen HS, Larbi JA, Janssen HHA (eds) Proceedings of the 7th Euroseminar on microscopy applied to building materials, June 29–July 2, 1999, Delft, The Netherlands, TU Delft-TNO, pp 345–353
Toumbakari EE, Van Gemert D, Tassios TP (1999b) Mechanical properties of multi-blend cementitious injection grouts for repair and strengthening of masonry structures. In: Swamy R (ed) Infrastructure regeneration and rehabilitation, Sheffield
Toumbakari EE, Van Gemert D, Tassios TP (1999c) Multi-blend cementitious injection grouts for repair and strengthening of masonry structures. In: Proceedings of international conference on creating with concrete volume: "Specialist techniques and materials for concrete construction", Dundee
Toumbakari EE, Van Gemert D, Tassios TP, Vintzileou E (2005) Experimental investigation and analytical modeling of the effect of injection grouts on the structural behaviour of three-leaf masonry walls. In: Modena c, Lourenco P, Roca P (eds) Proceedings of 4th international seminar on Structural analysis of historical constructions, 10–13 Nov Padova, Italy, Taylor and Francis Group, London, pp 707–717
Toumbakari EE, Vintzileou E, Tassios TP, Van Gemert D (2007) Shear behaviour of masonry units and lime-pozzolan-cement grout interfaces. In: Proceedings of 10th NAMC The Masonry Society, St Louis, Missouri, pp 767–776
Uranjek M, Žarnic R, Bokan-Bosiljkov V, Bosiljkov V (2014) Seismic resistance of stone masonry building and effect of grouting. Gradevinar 66(8):715–726. https://doi.org/10.14256/JCE.1031.2014
Valluzzi MR (2000) Comportamento meccanico di murature storiche consolidate con materiali e tecniche a base di calce. Ph.D. thesis, University of Trieste, p 276
Valluzzi MR, da Porto F, Modena C (2004) Behaviour and modelling of strengthened three-leaf stone masonry walls. Mater Stuct 37:184–192
Van Gemert D (1988) The use of grouting for the consolidation of historic masonry constructions. Advantages and limitations of the method. In: Lemaire RM, Van Balen K (eds) Stable-unstable? Structural consolidation of ancient buildings. Leuven University Press, pp 265–276
Van Rickstal F (2000) Grout injection of masonry, scientific approach and modeling. Katholieke Universiteit Leuven, Mei, p 2000
Vintzileou E (2011) Three-leaf masonry in compression, before and after grouting: a review of literature. Int J Archit Heritage 5:513–538
Vintzileou E, Tassios TP (1995) Three leaf stone masonry strengthened by injecting cement grouts. J Struct Eng ASCE 121(5):848–856
Vintzileou E, Miltiadou-Fezans A (2008) Mechanical properties of three-leaf stone masonry grouted with ternary or hydraulic lime-based grouts. Eng Struct 30(8):2265–2276
Vintzileou E, Mouzakis C, Adami CE, Karapitta L (2015) Seismic behavior of three-leaf stone masonry buildings before and after interventions: Shaking table tests on two-storey masonry model. Bull Earthq Eng. https://doi.org/10.1007/s10518-015-9746-x

Yüzer N, Oktay D, Ulukaya S, Gökyğit-Arpaci EY (2015) Mechanical behaviour of two-wythe brick masonry walls injected with hydraulic lime grout. In: Proceedings of international conference REHAB2015, 22–24 July 2015, Porto, Portugal

# Chapter 2
# Penetrability

**Abstract** This Chapter deals with the first component of injectability, i.e., the *penetrability* of the grains of the grout through the effective (nominal minimum) width $W_{nom}$ of masonry discontinuities (voids, microcracks, joints). As it is known, grouting is intended to fill voids, fissures and open joints of the masonry as a system, producing a "dendrite" (a three-dimensional skeleton), directly contributing to the strength of the masonry as a whole. However, to do so, the grout should be able to pass through the "narrowest" possible width of such discontinuities, in order to reach the maximum possible internal volume of masonry and open joints, avoiding most of possible blockages. In the specific case of three-leaf masonries, the most decisive result of the grouting is expected to be the strengthening of the bond along the interfaces between the external layers and the infill; the rather small voids, as well as pre-existing fissures along these interfaces, have to be penetrated. In this chapter the penetrability of hydraulic grouts is discussed, and relationships between (i) two characteristic diameters of the grains of the solid phase of the grout and (ii) the nominal minimum (effective) width of fissures and voids of the structure to be injected are proposed. Furthermore, the beneficial role of replacing part of the cement or hydraulic lime with ultrafine materials in order to improve penetrability is presented, and related criteria are proposed. Extensive experimental verifications of the proposed quantitative models are finally offered.

## 2.1 Introduction

First it is appropriate to describe the possible modes of grout-penetration in a heterogeneous medium like masonry, containing discontinuities (i.e., voids, fissures and open joints) which are the cause of low strength of masonry. A relatively low hydraulic pressure[1] is applied, and the grout is conveyed mainly through mortars filling the joints or infill materials of masonry. Its flow will be effectuated only through some particular routes of discontinuities that communicate to each other, directly or after

[1] Normally, this pressure is not higher than say 0.1 (MPa), because of the limited tensile strength of masonry, as well as in order to avoid segregation effects.

 
A. Miltiadou-Fezans and T. P. Tassios, *Mix-Design and Application of Hydraulic Grouts for Masonry Strengthening*, Springer Tracts in Civil Engineering,
https://doi.org/10.1007/978-3-030-85965-7_2

overpassing local small obstacles. The flow routes[2] followed by grout are therefore forming a kind of "dendrite" (a three-dimensional skeleton), that after setting contributes to the strength of the masonry as a whole. However, to do so, the grout should be able to pass through the "narrowest" possible width of such discontinuities, in order to reach the maximum possible internal volume of masonry and open joints, avoiding most of possible blockages of the grout.

Subsequently, the first step towards the rational design of hydraulic grouts is to ensure sufficient penetrability. Thus, the necessary grain size characteristics of the solid phase of the grout should be selected on the basis of the "effective minimum width" (or nominal width) of the discontinuities of the masonry. Three categories of such discontinuities (voids and fissures) should be considered: (a) In one leaf masonries, fine cracks due to block-to-mortar de-bonding, as well as relatively large voids left between blocks, (b) In three-leaf masonries, very large voids within the infill, as well as rather small voids or pre-existing cracks along the interfaces between external leaves and infill; filling of these fine cracks may very much improve the bond along the interfaces between the external layers and the infill, thus improving the integrity and resistance of masonry, (c) In all cases, generalized cracks produced by middle intensity earthquakes or by differential settlements or other causes.

(Note that in the case of two-leaf masonry, discontinuities similar to three-leaf masonries may be encountered).

For such discontinuities, to be successfully penetrated by the grout, a rough estimation of the "effective maximum" grain size of the solid ingredients of the grout should first be made.

In this chapter are presented the basic methods for testing penetrability and the criteria and procedures to be followed in order to make a first selection of suitable grain size distribution of solid materials of the grout as a function of the "effective minimum width of voids" of the milieu to be injected. Relationships between two characteristic diameters of the grains of the solid phase of the grout and the nominal minimum width of fissures and voids of the structure to be injected are proposed, based on the work of Miltiadou-Fezans and Tassios (2003, 2013). Furthermore, the beneficial role of replacing part of the cement or hydraulic lime with ultrafine materials in order to improve penetrability is also presented, and related criteria are proposed. Thus, a procedure is suggested to be followed in order to make a first selection of suitable grain size distribution of solid materials of the grout.

Based on these criteria and using commercial materials, one can prepare hydraulic grouts injectable in voids and cracks of a nominal minimum width of 0.1–0.2 mm. For lower nominal minimum widths (e. g., in cases of an infill with very fine silt or clay) the penetrability of hydraulic grouts cannot be satisfactory, at least with commercially available materials (Paillère et al. 1986, 1989; Miltiadou 1990; Miltiadou-Fezans 1998; Binda et al. 1993a, b, 1994).

[2] Note however that for a given fluidity, many other flow routes could also be followed should the grout be more penetrable, i.e., if its solid phase were finer.

## 2.2 Literature Survey

### 2.2.1 *Significance of Penetrability and Relevant Tests for Its Control*

Occasionally, literature may not be sufficiently clear about this issue; "penetrability" does not seem to be always distinguished from "fluidity" of the grout—a very important property which, however, may not be manifested at all if larger grains of the grout produce a blockage. This occasional underestimation of penetrability, may however be explained by the fact that if for other reasons (presented in Sect. 1.2) low cement content binary or ternary grouts (i.e., association of cement with ultrafine materials) or ultrafine hydraulic lime grouts are directly used, most of penetrability problems may frequently be overcome without additional efforts.

A literature survey on penetrability follows:

Referring to the case mentioned in Sect. 1.2 (1), guidance is offered by Gustafson and Stille (1996), regarding grouting of fractured rocks. The authors introduce the concept of an "effective aperture that governs the penetration radius" of cement-based grouts. Along the same lines, Eriksson (2002) introduced a method to quantify penetrability by means of standardized "filters"; for a given grout, a "minimum" and a "critical" aperture were thus determined, corresponding to a start of dripping and a constant maximum discharge, respectively [e.g., 60 and 100 μm for a special cement grout tested (from Eriksson et al. 2004)]. A similar penetrability-meter for cementitious grouts was also presented by Axelsson and Gustafson (2010), whereas Axelsson et al. (2009), among other factors, reiterate the importance of the ratio between the opening of the fissure and the maximum grains in the grout; and this is the basic approach to be followed in the present Chapter as well. Eriksson et al. (2004) offer also some guidance related to the risks due to possible oversized solid particles.

On the other hand, the specific problem of three-leaf masonry was successfully investigated by several authors, following two approaches.

#### 2.2.1.1 Simulating Infill Material in Cylinder Specimens

The conditions of the infill are experimentally reproduced, be it in an inevitably approximated way: within a plexiglass cylinder (d = 120–150 mm, h = 300 mm) a simulated infill is constructed, using old materials taken from the infill of the existing masonry and placed in the cylinder trying to reproduce its composition and voids in situ (Fig. 2.1a). Subsequently a candidate grout is injected, and several observations are made regarding its penetrability, fluidity and stability, as well as the final strength induced to this simulated infill.

This interesting technique for checking penetrability of grouts used for strengthening of infills of three-leaf masonries, was first introduced by Binda et al. (1993b,

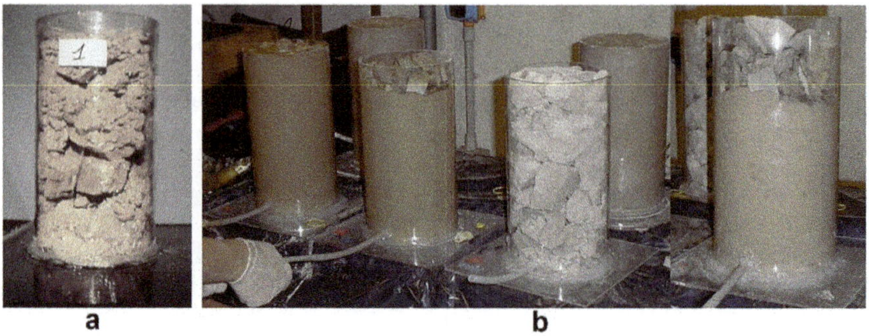


**Fig. 2.1** Penetrability test on plexiglass cylinders constructed to simulate infill material: **a** example of cylinder introduced by L. Binda (Courtesy of A. Elide Saisi), **b** examples of cylinders used by M. R. Valluzzi. Adapted from Valluzzi (2004), courtesy of the Author

1994, 1997) and subsequently followed by other authors too, e.g., Laefer et al. (1998), Binda et al. (2003).

Valluzzi (2000, 2004) and Valluzzi et al. (2003) followed a quasi similar procedure: they have filled plexiglas cylinders (d = 152 mm, h = 300 mm) with small pieces of stones and mortar (Fig. 2.1b), used in the infill of Laboratory made three-leaf walls to be tested before and after grouting. Various sizes of stones were used, resulting in 40–50% of voids of the mix. Based on this test, the penetrability of several hydraulic lime grouts was assessed, by measuring the rise of the grout along the height of the cylinder, the time for filling the cylinder and the volume of the grout.

An alternative technique was previously used by Miltiadou (1990) and Miltiadou et al. (1993); both cylindrical (d = 250, h = 500 mm) (Fig. 2.2a, b) and rectangular (250 x 250 x 200 mm) (Fig. 2.2c) infill specimens (consisting of pieces of stones and mortar) were prepared and tested in compression or shear (Casagrande test) respectively.

**Fig. 2.2** **a** Cylindrical specimens consisting of mortar and stones, simulating infill material with various voids percentage. **b** Cylindrical specimen during compression test before grouting. **c** Rectangular specimens for shear tests. **d** Rupture surface after direct shear test of an injected rectangular specimen. Adapted from Miltiadou (1990)

Metallic moulds were used instead of plexiglass ones; series of specimens were constructed using two types of stones and mortar, in an adequate content to simulate voids of 32 and 40%.

The specimens were tested in compression or shear before and after grouting or directly after grouting simulating a preventive intervention.

Various grout compositions were used: (a) binary compositions containing 75% cement (C10) and 25% pozzolan (densified silica fume or Santorini earth), and (b) a ternary solid phase composition consisting of 50% cement (C10), 22.5% pozzolan (densified silica fume or Santorini earth) and 27.5% hydrated lime in powder. Through this procedure particular problems of injectability were also examined, referring to simulated infills.

Moreover, in some cases the cylindrical specimens, after been injected and tested in compression, were *injected again* with the same grout composition and subsequently cut in slices. The careful examination of the penetration capacity of grouts injected gave important information for their penetrability capacity (Fig. 2.3).

It is interesting to note that, even very small width cracks opened in the pieces of stones due to the compression test after grouting, were also filled, together with many voids and cracks in the mortar itself.

Similar cylinders were also used by Kalagri et al. (2010).

#### 2.2.1.2 Sand Column Test to Simulate an Effective $W_{nom}$

The second approach (applicable to any type of masonries) is based on the concept of an "effective minimum opening $W_{nom}$" to be penetrated. In the case of granular materials, available knowledge allows for an approximate determination of such a decisive numerical value or for a direct experimental solution of the problem (see Mutman and Kavak 2011). The same holds true in the case of solid walls or monoliths with a network of cracks; statistical techniques (e.g., Gustafson and Stille 1996) may help in estimating $W_{nom}$. For ordinary masonries however, additional investigations are needed (see Chap. 5 of the present book) to this end. According to Bras and Henriques (2012), the variability of voids within masonry requires an ability to fine tune the rheological properties of the grout. Furthermore, it is recognized that "actual" conditions of the milieu to be grouted, are hardly reproductible in laboratory conditions. Thus, due to the multiple factors influencing the penetration of a grout (see Sect. 2.1 above), it seems that the idea to completely simulate the pores, voids, fissures and interfaces of masonry by means of an attempt to construct a true replica in a mould, is too ambitious; instead, only a "substitute milieu" could be examined reproducing mainly the effective minimum opening $W_{nom}$; thus, the penetrability is examined by studying the most decisive stage of the entire phenomenon. An appropriately selected granular mixture is used and grouted under laboratory conditions.

To this end, the so called "sand-column test" was conceived by Paillère and Rizoulières (1978) in the Laboratoire Central des Ponts et Chaussées (Paris).

**Fig. 2.3** **a** Section of a cylinder after a second application of grouting. Dark grey colour corresponds to the first grout; light grey to the second one. **b** Detail of the circled area showing the high penetrability of both grouts; even very small width cracks, opened in the pieces of stones due to the compression test after grouting, were also filled. **c**, **d** Similar details of other injected cylinders. Reworked from Miltiadou-Fezans and Tassios (2013)

The test was developed initially for the control of injectability of polymers for the repair of concrete structures. This test is a procedure to determine the penetrability of a liquid in a capillary network, as well as to measure its final bonding capacity (by measuring the splitting strength of cylindrical mortar specimens produced by the injection of a "sand column").

The principle of the test consists of injecting the liquid, (at low and constant pressure of 0.075 MPa), into a transparent plexiglass tube filled with graded sand, and kept in a vertical position. On the bottom and the top of the column, a metallic filter-net is placed.

The column is injected from its lower end; the time the grout takes to reach the various reference marks drawn along the tube is measured (Fig. 2.4). As reported in Paillère et al. (1989), when investigating the injectability of hydraulic grouts, the time $T_{36}$ the grout takes to reach the top of the column, a height of 36 cm, is measured. A hydraulic grout is considered to be injectable in the sand column, if it has the capacity not only to reach the upper part of the column (360 mm), but also when there is a continuous flow of grout from the sand column in the measuring vessel (see n° 10 in Fig. 2.4), at least of 20 ml. In cases in which the height of 36 cm is not reached or when after the 36 cm had been injected there is no continuous flow of grout to the measuring vessel, the injection of such grout is not satisfactory; the grout is characterized as non-injectable.

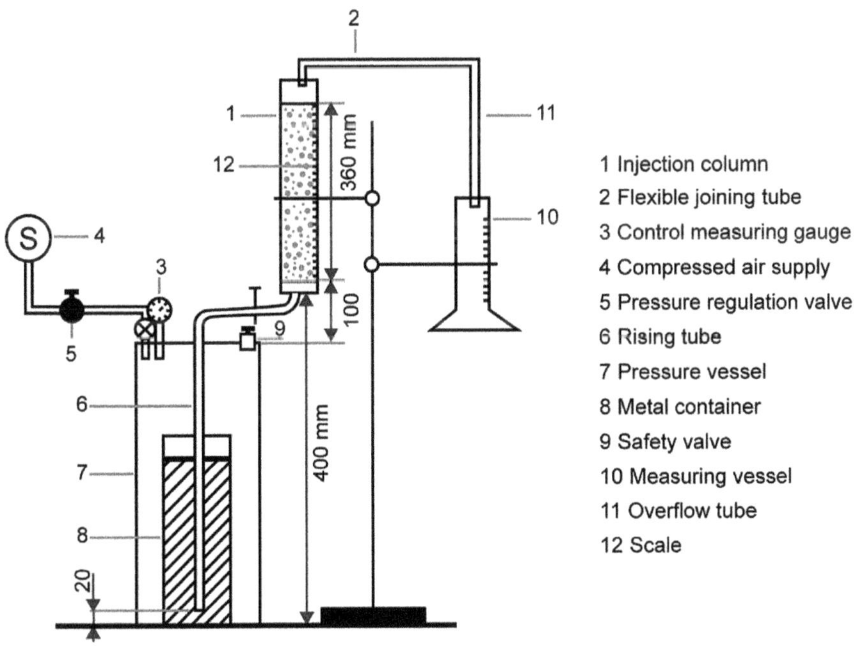


**Fig. 2.4** Schematic representation of the sand column injectability test

The sand column injectability test is described in the recommendations of RILEM TC-52 RAC, in the French standard NF P 18-891, as well as in the European standard EN 1771. This test was also used in investigations on injections undertaken by ICCROM (Ferragni et al. 1982, 1984) and by Miltiadou (1990), Toumbakari (2002), Miltiadou-Fezans et al. (2005, 2007), Kalagri et al. (2010), Biçer-Şimşir et al. (2010), Papayianni and Pachta (2015), Yüzer et al. (2015) and other researchers.

This test is particularly interesting, as it combines the simultaneous control of fluidity and penetrability. Furthermore, the use in this test of different gradings of sand, permits to simulate in a reproductible and fully controlled way voids and fissures of *different widths.* The test may be of paramount practical importance, since taking into account some actual parameters (be it with their "nominal" values), permits the systematic comparative study of multitude of grouts in a series of columns corresponding to different $W_{nom}$.

With this idealization of $W_{nom}$ in mind, the sand-column test does not simulate masonry itself, but offers the possibility to check experimentally the penetrability of a given grout through an effective fine void of a masonry characterized by the same nominal value of crack width. One of the advantages of this approach is that it allows a further quantification of the properties of the grout, as it will be attempted in the present chapter too.

The same approach with different practical applications is applied by other authors as well, e.g., Van Rickstal (2000), Bras (2011), Uranjek et al. (2014), Jorne et al. (2015), Van Gemert et al. 2015, Luso and Lourenço (2016, 2017), who have used plexiglass cylinders filled either with crashed bricks sieved to obtain different grain size distributions or with various grading combinations of sand, respectively, in order to simulate several masonry "permeabilities" (Fig. 2.5).

In what follows, further literature survey is carried out, regarding the numerical handling of the penetrability properties using the sand column technique.

### 2.2.2 *Maximum Grain Size Criterion*

As aforementioned, for a granular suspension (such as a hydraulic grout) to be able to penetrate, the grain size distribution of its solid phase should be compatible with the characteristic dimensions of the discontinuities (voids, fissures, interfaces, etc.) to be injected. In the field of soil grouting, most of the authors, (Littlejohn 1983; Mitchell 1970; Léonard 1961; Johnson 1958), formulated penetrability conditions in terms of (i) the ratio between the size of the larger grains (mostly $d_{85}$) of the solid phase of the grout and (ii) the size of the smaller grains (mostly $D_{15}$) of the soil to be injected (where 85% of the grout grains are smaller than $d_{85}$, whereas 15% of the soil grains are smaller than $D_{15}$).

In the field of the grouting of structures or fissured rocks, penetrability conditions are frequently expressed in terms of the ratio between the size of the larger grains of the grout and a "representative" diameter of orifices or a width of fissures or

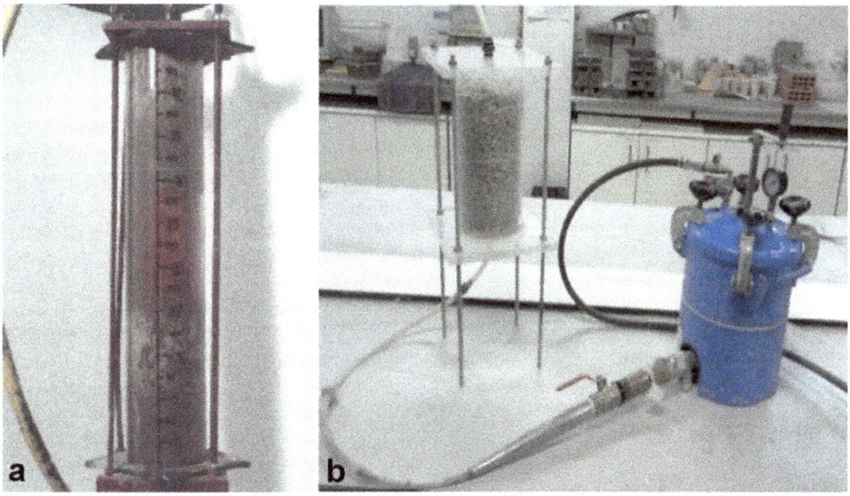


**Fig. 2.5** **a** Penetrability test on plexiglass cylinders filled with crashed-bricks sieved to obtain different grain size distributions. Adapted from Van Rickstal (2000), courtesy of the Author. **b** Crushed limestone sands and crushed bricks were used to simulate various types of porous media. Reproduced from Jorne et al. (2015)

discontinuities to be injected (Littlejohn 1983; Mitchell 1970; Hutchinson 1981; Cambefort 1977; Papadakis 1959).

Table 2.1 shows some of those ratios proposed in literature (Miltiadou-Fezans and Tassios 2013). In order to facilitate comparisons, in the last column of Table 2.1 all relationships are presented under the same format (Eq. 2.1).

$$d < \frac{W_{nom}}{n} \tag{2.1}$$

where

| | |
|---|---|
| $d$ | diameter characteristic of the size of the larger solid grains of the grout, frequently expressed as $d_{85}$ (or $d_{100}$); |
| $n$ | numerical factor, considerably larger than unity; |
| $W_{nom}$ | nominal lower value of the aperture of fissures or discontinuities to be injected. When such a value is not explicitly indicated by the author, as in the case of granular media, we have assumed that the approximation $W_{nom} \sim 0.15 \cdot D_{15}$ can be used instead; this value corresponds to the diameter of the smallest path passing through grains of the same size $D_{15}$, according to Dantu (1961). |

The factor "*n*" expresses the practical need of the grout grains to be "considerably" smaller than the characteristic aperture to be penetrated. This is due to several phenomena such as:

Table 2.1 Grain penetrability conditions, according to literature

| Author | Relationships | Grouted medium | Format: $d < W_{nom}/n$ |
|---|---|---|---|
| Johnson (1958) | $D_{15} / d_{85} > 25$ | Fine granular soil | $d_{85} < W_{nom} / 3 .75$ |
| Mitchell (1970) | $W_{nom} / d_{100} > 3$ | Fissured medium | $d_{100} < W_{nom} / 3$ |
| Littlejohn (1983) | $D_{15} / d_{85} > 25$<br>$W_{nom} / d_{100} > 5$ | Fine granular soil<br>Fissured medium | $d_{85} < W_{nom} / 3.75$<br>$d_{100} < W_{nom} / 5$ |
| Hutchinson (1981) | $W_{nom} / d_{max} > 3$ | Fine granular soil | $d_{max} < W_{nom} / 3$ |
| Cambefort (1977) | $W_{nom} / d_{100} > 1.5–2.0$ | Fissured medium | $d_{100} < W_{nom} / 1.5–2.0$ |
| Léonard (1961) | $D_{15} / d_{85} > 5–20$ | Fine granular soil | $d_{85} < W_{nom} / 0.75–3.0$ |
| Papadakis (1959) | $D_{min} / d_{max} > 10–20$ | Fine granular soil | $d_{100} < W_{nom} / 1.5–3.0$ |
| Paillère and Guinez (1984) | $W_{nom} / d_{100} > 1.5–2.3$ | Tests in "sand column" | $d_{100} < W_{nom} / 1.5–2.3$ |

Notation: $D_{15}$ = diameter of the soil grain, corresponding to 15% passing, $d_{85}$ = diameter of the grout grain, corresponding to 85% passing, $d_{100}$ = "maximum" diameter of the grout grains
Adapted from Miltiadou-Fezans and Tassios (2013)

- Frictions due to the irregular form of the grains.
- Electrostatic connections between particles.
- The agglomeration due to immediate hydration of the fines.

Table 2.1 shows that this factor "*n*", roughly varying from 1 to 5, may also depend on the grouted media and the conditions of application of the grouting. Despite the extent of the relevant literature, available papers very often do not offer but empirical and partly qualitative information.

To our knowledge, the first systematic works studying in a more precise way the relationship that could exist between the grading of the solid phase of the grout, and the penetrability of hydraulic grouts were those of Paillère and Guinez (1984), Paillère et al. (1989), Miltiadou et al. (1990) and Miltiadou (1990).

## 2.2.3 *Criterion for the Grading of the Coarser Grains*

Paillère and Guinez (1984), using three different grading curves of sands (0.63–1.25 mm, 1.60–2.50 mm and 2.50–4.00 mm), have shown that it is possible to obtain perfectly injectable cement grouts, when the pores' dimension is 1.5–2.3 times larger than the dimension of the coarser cement grains (Table 2.1).

Furthermore, they have associated the cement grouts penetrability limits into a well-defined granular "milieu", first with the percentage of cement grains larger than 32 μm and second with their grain size distribution. The authors have shown that the same sand column injectability test, initially developed for polymers, can be used for the evaluation of the penetrability of hydraulic grouts as well.

On the basis of the aforementioned research work, Paillère et al. (1989) and Miltiadou (1990), working more specifically on the injection of very fine cracks (the sand used in the column was 0.63–1.25 mm and 1.60–2.50 mm, corresponding to fine fissures of a width equal to 0.1–0.2 mm and 0.2–0.4 mm, respectively), have confirmed and refined the aforementioned results.

They have also shown that the maximum diameter of the coarser grains of the solid phase of the grout, or, in a more general way, other pertinent global parameters (like Blaine specific surface or modulus of fineness representing the totality of the grading curve), do not constitute an adequate criterion of the penetrability of cement grouts: They have proven that not only the maximum diameter but also the upper part of the entire grading curve of the cement must be taken into account; specific grading criteria were proposed to be respected in order to obtain grouts of high penetrability.

In Fig. 2.6 and Table 2.2 are presented the grading criteria for all three types of sand column proposed by Miltiadou (1990), for sand 0.63–1.25 mm and 1.60–2.50 mm, and by Paillère and Guinez (1984), for sand 2.50–4.00 mm, corresponding respectively to three nominal crack widths (or "voids") to be injected.

In Table 2.2 the value $W_{nom}$ appears, indicating a representative diameter of the orifices of the sand to be injected; this $W_{nom}$ value was approximately calculated on the basis of the diameter $D_{15}$ of the sand grains in the column, by means of the expression $W_{nom} = 0.15 \cdot D_{15}$ (Dantu 1961).

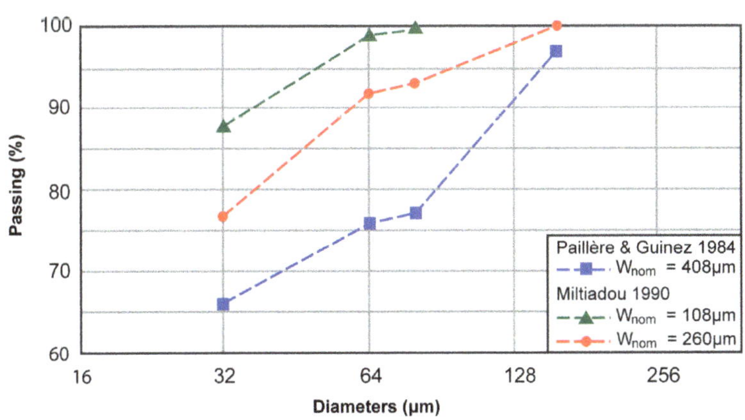


**Fig. 2.6** Penetrability criteria expressed as the lowest acceptable "position" of the upper part of the grading curves (of the coarser part of the cements), for several nominal values $W_{nom}$ of the "voids" to be grouted. Adapted from Miltiadou-Fezans and Tassios (2013)

**Table 2.2** Penetrability criteria expressed as the lowest acceptable "position" of the upper part of the grading curves (of the coarser part of the cements), for several nominal values $W_{nom}$ of the "voids" to be grouted

| Grading of the sand of the column $D_{min}/D_{max}$ in mm | Corresponding nominal crack width | Grading criterion for the solid phase of injectable grouts | | | |
|---|---|---|---|---|---|
| | | Upper limits of percentage of grains larger than | | | |
| | | 160 μm | 80 μm | 64 μm | 32 μm |
| 0.63/1.25 | ($W_{nom} = 108$ μm) | 0 | 0 | < 1 | < 12 |
| 1.60/2.50 | ($W_{nom} = 260$ μm) | 0 | < 7 | < 8 | < 23 |
| 2.50/4.00 | ($W_{nom} = 408$ μm) | 3 | < 23 | < 24 | < 34 |

Adapted from Miltiadou-Fezans and Tassios (2013)

Despite their experimental validity, the above-described criteria are connected only to three nominal values of orifices to be grouted, associated with the corresponding "sand columns" used during the pertinent experiments. For a more general approach, two criteria of broader applicability will be sought in what follows.

## 2.2.4 *Means to Improve Penetrability*

### 2.2.4.1 Use of Ultrafine Materials to Substitute Part of Basic Binder

Taking into account the aforementioned grading criteria, it is possible to improve penetrability by changing certain properties of commonly used cement grouts, when substituting part of the cement with ultrafine materials (Paillère et al. 1986, 1989; Miltiadou 1990). It has to be noted that, other pertinent experimental results, concerning fluidity and/or stability characteristics of pastes and grouts containing ultrafine materials (mainly fly ashes or silica fumes) were also already reported in the literature (Ranisch et al. 1989; Buil et al. 1987; Domone and Tank 1986; Aïtcin et al. 1984; Hobbs 1980), without however any specific reference to their consequences on penetration capacity of grouts studied).

Four different ultrafine materials contained grains no coarser than 12 μm were used by Miltiadou (1990) [one lime (L), two artificial pozzolans: silica fume (SF) and densified silica fume (DSF), and one natural pozzolan: Santorini earth (SE)]. The ultrafine materials were used to substitute a certain percentage of cement, thus preparing binary or ternary grout compositions, with or without superplasticizer. The penetrability, fluidity and stability characteristics were tested subsequently.

Based on sand column tests (sand diameter: 0.63/1.25 mm, $W_{nom} = 108$ μm) of such binary or ternary grouts, the author has observed that the following empirical "Eq. 2.2" should be respected in order to produce a penetrable blend, made of normal cements and ultrafines.

$$f_{vp} + \frac{R_{32}}{r_{32}} \geq 1 \tag{2.2}$$

where

$f_{vp}$ minimum volume percentage of ultrafines to replace a part of the cement to fullfil penetrability requirements (so that the blend material retained on 32 μm ($R_{32}$) be kept smaller than 12%).

$R_{32}$ volume percentage of blend material retained on 32 μm (suggested max value of $R_{32} = 12\%$, based on experimental results).

$r_{32}$ volume percentage of initial cement grains "retained" on 32 μm.

#### 2.2.4.2 Mixing Type and Procedure

Besides these quantitative considerations, it should be taken into account that one of the most important issues that drastically affects the penetration capacity of the hydraulic grouts, is the mixing type and procedure used for their preparation. In fact, it is very important to perfectly disperse all the grains of the solid phase, in order to ensure the wetting of their entire surfaces and their homogeneous distribution in the water, since interparticle forces are drastically affecting the penetrability, fluidity and stability of the suspensions. As these forces tend to produce flocculation of the grout, several mechanical means such as high turbulence (HT) and vibration mixing or ultrasound dispersion (US) may "break" the flocs, and establish a grout possessing considerably better injectability characteristics or pastes with a better fluidity and stability (Papadakis 1957; Bombled 1974; Roy and Asaga 1979; Legrand 1982; Tattersall and Baker 1988; Miltiadou 1990; Paillère et al. 1989; Hu 1995; Toumbakari et al. 1999; Fernandez-Altable and Casanova 2006; Bras and Henriques 2009).

The significance of using high turbulence mixing was once again confirmed by Macchi (2019), in laboratory control and in-situ application of a natural hydraulic lime and pozzolan grout composition (Miltiadou-Fezans et al. 2008) applied in a major monument in Italy, Cappella Guariniana della Sindone in Turin.

However, a broader presentation of this matter is beyond the scope of the present Chapter. What is important is to reiterate some basic research results concerning the appropriate mixing methods to be used for the preparation of hydraulic grouts. In fact, the injectability studies of hydraulic grouts, containing either plain constituent materials (cement or hydraulic lime) or various binary or ternary mixes combining them with ultrafine materials, have shown that the high turbulence mixing can be sufficient when the ultrafines, associated to the hydraulic lime or cement, are lime, natural pozzolans or silica fume; but this type of mixing is not capable to deflocculate the densified silica fume. This can be succeeded only if an ultrasound dispersion is used, as is has been shown by Paillère et al. (1989), Miltiadou (1990), and Toumbakari et al. (1999). Moreover, in the most of cases ultrasound mixing was proved to be the most satisfactory one, giving better results, concerning both penetrability and stability (Miltiadou 1990; Toumbakari et al. 1999). However, in the case of binary grouts consisting of hydraulic lime and natural pozzolan, experimental results have shown

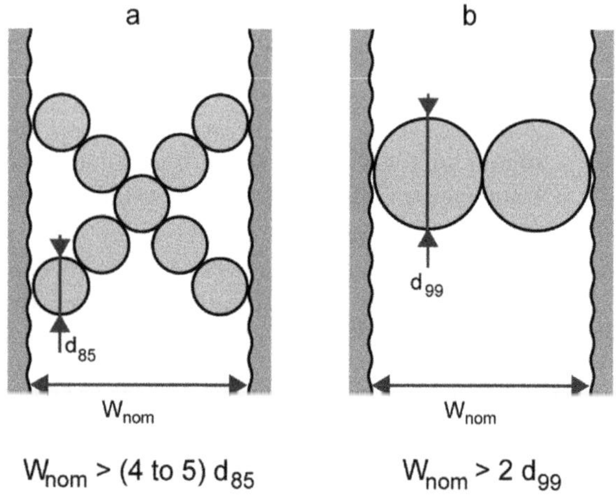


**Fig. 2.7** Schematic derivation of the maximum grains able to flow through a void (of a width equal to $W_{nom}$): **a** wall flocculation effects of relatively "frequent" large grains, should be avoided, **b** friction blocking of relatively "rare" large grains should be excluded. Adapted from Miltiadou-Fezans and Tassios (2013)

that high turbulence mixing may be a slightly better solution (Miltiadou-Fezans and Tassios 2012).[3]

In what follows, either high turbulence or ultrasound dispersion was used.

## 2.3 Simplified Model and Additional Experimental Data

Rough engineering judgement will be used in order to conceive approximate but physically meaningful criteria regarding penetrability. To this end, an estimator of the "most frequent" size of coarser grains will be first adopted—let it indicatively be the diameter $d_{85}$ (corresponding to the 85% passing).

A smooth flow of these grains through the nominal width $W_{nom}$ will be required; "wall flocculation blocking effects" (Fig. 2.7a) should be avoided. To this end, at least five flocculated grains should be considered, since, if only three grains were considered in each diagonal, the "cross" of grains thus formed would lack flexibility, and would be more easily blocked. Thus, roughly speaking, it is found that $W_{nom} > (4\ to\ 5)d_{85}$. On the other hand, although only very few grains have a size "$d_{99}$", it should also be ensured that they will not produce any "friction blocking" (Fig. 2.7b). Consequently, it should also be required that $W_{nom} > 2\ d_{99}$.

Oversimplified as they may seem, these two equations may be profitably used as a basis for further calibration by means of experiments, carried out under appropriate conditions.

Note: In the preceding analysis, only isolated grains of the solid phase were considered. It has however to be mentioned, that possible flocculation of grains may generate a larger "effective grain" size. That is why, the numerical relationships sought in this paragraph cannot be independent of the mixing procedure which is

[3] See also Sect. 3.4. Effects of mixing method on fluidity.

an extremely important factor in order to produce the necessary deflocculation of agglomerates, and ensure the dispersion, wetting and homogeneous distribution of all the grains of the solid phase of the grout. Besides, mixing procedure influences drastically the fluidity and stability of the grouts; as aforementioned both, very high turbulence and ultrasound dispersion will be considered in this book.

A first calibration of the above equations was based on the experimental penetrability results of seventeen commercial cements, covering a large spectrum of grading curves, obtained by Paillère et al. (1989) and Miltiadou (1990). The grouts were prepared using only cement and water. Based on the results of preliminary studies, a value of W/C equal to 0.75 was selected for the final comparative experimental investigation; beyond this water percentage, extensive bleeding and sedimentation of the grout was observed, without any further improvement of its fluidity.

Furthermore, in order to better identify the influence of the grading parameter alone, no superplasticizer was used. This was also imposed by the fact that for such water-to-solid ratios, the addition of a superplasticizer results very quickly in grout segregation. The mixing of the grouts was carried out using a very high turbulence screw mill, having a power of 600 W at a constant speed of 8000 rpm, for 6 min.

The penetrability of these grouts was evaluated by means of sand column tests, using two different gradings of sand, corresponding respectively to a $W_{nom}$ of 108 μm and 260 μm.

In Table 2.3 are presented the grading characteristics ($d_{85}$, $d_{95}$ and $d_{99}$) of the upper part of the grading curve of the cements tested, as well as the ratio of the $W_{nom}$ corresponding to the examined sand column to the aforementioned $d_{85}$, $d_{95}$ and $d_{99}$; bolded values indicate that the grout was able to penetrate in the sand column.

A first rough conclusion is that $W_{nom}/d_{85}$ ratios lower than 4 or 5, did not correspond to an easy penetration of the grouts.

A further refinement of this calibration is attempted using grouting penetrability results of compositions with a solid phase other than cement; grouts composed by hydraulic lime or ready mixes (based mainly on hydraulic lime) and grouts composed only by pozzolanic materials or lime, were tested. In Fig. 2.8 are presented the grading curves of the solid phase of eight hydraulic grouts; seven of them are based on hydraulic lime and one of them (Ledan TB1) is containing also cement (Kalagri et al. 2010). In Figs. 2.9 and 2.10 the grading curves of five white cements (C18–C22) and nine solid phases containing only natural pozzolanic materials or lime, are presented respectively.

Using each one of the above materials, grouts were prepared for various water to solids ratios having a fluidity factor[4] higher than the limit value corresponding to the $W_{nom}$ of the sand columns examined each time (Miltiadou-Fezans and Tassios 2012 and Chap. 3 of this book), so that to ensure sufficient fluidity. Thus, the grading characteristics were the only factor influencing the penetrability of the grouts studied. The grouts were prepared using an ultrasound dispersion mixer (20 kHz), assisted by a simultaneous mechanical stirring at 300 rpm. The mixing time was 3 min for the hydraulic lime-based grouts, pozzolan or lime grouts, and 2 min for cement grouts.

[4] For its definition, see Chap. 3.

**Table 2.3** Penetrability of seventeen normal commercial cements

| Cement n° | Grading characteristics of cements tested (μm) | | | Sand column 0.63/1.25, $D_{15} = 0.723$, $W_{nom} = 0.108$ (mm) | | | Sand column 1.60/2.50, $D_{15} = 1.735$, $W_{nom} = 0.260$ (mm) | | |
|---|---|---|---|---|---|---|---|---|---|
| | $d_{85}$ | $d_{95}$ | $d_{99}$ | $\frac{W_{nom}}{d_{85}}$ | $\frac{W_{nom}}{d_{95}}$ | $\frac{W_{nom}}{d_{99}}$ | $\frac{W_{nom}}{d_{85}}$ | $\frac{W_{nom}}{d_{95}}$ | $\frac{W_{nom}}{d_{99}}$ |
| C1 | 99.0 | 136.0 | 155 | 1.08 | 0.79 | 0.67 | 2.62 | 1.91 | 1.67 |
| C2 | 85.0 | 128.0 | 190 | 1.27 | 0.84 | 0.56 | 3.06 | 2.03 | 1.37 |
| C3 | 52.1 | 96.0 | 154 | 2.07 | 1.12 | 0.70 | 4.99 | 2.70 | 1.68 |
| C4 | 46.0 | 110.0 | 145 | 2.34 | 0.98 | 0.74 | **5.65** | **2.36** | **1.79** |
| C5 | 38.8 | 72.1 | 141 | 2.78 | 1.50 | 0.76 | **6.70** | **3.60** | **1.84** |
| C6 | 38.8 | 56.0 | 62 | 2.78 | 1.92 | 1.74 | **6.70** | **4.64** | **4.20** |
| C7 | 35.8 | 56.0 | 80 | 3.02 | 1.92 | 1.35 | **7.26** | **4.64** | **3.25** |
| C8 | 35.4 | 56.0 | 86 | 3.05 | 1.92 | 1.25 | **7.34** | **4.64** | **3.02** |
| C9 | 34.0 | 88.0 | 142 | 3.18 | 1.22 | 0.76 | **7.64** | **2.95** | **1.83** |
| C10 | 30.8 | 42.0 | 48 | 3.51 | 2.57 | 2.25 | **8.44** | **6.19** | **5.40** |
| C11 | 30.8 | 46.0 | 86 | 3.51 | 2.34 | 1.25 | **8.44** | **5.65** | **3.02** |
| C12 | 29.9 | 43.0 | 76 | 3.61 | 2.51 | 1.42 | **8.70** | **6.04** | **3.42** |
| C13 | 29.4 | 44.0 | 69 | **3.67** | **2.45** | **1.56** | **8.84** | **5.91** | **3.76** |
| C14 | 29.0 | 49.0 | 59 | **3.72** | **2.20** | **1.83** | **8.96** | **5.30** | **4.40** |
| C15 | 26.8 | 43.0 | 61 | **4.03** | **2.51** | **1.77** | **9.70** | **6.05** | **4.26** |
| C16 | 26.4 | 43.0 | 60 | **4.09** | **2.51** | **1.82** | **9.85** | **6.05** | **4.26** |
| C17 | 22.0 | 31.0 | 54 | **4.91** | **3.48** | **2.00** | **11.80** | **8.38** | **4.81** |

Bolded values indicate positive penetrability results (reworked from Miltiadou 1990)
Adapted from Miltiadou-Fezans and Tassios (2013)

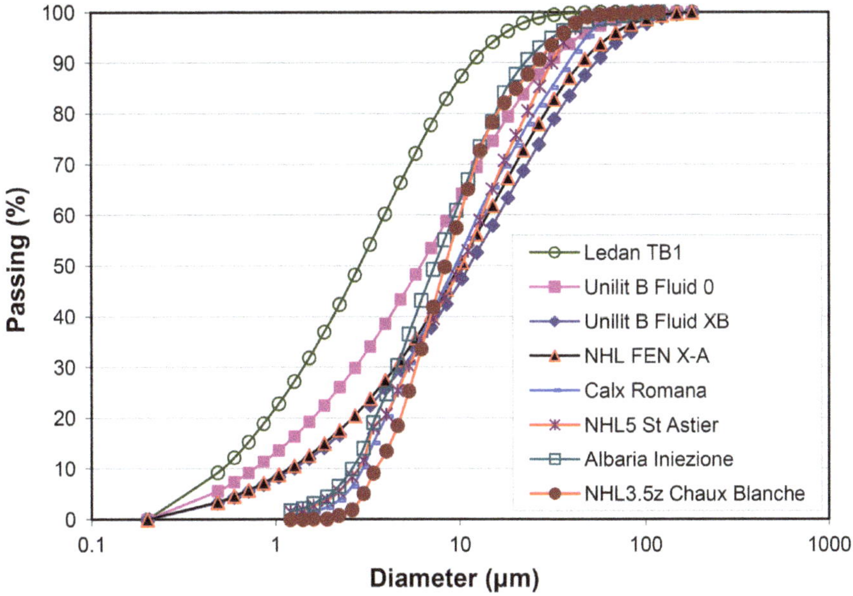


**Fig. 2.8** Grading curves of the eight grout solid phases, based mainly on natural hydraulic lime. Adapted from Miltiadou-Fezans and Tassios (2013)

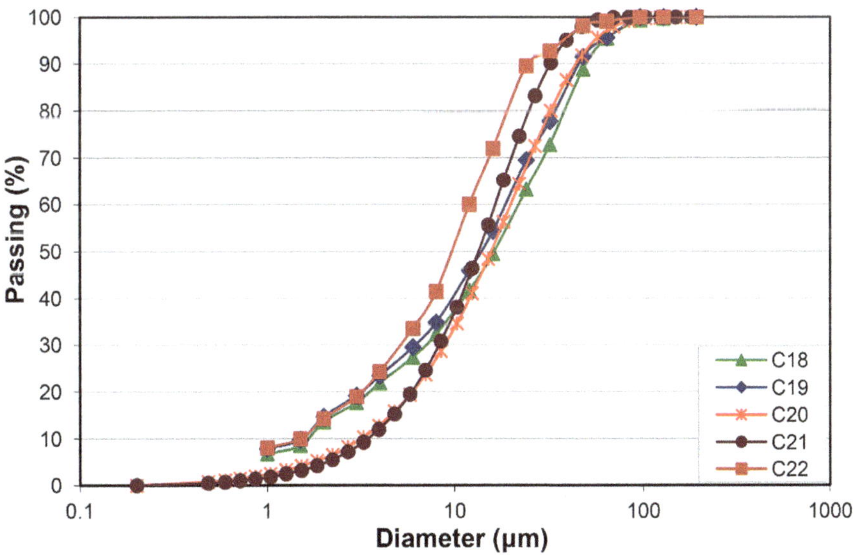


**Fig. 2.9** Grading curves of five white cements C18–C22. Adapted from Miltiadou-Fezans and Tassios (2013)

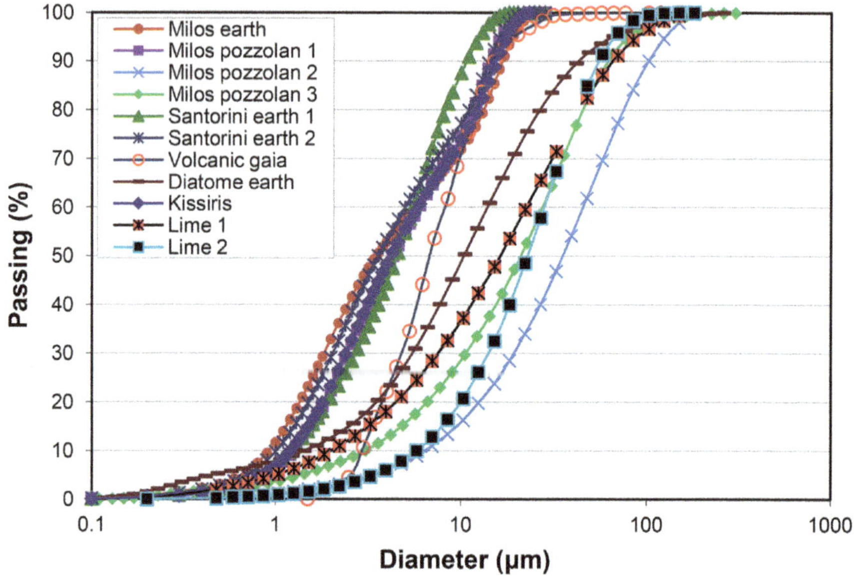


**Fig. 2.10** Grading curves of nine pozzolanic materials and two hydrated limes. Adapted from Miltiadou-Fezans and Tassios (2013)

The penetrability of these grouts was evaluated by means of sand column tests, using four different gradings of sand, corresponding respectively to a $W_{nom}$ of 108 μm, 140 μm, 175 μm and 205 μm. In Tables 2.4 and 2.5 are presented the characteristics ($d_{85}$, $d_{95}$ and $d_{99}$) of the upper part of the grading curves of the materials tested, as well as the ratio of the $W_{nom}$ to the aforementioned $d_{85}$, $d_{95}$ and $d_{99}$ values; bolded values indicate that the grout was able to penetrate to the sand column.

## 2.4 Calibration

Based on the data of Tables 2.1, 2.3, 2.4 and 2.5 the rounded figures presented in Table 2.6 may now be derived.

In suggesting these values, extreme values found in literature or in the aforementioned tests were not retained. For design purposes, the most frequent values are of practical importance; besides, the expected scattering will be reconsidered in Sect. 2.5 of this Chapter.

On Fig. 2.11 the proposed double penetrability criterion (suggested n-values, Table 2.6) is illustrated for three $W_{nom}$-categories, together with respective experimental findings (Fig. 2.6). The proposed criterion seems to be confirmed. In fact, the lowest acceptable grading curves, suggested for sand 2.50–4.00 mm by Paillère and Guinez (1984) and for sand 0.63–1.25 mm and 1.60–2.50 mm by Miltiadou (1990),

**Table 2.4** Penetrability of eight solid phases, based mainly on natural hydraulic lime and five white cements

| Grout solid phase | Grading characteristics of grout solid phase (μm) | | | Sand column 0.63/1.25, $D_{15} = 0.723$, $W_{nom} =$ 0.108 (mm) | | | Sand column 0.80/1.60, $D_{15} = 0.93$, $W_{nom} =$ 0.140 (mm) | | | Sand column 1.00/2.00, $D_{15} = 1.16$, $W_{nom} =$ 0.175 (mm) | | | Sand column 1.256/2.50, $D_{15} = 1.37$, $W_{nom} =$ 0.205 (mm) | | |
|---|---|---|---|---|---|---|---|---|---|---|---|---|---|---|---|
| | $d_{85}$ | $d_{95}$ | $d_{99}$ | $\frac{W_{nom}}{d_{85}}$ | $\frac{W_{nom}}{d_{95}}$ | $\frac{W_{nom}}{d_{99}}$ | $\frac{W_{nom}}{d_{85}}$ | $\frac{W_{nom}}{d_{95}}$ | $\frac{W_{nom}}{d_{99}}$ | $\frac{W_{nom}}{d_{85}}$ | $\frac{W_{nom}}{d_{95}}$ | $\frac{W_{nom}}{d_{99}}$ | $\frac{W_{nom}}{d_{85}}$ | $\frac{W_{nom}}{d_{95}}$ | $\frac{W_{nom}}{d_{99}}$ |
| Unilit B Fluid XB | 46.8 | 76.6 | 126.7 | 2.31 | 1.41 | 0.85 | 2.99 | 1.83 | 1.10 | 3.74 | 2.28 | 1.38 | **4.38** | **2.68** | **1.62** |
| C18 | 44.1 | 62.8 | 92.7 | 2.45 | 1.72 | 1.17 | 3.17 | 2.23 | 1.51 | 3.97 | 2.79 | 1.89 | **4.65** | **3.26** | **2.21** |
| C19 | 40.4 | 61.7 | 88.7 | 2.67 | 1.75 | 1.22 | 3.47 | 2.27 | 1.58 | 4.33 | 2.84 | 1.97 | **5.07** | **3.32** | **2.31** |
| C20 | 37.5 | 55.7 | 81.5 | 2.88 | 1.94 | 1.33 | **3.73** | **2.51** | **1.72** | **4.67** | **3.14** | **2.15** | **5.47** | **3.68** | **2.52** |
| NHL FEN X-A | 35.8 | 64.1 | 111 | 3.02 | 1.68 | 0.97 | 3.91 | 2.18 | 1.26 | **4.89** | **2.73** | **1.58** | **5.73** | **3.20** | **1.85** |
| Calx Romana | 31.2 | 48.2 | 64.4 | 3.46 | 2.24 | 1.68 | 4.49 | 2.90 | 2.17 | **5.61** | **3.63** | **2.72** | **6.57** | **4.25** | **3.18** |
| C21 | 28.1 | 39 | 54.4 | 3.84 | 2.77 | 1.99 | **4.98** | **3.59** | **2.57** | **6.23** | **4.49** | **3.22** | **7.30** | **5.26** | **3.77** |
| NHL5 St Astier | 26.8 | 38.3 | 51.4 | 4.03 | 2.82 | 2.10 | 5.22 | 3.66 | 2.72 | **6.53** | **4.57** | **3.40** | **7.65** | **5.35** | **3.99** |
| Unilit B Fluid 0 | 23.4 | 43.9 | 79.7 | 4.62 | 2.46 | 1.36 | **5.98** | **3.19** | **1.76** | **7.48** | **3.99** | **2.20** | **8.76** | **4.67** | **2.57** |
| C22 | 21.9 | 38.7 | 62.2 | **4.93** | **2.79** | **1.74** | **6.39** | **3.62** | **2.25** | **7.99** | **4.52** | **2.81** | **9.36** | **5.30** | **3.30** |
| NHL3.5z Chaux Blanche | 20.2 | 34.3 | 48.2 | **5.35** | **3.15** | **2.24** | **6.93** | **4.08** | **2.90** | **8.66** | **5.10** | **3.63** | **10.15** | **5.98** | **4.25** |
| Arbaria Iniezione | 18.1 | 31.6 | 68.2 | **5.97** | **3.42** | **1.58** | **7.73** | **4.43** | **2.05** | **9.67** | **5.54** | **2.57** | **11.33** | **6.49** | **3.01** |
| Ledan TB1 | 9.3 | 16.4 | 27.6 | **11.61** | **6.59** | **3.91** | **15.05** | **8.54** | **5.07** | **18.82** | **10.67** | **6.34** | **22.04** | **12.50** | **7.43** |

Bolded values indicate positive penetrability results
Adapted from Miltiadou-Fezans and Tassios (2013)

**Table 2.5** Penetrability of nine pozzolanic materials and two hydrated limes

| Grout solid phase | Grading characteristics of grout solid phase (μm) | | | Sand column 0.63/1.25, $D_{15}$= 0.723 $W_{nom}$= 0.108 (mm) | | | Sand column 0.80/1.60, $D_{15}$ = 0.93 $W_{nom}$ = 0.140 (mm) | | | Sand column 1.00/2.00, $D_{15}$ = 1.16 $W_{nom}$ = 0.175 (mm) | | | Sand column 1.256/2.50, $D_{15}$ = 1.37 $W_{nom}$ = 0.205 (mm) | | |
|---|---|---|---|---|---|---|---|---|---|---|---|---|---|---|---|
| | $d_{85}$ | $d_{95}$ | $d_{99}$ | $\frac{W_{nom}}{d_{85}}$ | $\frac{W_{nom}}{d_{95}}$ | $\frac{W_{nom}}{d_{99}}$ | $\frac{W_{nom}}{d_{85}}$ | $\frac{W_{nom}}{d_{95}}$ | $\frac{W_{nom}}{d_{99}}$ | $\frac{W_{nom}}{d_{85}}$ | $\frac{W_{nom}}{d_{95}}$ | $\frac{W_{nom}}{d_{99}}$ | $\frac{W_{nom}}{d_{85}}$ | $\frac{W_{nom}}{d_{95}}$ | $\frac{W_{nom}}{d_{99}}$ |
| Milos pozzolan 2 | 86.4 | 125.4 | 164.5 | 1.25 | 0.86 | 0.66 | 1.62 | 1.12 | 0.85 | 2.03 | 1.40 | 1.06 | 2.37 | 1.63 | 1.25 |
| Milos pozzolan 3 | 52.1 | 78.6 | 131.8 | 2.07 | 1.37 | 0.82 | 2.69 | 1.78 | 1.06 | 3.36 | 2.23 | 1.33 | **3.93** | **2.61** | **1.56** |
| Lime 2 | 47.5 | 66.9 | 93.8 | 2.27 | 1.61 | 1.15 | 2.95 | 2.09 | 1.49 | **3.68** | **2.62** | **1.87** | **4.32** | **3.06** | **2.19** |
| Diatome earth | 32.7 | 64.3 | 145.7 | 3.30 | 1.68 | 0.74 | 4.28 | 2.18 | 0.96 | 5.35 | 2.72 | 1.20 | **6.27** | **3.19** | **1.41** |
| Lime 1 | 32.7 | 88.6 | 138.6 | 3.30 | 1.22 | 0.78 | 4.28 | 1.58 | 1.01 | 5.35 | 1.98 | 1.26 | **6.27** | **2.31** | **1.48** |
| Milos earth | 15 | 19.0 | 22.0 | **7.20** | **5.68** | **4.91** | **9.33** | **7.37** | **6.36** | **11.67** | **9.21** | **7.95** | **13.67** | **10.79** | **9.32** |
| Volcanic gaia | 13.4 | 19.1 | 29.2 | **8.06** | **5.65** | **3.70** | **10.45** | **7.33** | **4.79** | **13.06** | **9.16** | **5.99** | **15.30** | **10.73** | **7.02** |
| Kissiris | 13.8 | 17.3 | 21.0 | **7.83** | **6.24** | **5.14** | **10.14** | **8.09** | **6.67** | **12.68** | **10.12** | **8.33** | **14.86** | **11.85** | **9.76** |
| Milos pozzolan 1 | 13.0 | 16.5 | 20.0 | **8.31** | **6.55** | **5.40** | **10.77** | **8.48** | **7.00** | **13.46** | **10.61** | **8.75** | **15.77** | **12.42** | **10.25** |
| Santorini earth 2 | 12.5 | 16.0 | 19.5 | **8.64** | **6.75** | **5.54** | **11.20** | **8.75** | **7.18** | **14.00** | **10.94** | **8.97** | **16.40** | **12.81** | **10.51** |
| Santorini earth 1 | 9.5 | 12.5 | 15.2 | **11.37** | **8.64** | **7.11** | **14.74** | **11.20** | **9.21** | **18.42** | **14.00** | **11.51** | **21.58** | **16.40** | **13.49** |

Bolded values indicate positive penetrability results
Adapted from Miltiadou-Fezans and Tassios (2013)

**Table 2.6** Suggested values for $W_{nom}/d_{85}$ and $W_{nom}/d_{99}$ on the basis of literature survey and present experimental investigation

| Ratio | Table 2.1 | Table 2.3 | Table 2.4 | Table 2.5 | Suggested rough values |
|---|---|---|---|---|---|
| $W_{nom} / d_{85} >$ | 3.00–3.75 | 3.7–5.3 | 3.7–6.0 | 3.7–4.0 | 4.0–5.0 |
| $W_{nom} / d_{99} >$ | 1.5–3.0 | 1.6–1.8 | 1.7–2.6 | 1.6–1.9 | 1.5–2.0 |

Adapted from Miltiadou-Fezans and Tassios (2013)

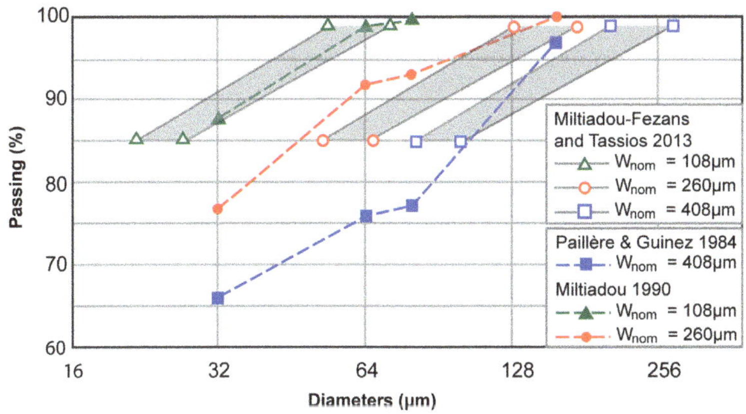


**Fig. 2.11** Experimental verification of the proposed double penetrability criterion; grey areas represent the proposed criterion. Adapted from Miltiadou-Fezans and Tassios (2013)

regarding the coarser part of cement of injectable grouts, seem to compare well with the double criterion, derived here above on the basis of experimental results with various solid phases of hydraulic grouts (grey and white cements, hydraulic limes, pozzolanic materials, limes, etc., Tables 2.4 and 2.5).

In conclusion, the rough model of Fig. 2.7 seems to be practically confirmed; the following expressions are therefore retained.

$$d_{85} < \frac{W_{nom}}{n} \quad with\ n = 4\ to\ 5, \quad d_{99} < \frac{W_{nom}}{n'} \quad with\ \ n' = 1.5\ to\ 2.0 \qquad (2.3a)$$

As an application, Table 2.7 offers five numerical examples of this criterion in selecting the coarser grain sizes of the solid phase of the grout to be used in grouting media, corresponding to given nominal size of voids or fissures.

**Additional evidence on Wnom**

As an indicative confirmation of the practicality of the values of $W_{nom}$ between 0.1 and 0.5 mm, as indicated in Table 2.7, a comparison is made here with the data of the "Association Française des Travaux en Souterrain—AFTES (1991) Recommendations on Grouting for Underground Works" (in "Tunneling and Underground Space Technology", Vol. 6, Number 4, 1991, p. 406).

**Table 2.7** Examples of penetrability conditions

| Nominal width of the pores or cracks, $W_{nom}$ (μm) | Maximum allowable values of characteristic diameters of the solid phase of the grout | |
|---|---|---|
| | $d_{85}$ (μm) | $d_{99}$ (μm) |
| 100 | 20–25 | 50–67 |
| 200 | 40–50 | 100–133 |
| 300 | 45–60 | 150–200 |
| 400 | 80–100 | 200–267 |
| 500 | 100–125 | 250–333 |

Reworked from Miltiadou-Fezans and Tassios (2013)

On Table 3 of that document (p. 408), several grout-compositions are recommended for various values of permeability coefficient "k", where

$$k^{m/s} = 10^{-2} \times D_{15}^2 \text{ (where } D_{15} \text{ in mm)} \quad [\text{AFTES (1991), p. 406}] \tag{2.4}$$

For values $10^{-3} < k < 10^{-1}$ m/s, these Recommendations for strengthening purposes suggest the use of several grouts composed of fine solid materials. On the other hand, in Sect. 2.2.2 of this Chapter the expression

$$W_{nom} \sim 0.15 \times D_{15} \tag{2.5}$$

was proposed, based on available literature. Combining "Eqs. 2.4 and 2.5", one finds

$$W_{nom} \sim 1.5\sqrt{k}$$

where "$W_{nom}$" in (mm) and "k" in (m/s).

Consequently, the $W_{nom}$-values corresponding to the k-values, permitting the use of the aforementioned grouts of fine solid materials according to "AFTES (1991)", would result as presented in Table 2.8.

These $W_{nom}$-values are of the same order of magnitude as the nominal values of the widths of effective voids presented in Table 2.7 on the basis of Sects. 2.2–2.4.

Finally, it has also to be noted that the proposed criteria are recently evaluated as follows:

- Jorne et al. (2015): "The [injectability] criteria created by Miltiadou-Fezans and Tassios (2013) showed to be most suitable".

**Table 2.8** The $W_{nom}$-values corresponding to the k-values permitting the use of grouts of fine solid materials according to "AFTES (1991)"

| $k^{m/s}$ | $10^{-3}$ | $10^{-2}$ | $10^{-1}$ |
|---|---|---|---|
| $W_{nom}^{mm}$ | 0.05 | 0.15 | 0.5 |

- Markou and Litsiou (2019): "The groutability criteria proposed by Miltiadou-Fezans and Tassios (2013) exhibit acceptable efficiency".

## 2.5 Practical Means to Improve Penetrability of Hydraulic Grouts

### 2.5.1 *A Simple Analytical Model*

Following the first criterion recommended by "Eq. 2.3", Fig. 2.12 shows how this criterion is applicable, when a percentage equal to $f_{vp}$ of a given basic binder is substituted by ultrafines. The coarser grains (diameter $d_v$) of the ultrafine material have to be in any case much smaller than the $d_{85}$ of the basic binder (usually, cement or hydraulic lime). For this simplified modelling, a straight-line DAG will be assumed to describe the grading curve of the given (unsuitable) basic binder. The ultrafines

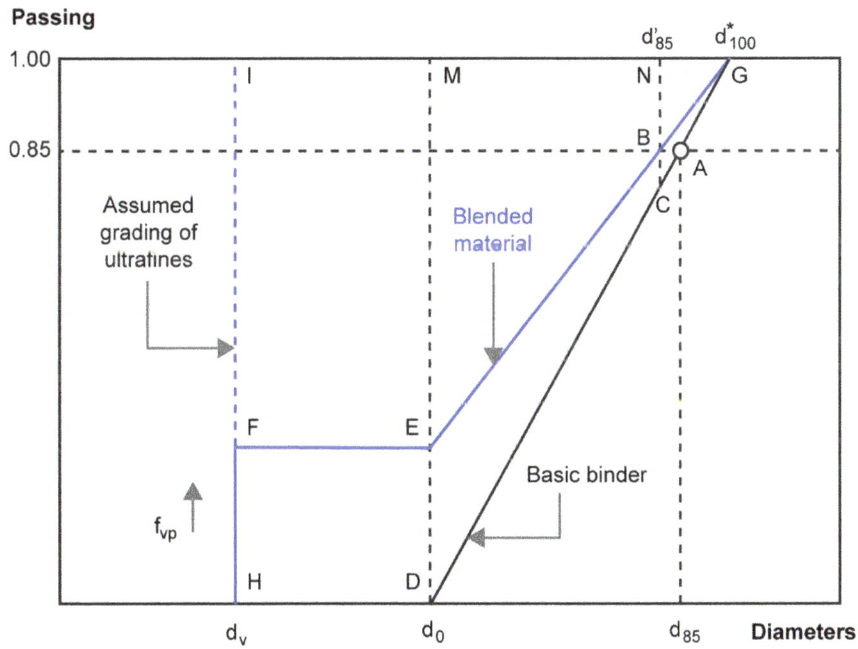


**Fig. 2.12** Thanks to a percentage "$f_{vp}$" of (quasi-monogranular) ultrafines replacing a part of the basic binder, the value of the characteristic diameter "$d'_{85}$" (point B) of the blended material is reduced, as compared to the value "$d_{85}$" (point A) of the basic binder (approximate grading curve DG). The value $d^*_{100}$ denotes the effective maximum grain size, after a linearization of the grading curve of the basic binder, such as to respect its $d_{85}$ value. Adapted from Miltiadou-Fezans and Tassios (2013)

will be considered as monogranular, with $d_v$ their maximum diameter; the vertical line HI in Fig. 2.12 represents their grading curve.

The "blended" solid phase, after the substitution of a percentage of the basic binder by ultrafines, will have a grading curve represented by the line HFEG. This blended material has now a lower characteristic diameter $d'_{85}$ at the level of 85% passing.

Simple geometry shows that:

$$\frac{AB}{GN} = \frac{CB}{CN}\left(= \frac{DE}{DM}\right)$$

where: $GN = d^*_{100} - d'_{85}$, $AB = d_{85} - d'_{85}$ and $\frac{DE}{DM} = f_{vp}$.

Consequently,

$$\frac{d_{85} - d'_{85}}{d^*_{100} - d'_{85}} = f_{vp} \text{ or } d'_{85} = \frac{d_{85} - f_{vp} \cdot d^*_{100}}{1 - f_{vp}}$$

or better, putting $f_{vp}^2 \sim 0$ (for, say, $f_{vp} < 0.25$)[5]:

$$d'_{85} = d_{85}\left[1 - \left(\frac{d^*_{100}}{d_{85}} - 1\right) \cdot f_{vp}\right] \tag{2.6}$$

where

$d'_{85}$ diameter of blended material at 85% passing
$d^*_{100}$ effective maximum grain size of basic binder
$f_{vp}$ volume percentage of ultrafines replacing a part of the basic binder to ensure penetrability.

By virtue of the first criterion of "Eq. 2.3", it should be

$$d'_{85} \leq \frac{W_{nom}}{n} \quad \text{with } n = 4 \text{ to } 5$$

The more conservative $n$-value will be used here ($n = 5$), to counterbalance the inaccuracy of "Eq. 2.6" in case of $f_{vp}$-values higher than 0.25. Thus, "Eq. 2.6" reshaped into an inequality, becomes:

$$\frac{W_{nom}}{n.d_{85}} + \left(\frac{d^*_{100}}{d_{85}} - 1\right) \cdot f_{vp} > 1 \tag{2.7}$$

Based on a statistical analysis of the 23 grading curves of binders presented in Figs. 2.8, 2.9 and 2.10, we may roughly put:

[5] It is however noted that in some cases $f_v$-values may be higher. That is an additional reason for a final trial-mix to be carried out.

$$\frac{d^*_{100}}{d_{85}} \approx 2(-0.5\ up\ to\ +1,0)$$

Consequently, "Eq. 2.7" could be further simplified as follows:

$$\min f_{vp} \approx 1 - \frac{W_{nom}}{n \cdot d_{85}}, \quad \text{with } n = 5 \tag{2.8a}$$

Despite its very approximate derivation, "Eq. 2.8a" may offer a good basis for practical applications,[6] as it will be further proved in the following.

### 2.5.2 Experimental Verification

Five normal cements were basically used by Miltiadou (1990), with large variety of grading curves (see cements C3, C6, C7, C8 and C10 in Table 2.3), in order to study the possibility to improve their injectability by substituting a part of the cement with ultrafines.

Three kinds of ultrafines were used: slaked lime powder (L), or ground natural pozzolanic material "Santorini" earth (SE), or densified silica fume (DSF). All ultrafines used had a maximum grain size roughly equal to 12 μm.

The aforementioned five cements were not suitable to penetrate the "fine discontinuities" of the sand column used 0.63/1.25 mm ($W_{nom} = 108$ μm); Table 2.9 shows that, in fact, the corresponding *n*-values are considerably lower than the recommended value 5.

For each of the previously mentioned ultrafines, its minimal percentage "$f_{vp}$" replacing a part of the basic binder was experimentally determined by the aforementioned author, in order for the blend to be successfully grouted in the sand column under consideration. A superplasticizer was also used (based on melamine formaldehyde resin), and ultrasounds were selected as mixing method. Table 2.10 shows these minimal (successful) $f_{vp}$ values needed, together with the corresponding *n*-values experimentally determined. In conclusion, it seems that "Eq. 2.8a" is experimentally confirmed; the *n*-value may indeed be taken equal to 5 for preliminary mix-designs.

**Table 2.9** Values $W_{nom}/d_{85}$ of grouts failing to penetrate the targeted sand column of $W_{nom} =$ 108 μm

| | Cement C3 $d_{85} = 52.1$ | Cement C6 $d_{85} = 38.8$ | Cement C7 $d_{85} = 35.8$ | Cement C8 $d_{85} = 35.4$ | Cement C10 $d_{85} = 30.8$ |
|---|---|---|---|---|---|
| $\frac{W_{nom}}{d_{85}} =$ | 2.07 | 2.78 | 3.02 | 3.05 | 3.50 |

Adapted from Miltiadou-Fezans and Tassios (2013)

[6] Besides, it is somehow interesting that it has the same format as the empirical "Eq. 2.2"; its denominator expressing characteristics of the initial binder, whereas its nominator is by definition "compatible" with properties of the blended material.

**Table 2.10** Minimal experimental percentages of ultrafines substituting a part of the five normal cements, so that the blend becomes injectable through a sand column 0.63/1.25 mm ($W_{nom}$ = 0.108 mm)

| Type of ultrafines | Required volume percentage $f_{vp}$ of ultrafines | | | | |
|---|---|---|---|---|---|
| | Cement 3<br>$d_{85} = 52.1$ | Cement 6<br>$d_{85} = 38.8$ | Cement 7<br>$d_{85} = 35.8$ | Cement 8<br>$d_{85} = 35.4$ | Cement 10<br>$d_{85} = 30.8$ |
| Lime (L) | 66 | 37 | 31 | 37 | 13 |
| Santorini earth (SE) | 64 | 35 | 30 | 36 | 13 |
| Densified silica fume (DSF) | 66 | 32 | 32 | 38 | 14 |
| Corresponding *n*-values in "Eq. 2.8a" | 5.9 | 4.3 | 4.3 | 4.8 | 4.0 |
| Average of *n*-values | ($n \cong 4.7 \pm 20\%$) | | | | |

Adapted from Miltiadou-Fezans and Tassios (2013)

However, despite the considerable scattering observed, it seems that n-values tend to increase with initial $d_{85}$-values, as it could be expected from the discussion on "Eq. 2.6": In fact, higher $d_{85}$-values would be needing higher percentages $f_{vp}$ of fines; thus, the approximation $f_{vp}^2 \sim 0$ may become more inaccurate, and the appropriate *n*-value should be correspondingly increased. Based on the data of Table 2.10, one could roughly suggest the following more precise expression.

$$n_{min} = 1 + 0.1\, d_{85}\ (\mu\text{m}) \tag{2.9}$$

where "$d_{85}$" denotes the 85% passing diameter of the *initial basic* binder. It may therefore be more advisable to select *n*-values as follows:

| $d_{85} \sim$[μm] | 30 | 40 | 50 |
|---|---|---|---|
| $n =$ | 4 | 5 | 6 |

In any event, such a preliminary selection of a percentage of ultrafines in the trial mix, will be subsequently checked by means of sand-column tests, and it will be appropriately corrected if necessary.

## 2.6 Conclusions Regarding the Penetrability Grading Criteria for the Grout's Solid Phase

1. Under appropriate stability and fluidity conditions (mixing method, use of superplasticizers), the penetrability of hydraulic grouts through structural media containing discontinuities, such as masonries, may be ensured if the coarser particles of the solid phase of the grouts observe the following grading rules:

$$d_{85} < \frac{W_{nom}}{n}, \quad d_{99} < \frac{W_{nom}}{2} \tag{2.3b}$$

where $n$ = a numerical factor roughly equal to 5 ($\pm 1$).

2. If the fineness of the available binder does not allow the fulfilment of the first of the aforementioned rules, the replacement of a part of the basic binder by ultrafine materials, (finer than the $d_{85}$ of the cement or hydraulic lime grains: $d_{99}^{UF} < d_{85}^{B}$) of a minimal percentage in absolute volume equal to:

$$f_{vp} \cong 1 - \frac{W_{nom}}{n \cdot d_{85}} \tag{2.8b}$$

is able to ensure penetrability.

It is reminded that this percentage being expressed in "volume", the final "masses" of ultrafine materials will be appropriately calculated.

3. If the second of the rules of "Eq. 2.3b" is not fulfilled, the only remedy is to select a finer basic binder.
4. A successful preliminary grout mix-design may thus be carried out, which however should be finally checked by means of an appropriate sand column test, representative of the effective minimal pores diameter or crack widths to be grouted.

Experimental results using a great variety of solid phases has been taken into account for the formulation of the proposed penetrability grading criteria, including all the materials commonly used in the composition of hydraulic grouts (i.e. cements, hydraulic limes, pozzolanic materials, lime, etc.); thus the proposed criteria may be very useful for the mix-design of all categories of hydraulic grouts.

On the basis of the aforementioned criteria, hydraulic grouts of high penetrability were designed and applied in various historical structures, including important monuments of World Heritage, as the Acropolis of Athens and the Byzantine Monastery of Daphni (Miltiadou-Fezans et al. 2005, 2007, 2008).

## 2.7 Estimation of a $W_{nom}$-value

In Sects. 2.2.1 and 2.2.2 the concept of a "nominal" lower value of the width of the internal discontinuities of masonry was introduced. This ($W_{nom}$) value was proved to be very useful in facilitating the mix-design, in terms of the penetrability of the grout. In the subsequent Chap. 3, the same concept is effectively used in the design of the grout in terms of its fluidity.

However, the selection of such a representative characteristic of a masonry, such as a $W_{nom}$-value, needs of further discussion, presented in Chap. 5.

## References


Aïtcin PC, Ballivy G, Parizeau R (1984) The use of condensed silica fume in grouts. Innovative cement grouting. ACI Special Publication SP-83, pp 1–18

Association Francaise des Travaux en Souterrain (AFTES) (1991) Recommendations on grouting for underground works. Tunn Undergr Space Technol **6**(4):383–461

Axelsson M, Gustafson G (2010) The penetracone, a new robust field measurement device for determining the penetrability of cementitious grouts. Tunn Undergr Space Technol 25:1–8

Axelsson M, Gustafson G, Fransson A (2009) Stop mechanism for cementitious grouts at different water-to-cement ratios. Tunn Undergr Space Technol 24:390–397

Biçer-Şimşir B, Griffin I, Palazzo Bertholon B, Rainer L (2010) Lime- based injection grouts for the conservation of architectural surfaces. Stud Conserv. https://doi.org/101179/sic.2010.55.Supplement-1.3

Binda L, Baronio G, Tiraboschi C (1993a) Repair of brick-masonries by injection of grouts: experimental research. J Struct Eng 20(1):29–44

Binda L, Modena C, Baronio G (1993b) Strengthening of masonries by injection technique. In: Proceedings of 6° NaMC, vol I, Philadelphia, pp 1–14

Binda L, Modena C, Baronio G, Gelmi A (1994) Experimental qualification of injection admixtures used for repair and strengthening of stone masonry walls. In: 10th International brick/block masonry conference, vol 2, Calgary, pp 539–548

Binda L, Modena C, Baronio G, Abbaneo S (1997) Repair and investigation techniques for stone masonry walls. Constr Build Mater 11(3):133–142

Binda L, Baronio G, Tiraboschi C, Tedeschi C (2003) Experimental research for the choice of adequate materials for the reconstruction of the Cathedral of Noto. Constr Build Mater 17:629–639

Bombled JP (1974) Rhéologie des mortiers et des bétons frais, étude de la pâte interstitielle de ciment. Revue des matériaux de construction, n°688, Mai-Juin pp 137–155

Bras A (2011) Grout optimization for masonry consolidation. PhD thesis, Universidade Nova de Lisboa, Lisbon, Portugal

Bras A, Henriques FMA (2009) The influence of the mixing procedures on the optimization of fresh grout properties. Mater Struct 42:1423–1432

Bras A, Henriques FMA (2012) Natural hydraulic lime based grouts—the selection of grout injection parameters for masonry consolidation. Constr Build Mater 26:135–144

Buil M, Paillère A-M, Hamon JP (1987) Application d'un modèle de viscosité de suspensions concentrées polydispersées à la formulation de coulis d'injection à base de liants hydrauliques. In: Proceedings of the first international RILEM conference on from materials science to construction materials engineering, vol 1, Paris, Chapman and Hall Edit., pp 349–356

Cambefort H (1977) Principes et applications de l'injection. Annales de l'ITBTP, Paris, Supp. n° 353, Série: Sols et fondations, n° 144, 23 p

Dantu P (1961) Etude mécanique d'un milieu pulvérulent formé de sphères égales de compacité maxima. 5ème Congrès International de Mécanique des sols et des Travaux de Fondations, Paris, Dunod, Publication 61-3, 10 p

Domone PL, Tank SB (1986) Use of condensed silica fume in Portland cement grouts. In: Proceedings of the second international conference on fly ash, silica fume, slag and natural pozzolans in concrete, vol 2, Madrid, Spain, pp 1231–1260

EN 1771 (2004) Products and systems for the protection and repair of concrete structures. Test methods. Determination of injectability and splitting test

Eriksson M (2002) Prediction of grout spread and sealing effect. A probabilistic approach. Doctoral thesis. Royal Institute of Technology, Stockholm, Sweden

Eriksson M, Friedrich M, Vorschulze Ch (2004) Variations in the rheology and penetrability of cement-based grouts—an experimental study. Cem Concr Res 34:1111–1119

Fernandez-Altable V, Casanova I (2006) Influence of mixing sequence and superplasticizer dosage on the rheological response of cement pastes at different temperatures. Cem Concr Res 36:1222–1230

Ferragni D, Malliet J, Di Martino S, Forti M (1982) Essais de laboratoire sur des coulis à base de ciment. In: Proceedings of the international symposium on Mortiers, ciments et coulis utilisés dans la conservation des bâtiments historiques, Ed. ICCROM, Rome, 3-6 Nov 1981, pp 185–203
Ferragni D, Forti M, Malliet J, Mora P, Teutonico JM, Torraca G (1984) Injection grouting of mural paintings and mosaics. In: Proceedings of the international symposium on adhesives and consolidants, Paris, 2–8 Sept, pp 110–116
Gustafson G, Stille H (1996) Prediction of groutability from grout properties and hydrogeological data. Tunn Undergr Space Technol 11(3):325–332
Hobbs DW (1980) The effect of pulverized-fuel ash upon the workability of cement paste and concrete. Mag Concr Res 32(113):219–226
Hu C (1995) Rhéologie des betons fluids. Etudes et Recherches des Laboratoires des Ponts et Chaussées. Série ouvrages d'art OA16, LCPC, pp 1–202
Hutchinson MT (1981) Principles of grouting II. Cem Concr Assoc Training Centre J
Johnson SJ (1958) Cement and clay grouting of foundations: Grouting with clay-cement grouts. J Soil Mech Found Eng Div ASCE 84(1):1–12
Jorne F, Henriques FMA, Baltazar LG (2015) Injection capacity of hydraulic lime grouts in different porous media. Mater Struct 48:2211–2233. https://doi.org/10.1617/s11527-014-0304-9
Kalagri A, Miltiadou-Fezans A, Vintzileou E (2010) Design and evaluation of hydraulic lime grouts for the strengthening of stone masonry historic structures. Mater Struct 43:1135–1146, doi 0.1617/s11527-009-9572-11
Laefer D, Baronio G, Anzani A, Binda L (1998) Measurement of grout injection efficacy for stone masonry walls. In: Conv. 7NAMC, vol 1, Notre Dame, USA, pp 484–496
Legrand C (1982) La structure des suspensions de ciments. Le béton hydraulique, Presses de l'École Nationale des Ponts et Chaussées, Paris, pp 99–113
Léonard ZF (1961) Grouting: clay based and chemical. Eng 864–866
Littlejohn GS (1983) Chemical grouting. South African Institution of Civil Engineers, University of Witwatersrand Johannesburg, 4–6th July 1983
Luso E, Lourenço PB (2016) Experimental characterization of commercial lime-based grouts for stone masonry consolidation. Constr Build Mater 102:216–225
Luso E, Lourenço PB (2017) Experimental laboratory design of lime based grouts for masonry consolidation. Int J Archit Heritage. https://doi.org/10.1080/15583058.2017.1354095
Macchi G (2019) Giorgio Macchi progetti. Il poligrafo casa editrice, Oct 2019, p 371
Markou IN, Litsiou AE (2019) Efficiency of soil groutability criteria for cement suspension grouting. In: Proceedings of the XVII European conference on soil mechanics and geotechnical engineering, geotechnical engineering foundation of the future, Reykjavik. https://doi.org/10.32075/17ECSMGE-2019-0197
Miltiadou AE (1990) Étude des coulis hydrauliques pour la réparation et le renforcement des structures et des monuments historiques en maçonnerie. Thèse de Doctorat de l'Ecole Nationale des Ponts et Chaussées. Pub by LCPC in Collection Etudes et recherches des Laboratoires des Ponts et Chaussées, série Ouvrages d'art, OA8 ISSN 1161-028X, LCPC, Décembre 1991, Paris, France, p 278
Miltiadou AE, Paillère A-M, Serrano JJ, Denis A, Musicas N (1990) Formulation de coulis hydrauliques pour l'injection des fissures et cavités des structures en maçonnerie dégradées. In: Proc. Inter. Conf. Structural conservation of stone masonry, 31/10–3/11/1989, Athens, Greece, Pub. ICCROM, pp 299–312
Miltiadou AE, Durville J-L, Martineau F, Massieu E, Serrano J-J (1993) Etude mécanique de mélanges cailloux-mortier-influence de l'injection de coulis. Bull liaison Lab Ponts Chaussées-183-janv.-févr. 1993, Réf. 3677:75–84
Miltiadou-Fezans A (1998) Criteria for the design of hydraulic grouts injectable into fine cracks and evaluation of their efficiency. In: Biscontin G, Moropoulou A, Erdik M, Delgado Rodrigues J (eds) Compatible materials for the protection of European cultural heritage, PACT 55 1998, Technical Chamber of Greece, pp 149–163

Miltiadou-Fezans A, Tassios TP (2003) Penetrability of hydraulic grouts in structural strengthening. In: Barsony (ed) Anniversary volume honouring Peter Lenkei, Structural research. University of Pecs, Hungary

Miltiadou-Fezans A, Tassios TP (2012) Fluidity of hydraulic grouts for masonry strengthening. RILEM Mater Struct. https://doi.org/10.1617/s11527-012-9872-8

Miltiadou-Fezans A, Tassios TP (2013) Penetrability of hydraulic grouts. RILEM Mater Struct. https://doi.org/10.1617/s11527-012-0005-1

Miltiadou-Fezans A, Papakonstantinou E, Zambas K, Panou A, Frantzikinaki K (2005) Design and application of hydraulic grouts of high injectability for structural restoration of the column drums of the Parthenon Opisthodomos. In: Brebbia CA, Torpiano A (eds) Structural studies, repairs and maintenance of heritage architecture IX. WIT Press, Southampton, pp 461–472

Miltiadou-Fezans A, Kalagri A, Delinikolas N (2007) Design of hydraulic grout and application methodology for stone masonry structures bearing mosaics and mural paintings: the case of the Katholikon of Dafni Monastery. In: Arun G (ed) Proceedings of international symposium studies on historical heritage, Antalya, Turkey, 17–21 Sept 2007, pp 649–656

Miltiadou-Fezans A, Kalagri A, Kakkinou S, Ziagou A, Delinikolas N, Zarogianni E, Chorafa E (2008) Methodology for in situ application of hydraulic grouts on historic masonry structures. The case of the Katholikon of Dafni Monastery. In: D'Ayala D, Fodde E (eds) Proceedings of the 6th international conference on structural analysis of historic construction, vol II, 2–4 July, Bath, UK. CRC Press/Balkema, Taylor and Francis Group, pp 1025–1033

Mitchell KJ (1970) In-plane treatment of foundation soils. J Soil Mech Found Div Proc ASCE SM1 73–110

Mutman U, Kavak A (2011) Improvement of granular soils by low pressure grouting. Int J Phys Sci 6(17):4311–4322

NF P 18-891:1986. Produits spéciaux destinés aux constructions en béton hydraulique. Produit pour injections dans des structures en béton. Essai d'injectabilité à la colonne de sable en milieux sec et humide

Papadakis M (1957) Recherches sur le malaxage à haute turbulence des suspensions de ciment. Publication Technique No. 82-83. Extrait de la Revue des Matériaux de Construction, CERILH, pp 1–25

Papadakis M (1959) L'injectabilité des coulis et mortiers de ciments. Revue des matériaux de construction, 531, publication technique n°11 CERILH, p 48

Papayianni I, Pachta V (2015) Experimental study on the performance of lime based grouts used in consolidating historic masonries. Mater Struct 48:2111–2121 (2015). https://doi.org/10.1617/s11527-014-0296-5

Paillère AM, Rizoulières Y (1978) Réparation des structures en béton par injection de polymères. Bull liaison Lab Ponts Chaussées (96):17–23

Paillère AM, Guinez R (1984) Recherche d'une formulation de coulis à base de liants hydrauliques pour l'injection dans les fines fissures et les cavités. Bull liaison Lab Ponts Chaussées Paris (130):51–57

Paillère AM, Serrano JJ, Buil M (1986) Possibilités offertes par l'emploi d'ultrafines siliceuses dans les coulis d'injection à base de liants hydrauliques. Bull liaison Lab Ponts Chaussées Paris (141):23–25

Paillère AM, Buil M, Miltiadou A, Guinez R, Serrano JJ (1989) Use of silica fume and superplasticizers in cement grouts for injection of fine cracks. In: Proceedings of the third international conference on use of fly ash, silica fume, slag and natural pozzolans in concrete, vol 2, Trondheim, Norway, SP-ACI, pp 1131–1157

Ranisch EH, Rostasy FS, Herschelman F (1989) Properties of cement grouts with silica fume addition for the injection of post-tensioning ducts. In: Proceedings of the third international conference on the use of fly ash, silica fume, slag and natural pozzolans in concrete, vol 2, 18–23 June 1989, Trondheim, Norway, SP 114-56, pp 1159–1171

Roy DM, Asaga K (1979) Rheological properties of cement mixes: III. The effects of mixing procedures on viscometric properties of mixes containing superplasticizers. Cem Concr Res 9(6):731–739

RILEM TC-52 RAC. Recommended tests to measure the adhesion between resin based materials and concrete

Tattersall GH, Baker PH (1988) The effect of vibration on the rheological properties of fresh concrete. Mag Concr Res 40(143):79–89

Toumbakari EE (2002) Lime-pozzolan-cement grouts and their structural effects on composite masonry walls. Ph.D. thesis, Department of Civil Engineering, Katholieke Universiteit Leuven, p 364

Toumbakari EE, Van Gemert D, Tassios TP, Tenoutasse N (1999) Effect of mixing procedure on injectability of cementitious grouts. Cem Concr Res 29:867–872

Uranjek M, Zarnic R, Bokan-Bosiljkov V, Bosiljkov V (2014) Seismic resistance of stone masonry building and effect of grouting. Gradevinar 66(8):715–726. https://doi.org/10.14256/JCE.1031.2014

Valluzzi MR (2000) Comportamento meccanico di murature storiche consolidate con materiali e tecniche a base di calce. PhD thesis, University of Trieste, p 276

Valluzzi MR (2004) Consolidamento di murature in pietra. Inezioni di calce idraulica natural. Collana Scientifica "Recifere", Ed. Gruppo Editoriale Faenza Editrice S.p.A., p 128

Valluzzi MR, Da Porto F, Modena C (2003) Grout requirements for the injection of stone masonry walls. In: A new era of building. Proceedings of the conference of ICPCM, Cairo, Egypt, 18–20 Feb

Van Gemert D, Ignoul S, Brosens K, Toumbakari EE (2015) Consolidation and strengthening of historical masonry by means of mineral grouts: grout development. Restor Build Monuments 21(1):29–45

Van Rickstal F (2000) Grout injection of masonry, scientific approach and modeling. Ph.D. Thesis, departement of Civil engineering, Katholieke Universiteit Leuven, May 2000

Yüzer N, Oktay D, Ulukaya S, Gökyğit-Arpaci EY (2015) Mechanical behaviour of two-wythe brick masonry walls injected with hydraulic lime grout. In: Proceedings of international conference REHAB 2015, 22–24 July 2015, Porto, Portugal

# Chapter 3
# Fluidity

**Abstract** This Chapter refers to *fluidity*, i.e. the second component of injectability of the grout: The grout should be able to easily flow along the sinuous paths of internal interconnected discontinuities, up to a distance larger than the distance between consecutive grout tubes. To this end, appropriate water-to-solid ratio (W/S), superplasticizer content (SP) and mixing technique should be selected. A new practical (but scientifically significant) fluidity measurement is proposed (the Fluidity Factor Test FFT), and a "fluidity factor" is defined. It is proved that the follow-up of this factor as a function of the water-to-solids ratio may reveal fundamental characteristics of the grout-composition under design. The influence of the mixing method and superplasticizer on grout's fluidity is also experimentally studied. The chapter concludes with a case study to highlight the practical use of the proposed test.

## 3.1 Introduction

As aforementioned, grouting is intended to fill voids, fissures and open joints of the masonry as a system, producing a "dendrite" (a 3D skeleton), directly contributing to the strength, tightness and durability of the masonry. To this end, the grout should first be able to pass through the "narrowest" possible width of such discontinuities and, second, overcome flow-resistances, in order to reach and fill the maximum possible internal volume of masonry voids and open joints, avoiding most of possible blockages (Miltiadou-Fezans and Tassios 2013a). Thus, depending on the characteristics of the milieu to be injected, a suitable *fluidity* should also be decided; to this end, appropriate water-to-solid ratio (W/S), superplasticizer content (SP) and mixing technique should be selected. That is why, very often, the tests proposed to check penetrability are simultaneously checking fluidity as well.

In this chapter, the basic methods for testing fluidity and the criteria and procedures to be followed in order to make a first selection of suitable fluidity of the grout as a function of the "effective minimum width of voids" of the milieu to be injected are presented. A new practical fluidity measurement is proposed [the Fluidity Factor Test (FFT)] and a "fluidity factor" is defined accordingly. An empirical formula

 
A. Miltiadou-Fezans and T. P. Tassios, *Mix-Design and Application of Hydraulic Grouts for Masonry Strengthening*, Springer Tracts in Civil Engineering,
https://doi.org/10.1007/978-3-030-85965-7_3

estimating the lower fluidity factor values for grouts to be injectable through a given milieu having a given $W_{nom}$-value (the lower nominal value of crack widths to be injected) is proposed, based on the work of Miltiadou-Fezans and Tassios (2012); thanks to this expression, the necessary fluidity ensuring injectability in a given milieu can be estimated. The influence of the mixing method and superplasticizer on grout's fluidity is also discussed.

## 3.2 Literature Survey

### 3.2.1 *Significance of Fluidity*

The significance of the concept of fluidity becomes apparent through the basic rheological equation under quasi-laminar flow conditions:

$$V = \alpha \cdot \left(\frac{\Delta p}{\ell}\right)^{\beta} \cdot d_{ch}^{\gamma} \cdot \eta^{-\delta} \tag{3.1}$$

where

| | |
|---|---|
| $V$ | velocity of flow |
| $\Delta p$ | pressure loss along a length "$\ell$" |
| $\ell$ | total length of the grouted "channel" |
| $d_{\mathrm{ch}}$ | equivalent diameter of the "channel" (the void) |
| $\eta$ | viscosity of the grout |
| $\alpha, \beta, \gamma, \delta$ | constants, depending on the roughness of the walls and the form of the cross section of the channel, as well as on the rheological properties of the grout. |

Normally, a maximum grouting pressure ($p_{max} = \Delta p$) is imposed, depending on the tensile strength of the grouted body; on the other hand, the length "$\ell$" of flow from nozzle to nozzle may be estimated, roughly though. Thus, for a given equivalent diameter "d" of the grouted voids, and a practically expected grout discharge, an order of magnitude of required viscosity of the grout could be estimated for a given velocity.

The concept of fluidity, however, cannot be unequivocally expressed by just one parameter such as the viscosity of the grout. Moreover, the constants of "Eq. 3.1" can hardly be known—even under well-defined industrial conditions. They may even be considered as pseudo-constants, since they may also depend on the variables. Besides, the rheological behaviour of the given type of grout (Newtonian or Bingham fluid) plays also an important role to its flow capacity in a given milieu.

Regarding this fundamental property of the grouts, extended research has been undertaken for the rheological characterization of grouts, and the important role of yield stress and plastic viscosity in their flow behaviour is recognized. A remarkable research corpus is available in the literature, mainly for soil and rock grouting, pre-tensioning and post-tensioning concrete, preplaced-aggregate concrete, but also for masonry grouting. Among others, the work of Jorne et al. (2015), Mohammed et al. (2014), Baltazar et al. (2012), Miltiadou-Fezans and Tassios (2012), Bras et al. (2010, 2013), Fernàndez-Altable and Casanova (2006), Le Roy and Roussel (2005), Yahia and Khayat (2003), Saric-Coric et al. (2003), Valluzzi et al. (2003), Toumbakari (2002), Mirza et al. (2002), Van Rickstal (2000), Khayat and Yahia (1998), Viseur and Barrioulet (1998), Miltiadou (1990), Ranisch et al. (1989), Domone and Tank (1986), Aïtcin et al. (1984), Littlejohn (1982), Huang (1974), Ritchie (1965), Papadakis, (1957, 1959), should be mentioned here.

On the basis of the results of these research works, the following parameters were identified as very important in grout mix design: appropriate water to solids ratio (with or without superplasticizer), particle shape and size distribution, specific surface area, chemical characteristics of the solid phase, as well as the mixing method and temperature.

### 3.2.2 *Relevant Tests for the Control of Fluidity (The Flow Cone Test)*

For in Lab determination of the rheological characteristics of Newtonian and Bingham fluids, the plastic viscosity "$\eta$" and the yield stress "$\tau_0$" are measured, using an adequate type of viscometer (coaxial cylinder viscometer, rotational plate to plate rheometer, etc.).

However, as these characteristics can only be measured in the Laboratory, other more practical estimators of fluidity that can be also determined in the worksite are frequently used, such as the "flow time" through a cone (Şahmaran et al. 2008; Nguyen et al. 2006, 2011; Roussel and Le Roy 2005; Le Roy and Roussel 2005; Yahia and Khayat 2003; Valluzzi et al. 2003; Toumbakari 2002; Mirza et al. 2002; Van Rickstal 2000; Khayat and Yahia 1998; Viseur and Barrioulet 1998; Agullò et al. 1999; Miltiadou 1990).

The test procedure to measure the flow time is roughly the following: A thoroughly cleaned cone is attached to a stand and a graduated cylinder[1] is put underneath the nozzle of the cone; after closing its nozzle, a certain volume (depending on the standard used and the procedure adopted by the research team) of the tested material is poured into the cone. The orifice is opened and a stopwatch started; the time taken for a certain quantity of grout to run into the cylinder is measured. The time of flow may be recorded at set intervals of 100 ml up to 1000 ml (or 500 ml or other quantity, depending on the standard or the researcher).

[1] The adequate volume and diameter of the cylinder, as well as the distance of its upper part from the lower part of the nozzle, are given by the corresponding standard.

The grout is characterized as "more fluid" when the flow time is "shorter", and inversely, the "longer" the flow time the "less fluid" is the grout. To calibrate the equipment and the whole procedure, the flow time of a certain volume of water is usually used.

The test has been intensively used by the researchers; specific standards have been developed to ensure for its correct application in the cases of preplaced aggregates concrete, prestressing tendons and for geotechnical works. The most common standards are the following:

1. ASTM C939-02: Standard Test Method for Flow of Grout for Preplaced-Aggregate Concrete (Flow Cone Method); a flow cone with an orifice having 12.7 mm diameter is used, since this type of grouts contain in their solid phase cement and sand.
2. ASTM D6910-04: Standard test Method for Marsh Funnel Viscosity of Clay Construction Slurries; a Marsh cone with an orifice having 4.75 mm diameter is used.
3. NF P 18-358 (1985): Adjuvants pour bétons, mortiers et coulis—Coulis courant d'injection pour précontrainte—Mesure de la fluidité et de la réduction d'eau; a Marsh cone is used with an orifice having 10 mm diameter.
4. NF P 18-507 (1992): Additions pour béton hydraulique—Besoin d'eau, contrôle de la régularité—Méthode par mesure de la fluidité par écoulement au Cône de Marsh; a Marsh cone with an orifice having 8 mm diameter is recommended.
5. EN 445: 2007: Grout for prestressing tendons—Test methods; a Marsh cone with an orifice having 10 mm diameter is used.
6. EN 12715: 2001: Execution of special geotechnical work—grouting; a Marsh cone with an orifice having 4.75 mm is used.

In Fig. 3.1 and in Table 3.1 the main characteristics of the above standardized cones are presented.

It is observed that for finer and more fluid grouts, the diameter of the nozzle is understandably reduced considerably.

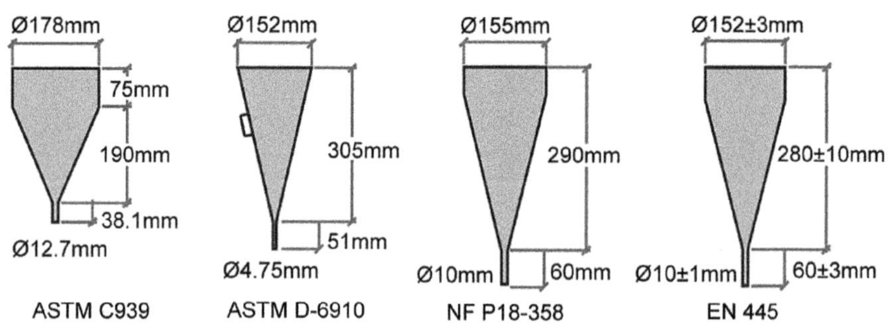


**Fig. 3.1** The most common standardized flow cones used for measuring the flow time of various types of grouts

**Table 3.1** The main characteristics of the most common standardized flow cones used for measuring the flow time of various types of grouts

| Norm | Flow time of water (s) | Volume inserted in the cone (ml) | Volume of grout flowed from cone (ml) | Nozzle internal diameter (mm) | Number of tests | Criteria |
|---|---|---|---|---|---|---|
| ASTM C939-02 | 8.0 ± 0.2 (1725 ± 5 ml of water, at 23.0 ± 0 °C) | > 1725 ± 5 | 1725 ± 5 | 12.7 | At least two tests (flow time should not differ by more than 2.49 s) | Flow time < 35 s |
| ASTM D 6910-04 | 26.0 ± 0.5 (introduce 1500 ml of water, measure the flow time of 946 ml, at 21 ± 3 °C) | 1500 | 946 | 4.75 | Several times to provide more reproducible results | |
| NF P 18-358 (1985) | | 1850 ± 50 | 1000 | 10 | Three tests after mixing ($t_0$) in an interval < 3 min | Fascicule N° 65* $t_0 \leq 25$ s $t_{30} \leq 25$ s |
| NF P 18-507 (1992) | | | | 8 | | |
| EN 445: 2007 | | 1700 ± 17 | 1000 | 10.0 | Tests are needed after mixing ($t_0$) and 30 min after mixing ($t_{30}$) | EN 447:2007 $t_0 \leq 25$ s $t_{30} \leq 25$ s $1.2\, t_0 \geq t_{30} \geq 0.8\, t_0$ |
| EN 12715: 2001 | 28.0 ± 0.5 (introduce 1500 ml, measure the flow time of 1000 ml, at 21 ± 3 °C) | 1500 | 1000 | 4.75 | | |

*Fascicule N° 65, Décret no 85-404 du 3 avril 1985. Exécution des ouvrages de génie civil en béton armé ou précontraint, Cahier des clauses techniques générales, Marchés Publics de Travaux, Ministère de l'Urbanisme, du Logement et des Transports, Ministère de l'Economie, des Finances et du Budget, France

The test however presents certain disadvantages, already mentioned in the literature (Agullò et al. 1999; Khayat and Yahia 1998). The fluid pressure acting on the nozzle is not constant, as it is practically depending on the volume of the grout each moment remaining in the cone. In other words, the pressure at the nozzle decreases during the test, starting from a certain value at the beginning of the test, and becoming equal to zero at the end of the test. Moreover, there is a non-negligible influence of the roughness of the interior surfaces of the cone on the flow time.

Nevertheless, the test has been frequently used by researchers, as it has the advantage to be simple and easy to be conducted both in laboratory and under work-site conditions, allowing for direct comparison of fluidity of grouts prepared in the lab with those prepared in situ; possible corrective measures may be taken before grouting application.

But the fact is that on the basis of the results of relevant research works (Nguyen et al. 2011; Le Roy and Roussel 2005; Roussel and Le Roy 2005; Agullò et al. 1999; Khayat and Yahia 1998, etc.), the flow time of a grout through a cone depends on a series of parameters: the volume of material placed in the cone, the volume of material for which the flow time is measured, the kind of cone's internal surfaces, the cone geometry, etc. The importance of the diameter of cone's orifice was also noted in the literature. That is why, Khayat and Yahia (1998) used a modified Marsh cone with a 4.56 mm nozzle diameter, whereas an 8 mm nozzle diameter is used by Agullò et al. (1999) and Nguyen et al. (2011).

Moreover, Le Roy and Roussel (2005) present an analytical modelling of the flow process in a Marsh cone, on the basis of the comparison of experimental results obtained using two Marsh cones differing by their nozzle diameter (10 and 8 mm) and those obtained using a viscometer. They have concluded that the Marsh Cone test may be considered adequate, mainly for grouts with Newtonian behaviour or grouts whose behaviour may be approximated by a Bingham law, *only* if they have very low plastic yield value, of the order of 20 Pa. For higher values flow time is strongly affected and "the Marsh cone accuracy diminishes very rapidly". Furthermore, they point out that the excessive debit obtained on low viscosity grouts with the nozzle diameters used in their tests (10 and 8 mm) is responsible for the discrepancy observed in their model. Thus, in order to avoid an overestimation of the viscosity, the use of a narrower nozzle is suggested in their conclusions, both for Newtonian and very diluted Bingham fluids, so that to be able to differentiate a correct grout from an over-watered one.

This conclusion is very important for masonry grouting, as the grouts to be used should have a very low viscosity and yield stress, in order to be able to penetrate and move in the fine cracks and discontinuities of the interior of masonry; this is confirmed by the literature results presented in Table 3.2.

On this basis, other authors, working on high penetrability grouts, have also considered that the measurement of flow time through a Marsh cone may have a better physical meaning when the nozzle diameter of the cone is smaller enough, for the instrument to be more sensitive. Miltiadou-Fezans et al. (1998, 2006a) have compared the results of flow time of only 100 ml of grout (out of 1000 ml inserted in the cone) using two Marsh cones, having a nozzle diameter of 4.75 mm and 3 mm respectively (Fig. 3.2). Moreover, a corelation with the sand column injectability test, corresponding to the $W_{nom}$ characterizing the structure to be injected, has been realized, since the sand column test controls penetrability and fluidity as well (see Sect. 2.2.1.2).

**Table 3.2** Plastic viscosity and yield stress of high injectability grouts for masonry strengthening

| | Plastic viscosity (mPa s) | Yield stress (Pa) |
|---|---|---|
| Miltiadou (1990) | 0.012–0.018 | 8.00–11.00 |
| Toumbakari et al. (1999) | 0.012–0.017 | 3.70–9.20 |
| Van Rickstal (2000) | 0.005–0.138 | |
| Eriksson et al. (2004) | 0.01–0.12 | 0.40–3.50 |
| Bras and Henriques (2012) | 0.10–0.15 | 0.42–1.04 |
| Baltazar et al. (2014) | 0.10–0.20 | 8.00–10.00 |
| Baltazar and Henriques (2013) | 0.04–0.15 | 0.52–32.00 |
| Jorne et al. (2015) | 0.10 | 0.63 |

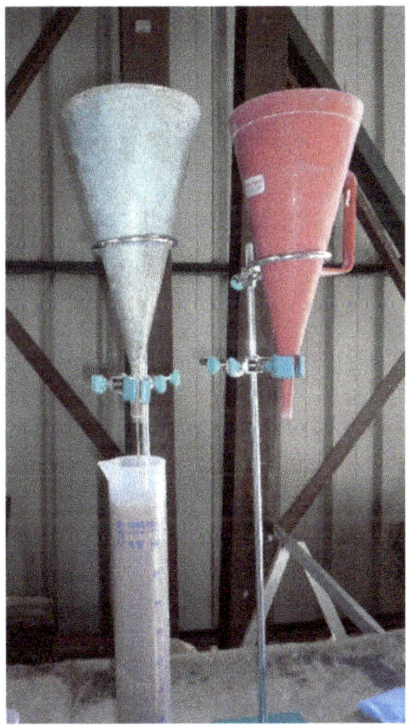

**Fig. 3.2** Two Marsh cones having a nozzle diameter of 3 mm (left) and 4.75 mm (right)

In Figs. 3.3 and 3.4 an injectability diagram is presented,[2] including the evolution of the basic grouting characteristics as a function of the W/S ratio: (a) the time $T_{36}$

[2] From the Research Reports regarding the design of the hydraulic grouts to be used (a) for the drums of the columns of Parthenon Opisthodomos (Miltiadou-Fezans et al. 1998) and (b) for the Katholikon of Daphni Monastery (Miltiadou-Fezans et al. 2006a).

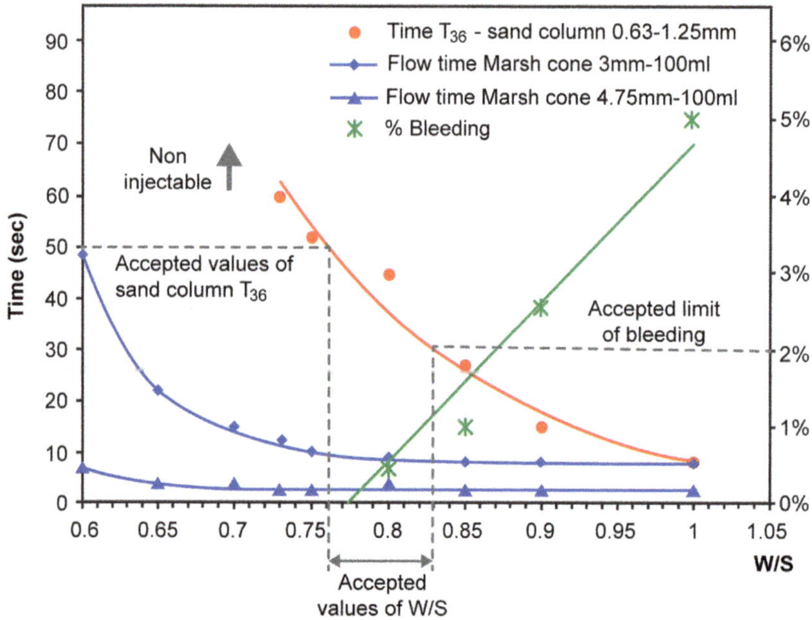


**Fig. 3.3** Injectability diagram of a White Cement C22 and Milos earth mixture at a ratio of 75/25, versus the W/S ratio

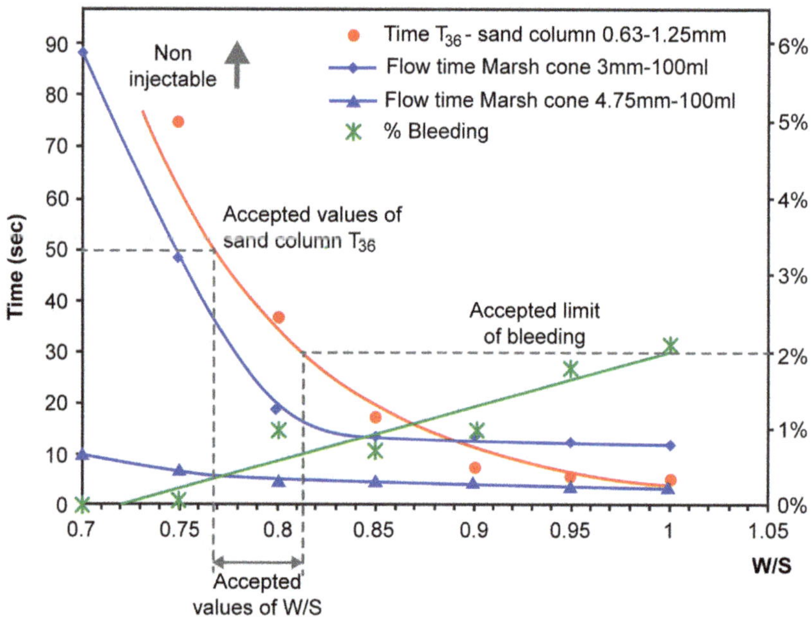


**Fig. 3.4** Injectability diagram of a 100% NHL5 grout, versus the W/S ratio

needed by an injectable grout to penetrate and fill the total height of 36 cm in a sand column (with a $W_{nom} = 108\ \mu m$ in Fig. 3.3 and 205 μm in Fig. 3.4, respectively), (b) the flow time of 100 ml of grout (out of 1000 ml inserted), using a cone with a 3.00 mm nozzle diameter, (c) the flow time of 100 ml of grout (out of 1000 ml inserted), using a cone of 4.75 mm nozzle diameter and (d) the percentage of bleeding of the grout, characterizing the stability of the suspension. It is very interesting to note that the curve corresponding to flow time of 100 ml using a 3 mm nozzle diameter Marsh cone and the sand column $T_{36}$ time, have similar qualitative characteristics. For low W/S ratio high values of Marsh cone flow time is obtained or the flow is not possible. Similarly, in the sand column test high values of $T_{36}$ is obtained or the grout is not injectable to the sand column. It is observed that a cone with a 4.75 mm orifice diameter gives results insensitive in the W/S variations. On the contrary a 3 mm orifice diameter offers the possibility to identify the rapid reduction of fluidity for low W/S values, the same way as the sand column test.

The following conclusions may be formulated:

- In the specific case of the extra fluid grouts to be injected in masonry elements, large diameter orifices of Marsh cones are not appropriate.
- The various disadvantages of the actual standard Marsh cone test, may be alleviated by measuring the time needed for only 100 $cm^3$ to flow out of an orifice of 3 mm diameter, whereas the entire quantity of grout in the cone is equal to 1000 $cm^3$. Thus, the following advantages are obtained: almost constant pressure on the nozzle, smaller influence of friction on the internal surface of the cone, and sensitivity versus variations.
- Because of the above favourable circumstances, a new test arrangement may yield a fluidity characteristic of the grout expressed in scientific terms, instead of the empirical flow time values of the standard Marsh cone. Thus, a better communicability among researches may also be achieved.

### 3.2.3 Fluidity Factor Test (FFT)

On the basis of the previous conclusions, and in order to improve the physical significance of the Marsh-cone test, Miltiadou-Fezans and Tassios (2012) have proposed the Fluidity Factor Test (FFT).

A Marsh cone with a 3 mm nozzle-diameter (appropriate for extra fluid grouts) is used, and 1000 $cm^3$ of grout are placed in the cone; the flow time $t_f$ is measured for a flow of only $Q = 100\ cm^3$ of grout to pass through (Fig. 3.5). Thus, the fluid pressure acting on the nozzle is practically kept constant during this rapid test, and the influence of the roughness of walls is minimised. The advantage of choosing a small grout-volume for reference-flow-time measurements was also indicated by Agullò et al. as early as in 1999.

The concept of a "fluidity factor" $F_l$ has been introduced:

$$F_l = \frac{Q}{A \cdot t_f} \tag{3.2}$$

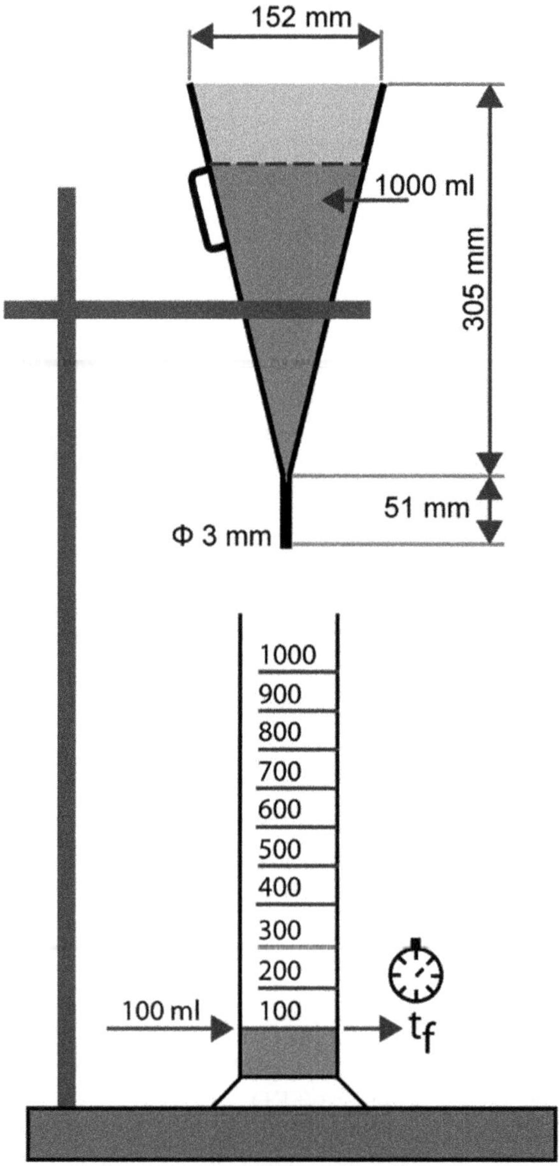


**Fig. 3.5** For the fluidity factor test (FFT), a 3 mm nozzle-diameter Marsh cone is used, a volume of 1000 $cm^3$ is placed in the cone, and the flow time $t_f$ of only Q = 100 $cm^3$ is measured

where “A” denotes the area of the cross section of the nozzle. Thus, “more fluid” grouts are characterized by higher $F_l$-values i.e., by higher “velocities” of flow (in units mm/s).

Furthermore, for each grout mix, it is strongly suggested to study a broader spectrum of water-to-solids ratio (W/S), in order to counterbalance scattering of measurements, and to exploit the possibility to collect additional rheological information on the grout mix under consideration. Thus, in Fig. 3.6, the following important characteristics may be identified, based on a rough bilinearisation of the experimental findings:

- $\left(\frac{W}{S}\right)_{o,f}$, the "absolute" *minimum* possible water-to-solids ratio.
- $\left(\frac{W}{S}\right)_{u,f}$, an ultimate value of water-to-solids ratio, which practically results in the *maximum* possible fluidity factor (for this particular composition of solids).
- max $F_l$, the maximum possible value of the fluidity factor.
- the cohesion index of the grout $C.I. = \left[\left(\frac{W}{S}\right)_{u,f} - \left(\frac{W}{S}\right)_{o,f}\right] / \max\ F_l$, which may be very useful in stability considerations (against bleeding, segregation or washout effects).

Figure 3.7 and Table 3.3 show the results of the application of the FFT method in the case of grout compositions of: (i) a Portland cement (C23) with Blaine Specific surface $S_A = 4100\ cm^2/g$, satisfying the penetrability grading rules (Sect. 2.6) for a $W_{nom} = 108\ \mu m$ (corresponding to the 0.63–1.25 mm sand column), and (ii) a Santorini earth (SE1), with max d < 20 μm and $S_A = 8596\ cm^2/g$.

Ultrasounds (US) were selected as mixing method. The US mixing consisted of an ultrasound dispersion at 20 kHz and a simultaneous mechanical stirring at 300 trs/min (Miltiadou 1990). The mixing time was 2 min for each ultrafine material (i.e., 2 min for Santorini earth, followed by 2 more minutes after the addition of the cement). As expected, rheological characteristics are approximately linear functions of the Santorini earth-content.

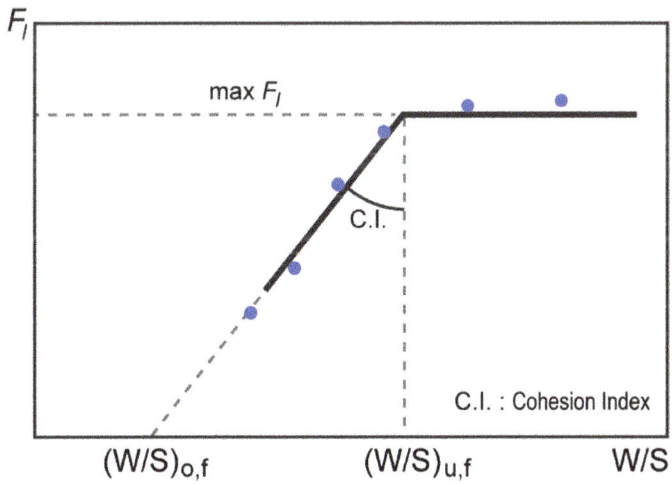


**Fig. 3.6** Several rheological characteristics of grout mixes with given composition of solids and various water percentages, assessed by means of the Fluidity Factor Test method

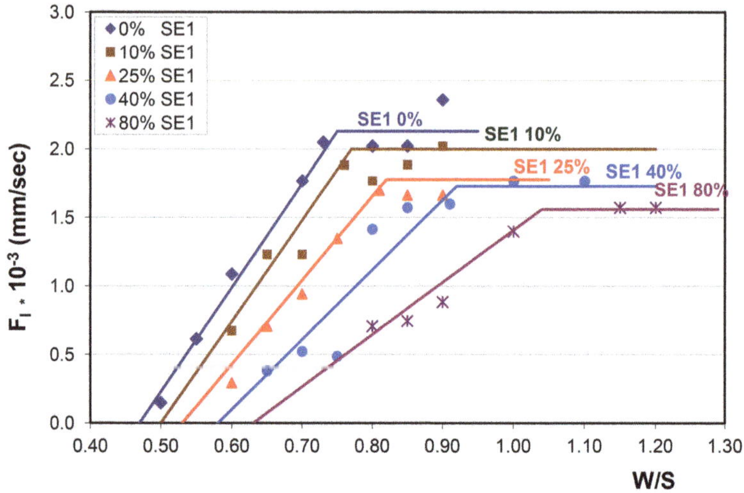


**Fig. 3.7** Fluidity factor measurements (FFT method) for several cement C23 and Santorini earth SE1 compositions (superplasticizer SP% = 0). Rough linearisations. Adapted from Miltiadou-Fezans and Tassios (2012)

**Table 3.3** Rheological characteristics of cement C23 and Santorini earth (SE1) compositions (SP% = 0), measured by means of the FFT method, and compared to limit (W/S)-values observed in the sand-column test and in stability tests

| SE1% | $S_A$* ($cm^2/g$) | Fluidity test (FFT) | | Penetration in the sand column | $10^{-3}$ max $F_l$ (mm/s) | C.I. (s/mm) (Fig. 3.6) | Initiation of bleeding | Initiation of segregation |
|---|---|---|---|---|---|---|---|---|
| | | $(W/S)_{o,f}$ | $(W/S)_{u,f}$ | $(W/S)_{sand\text{-}col}$ | | | $(W/S)_{0\text{-}bl}$ | $(W/S)_{segr}$** |
| 0 | 4100 | 0.47 | 0.75 | 0.70 | 2.13 | 131 | 0.55 | 1.48 |
| 10 | 4550 | 0.50 | 0.77 | 0.75 | 2.00 | 135 | 0.64 | 1.56 |
| 25 | 5225 | 0.53 | 0.82 | 0.75 | 1.78 | 163 | 0.72 | 1.67 |
| 40 | 5898 | 0.58 | 0.92 | 0.80 | 1.73 | 196 | 0.80 | 1.78 |
| 80 | 7697 | 0.63 | 1.04 | 1.10 | 1.56 | 264 | 0.89 | 2.08 |

*$S_A$: Calculated average fineness of the solid phase of the grout, on the basis of the measured fineness of its constituents, **$(W/S)_{segr}$: Limit water to solid phase ratio calculated using the "Eq. 4.41"
Adapted from Miltiadou-Fezans and Tassios (2012)

On Table 3.3 are also reported the values $(W/S)_{sand\text{-}col}$ for which the grout does penetrate in the aforementioned sand column[3], the experimental values $(W/S)_{0\text{-}bl}$ for which the grout starts to exhibit bleeding, and the calculated values $(W/S)_{segr}$ for which the grout is estimated that starts to exhibit segregation, obtained by means of the "Eq. 4.41" presented in Chap. 4, Sect. 4.5.4 (proposed by Miltiadou-Fezans and Tassios 2013b).

[3] References regarding the sand column test are given in Sect. 2.2.1.

It is interesting to observe that water-to-solids ratios $(W/S)_{u,f}$ leading to the maximum possible fluidity factor are somehow comparable to those enabling the grout to penetrate in the sand-column test, and they are reasonably higher of those water-to-solid values initiating bleeding.

These facts may be considered as another indication of the wealth of information and the practical usefulness of the proposed FFT-method.

A similar limit $(W/S)_u$-value ("saturation point") was indicated by Khayat and Yahia (1998), using the mini-slump method in the case of grouts of various associations of cement with fine materials of different nature, and gradually increasing superplasticizer percentage. On the other hand, Agullò et al. (1999) have also defined the maximum dosage of superplasticizer in cement pastes in terms of the saturation point, beyond which there is no significant decrease in flow time using the Marsh cone test.

Besides, the importance of a separate assessment of the cohesion of the grout was reiterated by Lombardi (1985), who has also proposed a practical method (the "plate cohesion meter") for the measurement of cohesion. In this connection, the Cohesiveness Index (C.I.) measured by means of the FFT-method (see Fig. 3.6) seems to be roughly proportional to the critical water-to-solid ratio $(W/S)_{segr}$ initiating segregation, as it was calculated using the aforementioned "Eq. 4.41". In fact, for the examined grouts blended with Santorini earth (SE1), we may say that roughly

$$C.I. \sim 90 \cdot \left(\frac{W}{S}\right)_{segr} \quad (\pm 10\%) \tag{3.3}$$

for $S_A$-values lower than 8000 $cm^2/g$.

Once again, the FFT-method seems to offer significant predictions both about fluidity and stability features of grouts.

It is worth to note that the same physically meaningful characteristics of a given mix regarding its fluidity, as described above, can also be recognized and assessed by means of other well established rheological test methods, such as e.g., the coaxial viscosimeter.

As it is known, coaxial viscosimeters are made of an inner and a rotating outer cylinder. By measuring the applied torque on one of the cylinders for various angular velocities, the plastic viscosity and the yield value of the tested material are calculated.

In Fig. 3.8 the results of such tests realized by Miltiadou (1990) are presented. They indicate that in fact, viscosity measurements disclose that there is a physically significant concept of a critical $(W/S)_u$-ratio, corresponding to a practical upper limit of fluidity (under stable conditions): It corresponds to the lowest possible limit of yield stress $\tau_o$ and to the lowest possible limit of viscosity $\eta$, in a way similar to that suggested by the proposed FFT method (see Figs. 3.6 and 3.7). This observation further confirms the physical significance of the FFT-method.

It may be said that the $F_l$ versus $W/S$ diagram (Fig. 3.6) is a mere *inversion* of the $\tau_o$ versus $W/S$ (or $\eta$ versus $W/S$) diagram (Fig. 3.8), offering similar information by means of much simpler means.

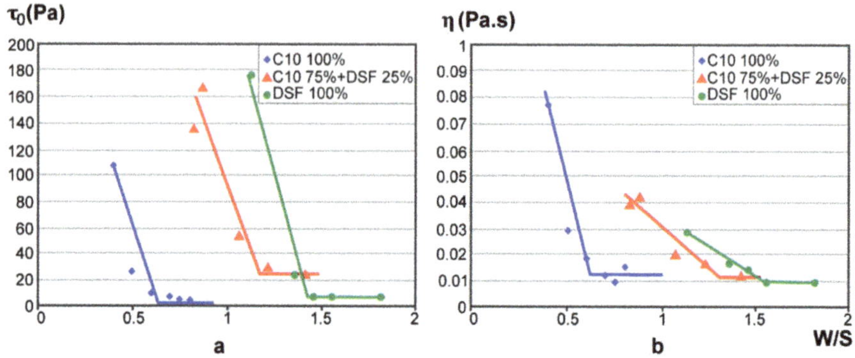


**Fig. 3.8** Evolution **a** of yield stress $\tau_0$ and **b** of viscosity $\eta$, as a function of the water content of grouts composed by cement C10 and densified silica fume (DSF) (SP% = 0). Adapted from Miltiadou-Fezans and Tassios (2012)

## 3.3 Further Significance of the Concept of Fluidity Factor—Acceptable Lower Fluidity Factor Values

As presented in Chap. 2, the sand column test is a very useful test modelling the part of the structure to be grouted and combining the checking of (i) the suitable fineness of the grains of the grout, and (ii) the appropriate fluidity of the suspension (Paillère and Rizoulières 1978). In the previous paragraphs, fluidity was considered independently of penetrability and was measured by means of viscosimeters or by the more practical Fluidity Factor Test (FFT).

In what follows, the importance of the property of fluidity per se is reiterated. To this end, a large number of grouts is considered, their solid phase being designed to penetrate a given sand column (of a given $W_{nom}$ = nominal lower value of the aperture of fissures or orifices to be injected); but it was experimentally proved that none of them was able to penetrate, unless its fluidity was higher than a specific lower limit of $F_l$-value. This limit was practically constant for all grouts of each $W_{nom}$-category.

Figures 3.9, 3.10, 3.11 and 3.12 show the values of the fluidity factor for injectable and non-injectable grouts that were tested, both using the FFT method and the sand column test. A large variety of grout compositions was examined (Miltiadou-Fezans 2000; Miltiadou-Fezans et al. 1998, 2001a, b, 2003, 2004a, b, c, 2006a, b, 2007a, b; Kalagri et al. 2010) with fineness ranging from 4000 to 10000 $cm^2/g$. The solid phase of the grouts consisted either of plain cement or plain hydraulic lime, either of various binary or ternary combinations of hydraulic lime or cement with lime or/and pozzolans.

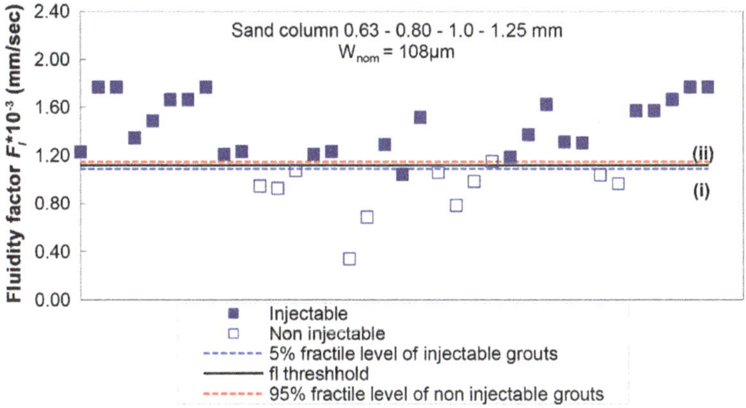


**Fig. 3.9** There is a lower fluidity-factor-value (approximately equal to $1.12 \times 10^3$ mm/s) for grout mixes to be injectable through this specific sand column with nominal value of widths of voids equal to 108 μm. Adapted from Miltiadou-Fezans and Tassios (2012)

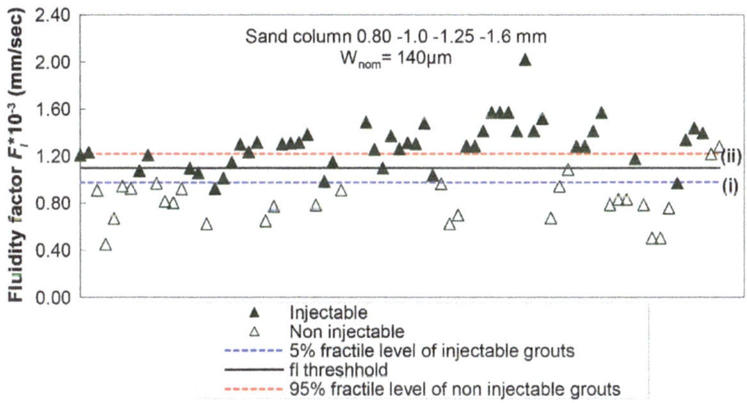


**Fig. 3.10** There is a lower fluidity-factor-value (approximately equal to $1.1 \times 10^3$ mm/s) for grout mixes to be injectable through this specific sand column with nominal value of widths of voids equal to 140 μm. Adapted from Miltiadou-Fezans and Tassios (2012)

Furthermore, the solid phase of each grout was selected so that to satisfy the penetrability grading criteria[4] for at least one of the four different $W_{nom}$-values: 108, 140, 175 and 205 μm. Thus, penetrability was ensured, and the role of fluidity was examined independently. For the preparation of the grouts the ultra sound (US) mixing method previously presented was selected. The mixing time was 2 min for

[4] These criteria were: $d_{99} < W_{nom}/2$ and $d_{85} < W_{nom}/n$, where n ~ 5 (±1), $d_{85}$ = diameter of the grout grain, corresponding to 85% passing and $d_{99}$ = "maximum" diameter of the grout grains. See also Sect. 2.6.

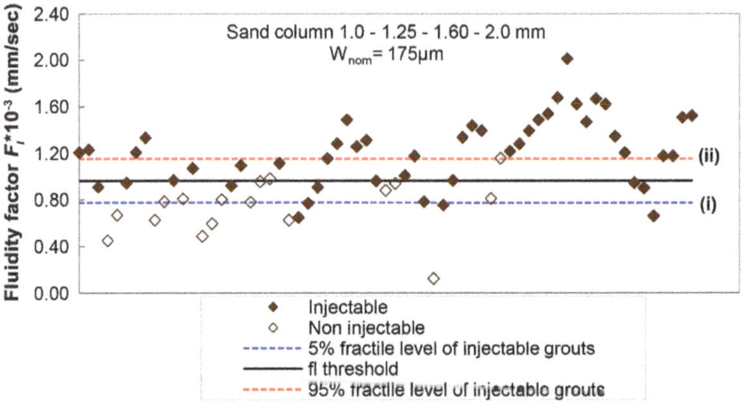


**Fig. 3.11** There is a lower fluidity-factor-value (approximately equal to 0.98 × $10^3$ mm/s) for grout mixes to be injectable through this specific sand column with nominal value of widths of voids equal to 175 μm. Adapted from Miltiadou-Fezans and Tassios (2012)

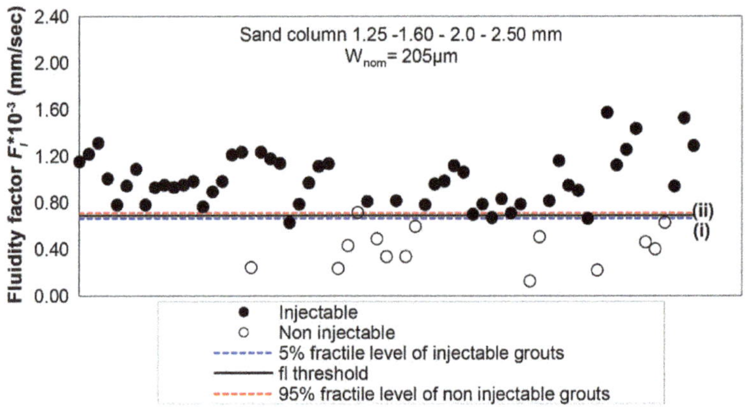


**Fig. 3.12** There is a lower fluidity-factor-value (approximately equal to 0.7 × $10^3$ mm/s) for grout mixes to be injectable through this specific sand column with nominal value of widths of voids equal to 205 μm. Adapted from Miltiadou-Fezans and Tassios (2012)

each ultrafine material, followed by 2 more minutes after the addition of cement or hydraulic lime.

Each grout was tested in the corresponding sand column test, and for every sand grading the grouts were classified in two categories: those that are injectable into the sand column without difficulties, and those that are not injectable at all. The $F_l$-threshold is defined as the average of (i) a 5%-fractile level of "injectable" (sand column test) and of (ii) a 95%-fractile level of "non injectable" grouts. However, the position of the corresponding right lines (i) and (ii) on Figs. 3.9, 3.10, 3.11 and 3.12 is approximate, and may be slightly influenced by a larger number of tests.

From these experimental results, an indicative lower value of the fluidity factor can be derived for each $W_{nom}$, as presented in Table 3.4, Fig. 3.13 and "Eq. 3.4".

Thus an "absolute" minimum value of required fluidity may be derived:

$$\min F_l \approx \left[1.2 - 45\,(W_{nom} - 0.1)^2\right]10^3[\text{mm/s}] \text{ for } 0.1 < W_{nom} < 0.2\ [\text{mm}] \quad (3.4)$$

Since, however, fluidity may be gradually reduced along the discontinuities of the grouted medium, actually selected fluidities, could be relatively higher, whenever possible, provided that stability is not jeopardized. Thus, a more uniform distribution of grout within most of the fissures along the medium will be ensured.

**Table 3.4** Lower fluidity factor values, for grouts to be injectable through sand-columns of several nominal values of widths of voids $W_{nom}$

| Grading of the sand of the column $D_{min}/D_{max}$ in mm | Corresponding nominal crack width | Lower threshold of fluidity factor in mm/s |
|---|---|---|
| 0.63/1.25 | $W_{nom} = 108\ \mu m$ | $1.12 \times 10^3$ |
| 0.80/1.60 | $W_{nom} = 140\ \mu m$ | $1.1 \times 10^3$ |
| 1.00/2.00 | $W_{nom} = 175\ \mu m$ | $0.98 \times 10^3$ |
| 1.25/2.50 | $W_{nom} = 205\ \mu m$ | $0.70 \times 10^3$ |

Adapted from Miltiadou-Fezans and Tassios (2012)

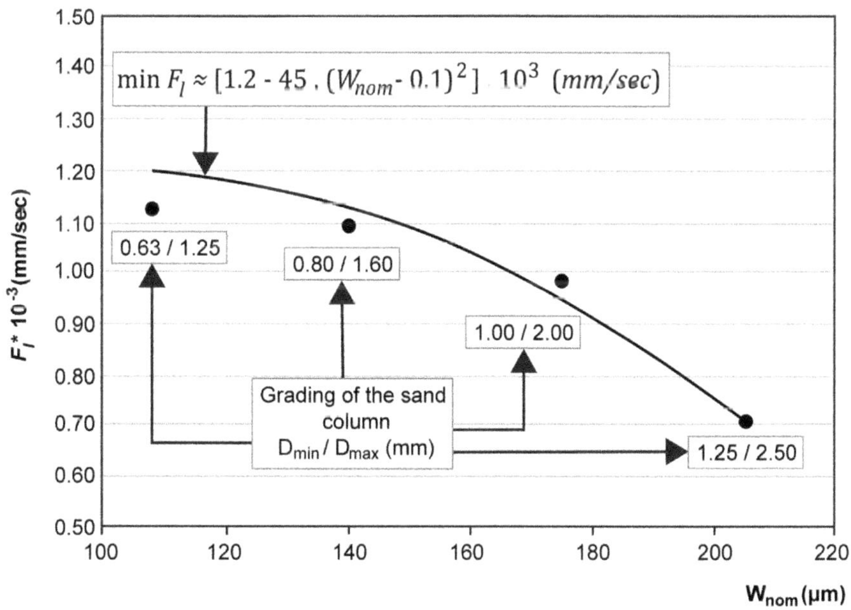


**Fig. 3.13** Lower fluidity factor values, for grouts to be injectable through sand-columns of several nominal values of widths of voids $W_{nom}$. Adapted from Miltiadou-Fezans and Tassios (2012)

## 3.4 Effects of Mixing Method on Fluidity

As it is known, the fluidity of a cement paste or hydraulic grout may be drastically affected by the mixing procedure, since interparticle forces are drastically affecting the rheology of grouts. As these forces tend to produce flocculation of the grains, several mechanical means such as high turbulence and vibration mixing or ultrasound dispersion may "break" the flocs, and establish a considerably better fluidity of the grout or cement paste (Mirza et al. 2013; Svermova et al. 2003; Toumbakari et al. 1999; Hu 1995; Miltiadou 1990; Paillère et al. 1989; Tattersall and Baker 1988; Legrand 1982; Roy and Asaga 1979; Bombled 1974; Papadakis 1957).

Indicatively, it is reported in Paillère et al. (1989) and Miltiadou (1990) that for a mix of cement and densified silica fume, (C/DSF = 0.50/0.50, with W/S = 1.00), the application of high turbulence (HT) mixing method was unable to produce a mix penetrating the "sand column" corresponding to $W_{nom} = 108\ \mu m$, despite the use of a 3.33% superplasticizer admixture. For the same mix, penetration through the aforementioned "sand column" became feasible, even with a lower SP-percentage equal to 2.33%, when the high turbulence (HT) mixing was combined with a short application of ultrasound dispersion (US). Moreover, the use of ultrasound dispersion alone accentuated this improvement: the penetration through the same sand column was achieved with only 0.83% SP-percentage.

Similarly, Fig. 3.14, based on results of Toumbakari (2002), shows approximate in-time values of rheological characteristics of a ternary grout (composed by 17.5% of lime, 52.5% of trass and 30% of cement, with a W/S = 0.85 and 1.2% of superplasticizer), after high turbulence or ultrasound mixings; the differences between the two mixing methods seem to be clear enough.

A comparative presentation of US and HT mixing is also shown in Fig. 3.15. Both mixing methods were applied on lime/pozzolan/cement (25/45/30) grouts, for various values of water to solids ratio, without superplasticizer. Apparently, when US mixing was applied, higher values of the fluidity factor were achieved (Miltiadou-Fezans and Tassios 2012).

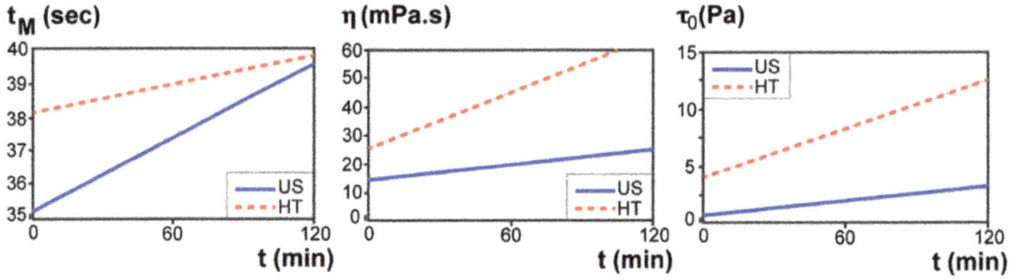


**Fig. 3.14** In-time rheological behaviour of a ternary grout, after a high turbulence (HT) mixing (rate 50 re. per sec, 9 min), or after an ultrasound mixing (US at 28 kHz, 6 min). Notation: $t_M$ = Marsh cone flow time ($d_{nozzle}$ = 14 mm), $\eta$ = apparent viscosity, $\tau_0$ = yield stress, t = time after mixing. Based on results from Toumbakari (2002). Adapted from Miltiadou Fezans and Tassios (2012)

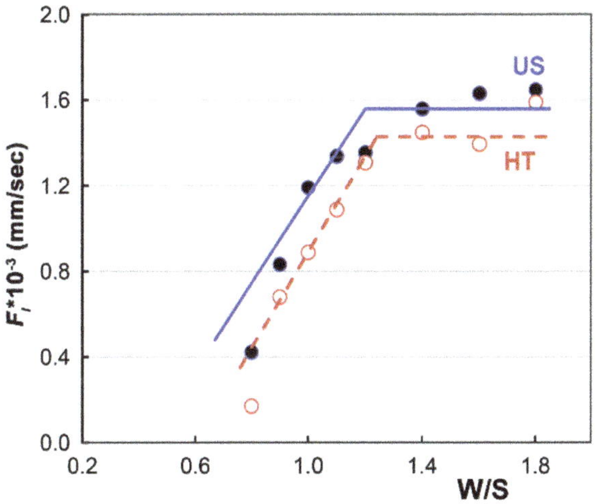


**Fig. 3.15** Comparison between US and HT mixing: fluidity factor values of lime/pozzolan/cement (25/45/30) grouts. Adapted from Miltiadou-Fezans and Tassios (2012)

Nevertheless, there are indications (see Fig. 3.16) that in the case of grouts consisting of pozzolan and hydraulic lime, US mixing does not always give better fluidity of the grout, as compared to HT mixing. The same seems true also in the presence of a superplasticizer.

In conclusion, it was proved that for practical purposes HT and US mixing may be considered as almost equivalent, except for the case of grouts containing densified silica fume. This is also indicatively confirmed in Table 3.5, presenting the injectability characteristics of a NHL5 (90%) and pozzolan (10%) grout measured (i) in the Laboratory using an ultrasound (US) mixing and (ii) in situ using a high turbulence (HT) mixing.

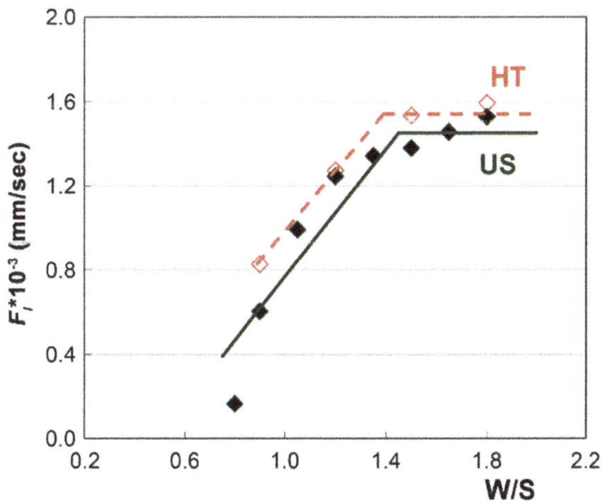


**Fig. 3.16** Comparison between US and HT mixing: fluidity factor values of hydraulic lime/pozzolan (90/10) grouts. Adapted from Miltiadou-Fezans and Tassios (2012)

**Table 3.5** Composition and injectability characteristics of a NHL5 (90%) and pozzolan (10%) grout measured (i) in the Laboratory using an ultrasound (US) mixing and (ii) in situ using a high turbulence mixing (HT)

| Grout composition | | |
|---|---|---|
| NHL5 | 90% | |
| Pozzolan | 10% | |
| Superplasticizer (1), (2) | 1% | |
| Water (1) | 80% | |
| Grout properties | In lab, US mixing | In situ, HT mixing |
| $T_{36}$ (s): sand column 1.25/2.50 mm (voids ~ 0.2–0.4 mm) | 19–22 | |
| Bleeding | < 1% | 1% |
| Fluidity Factor Fl × $10^{-3}$ > 0.7 (mm/s) | 0.88 | 0.89 |
| Flow time of 500 ml out of 1000 inserted in the Marsh cone with a 4.75 mm nozzle diameter: $t_{500\ ml,d=4.75}$ (s) | In lab | In situ |
| 0 min after mixing | 21 | 22 |
| 60 min after mixing (agitated) | 23 | 25 |
| Apparent density (gr/cm$^3$) | In lab | In situ |
| 0 min after mixing | 1.5050 | 1.4978 |
| 60 min after mixing (agitated) | 1.4986 | 1.4870 |

(1) % of the solid phase of the grout
(2) Superplasticizer based on polycarboxylic ether
Adapted from Miltiadou-Fezans et al. (2008)

## 3.5 Effect of Superplasticizers

The other important means to increase fluidity is known to be the use of appropriate superplasticizers (SP), resulting in electrostatic repulsive forces and consequently in the dispersion of solid particles (Paillère et al. 1990; Lapasin et al. 1980; Banfill 1980; Buil et al. 1989; Andersen and Roy 1987; Malhotra 1979; Daimon and Roy 1978, 1979; Rixom 1978). This is also confirmed by the results presented in Fig. 3.17 under the same conditions, as those in Fig. 3.15.

Expectedly, in the presence of a superplasticizer, an increase of fluidity is observed for the same W/S ratio, independently of the mixing method (see mean curves in Fig. 3.18).

However, the study of the several issues related to the use of such superplasticizers in grout mix design is beyond the scope of this work. What is basically needed is to reiterate the importance of only some aspects related to the use of superplasticizers; these aspects should be specifically examined when studying a mix-design:

- Chemical compatibility should be secured.

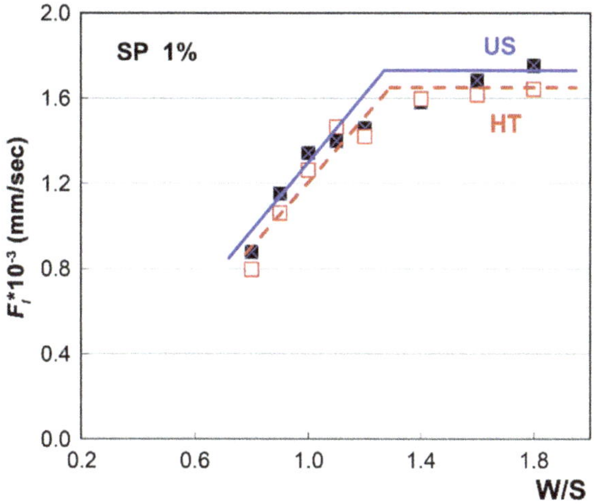


**Fig. 3.17** Fluidity factor values of lime/pozzolan/cement (25/45/30) grouts (superplasticizer 1%) for two mixing methods US and HT. Adapted from Miltiadou-Fezans and Tassios (2012)

**Fig. 3.18** Mean fluidity factor values of lime/pozzolan/cement (25/45/30) grouts (superplasticizer 0 and 1%) for US and HT mixing methods. Adapted from Miltiadou-Fezans and Tassios (2012)

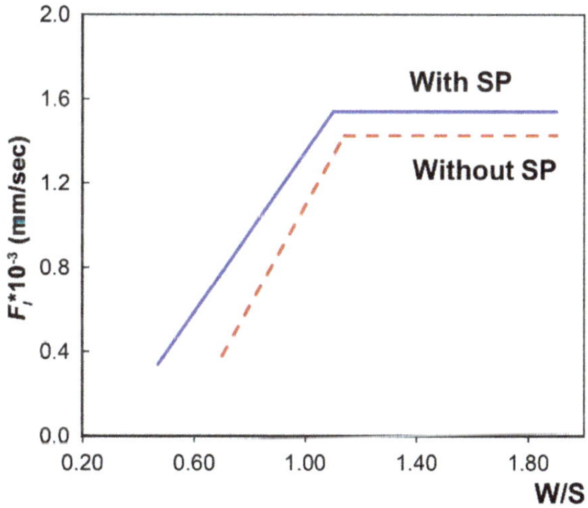


- Higher percentages of superplasticizers should be checked against possible instability (bleeding and segregation) side effects.
- Check that there is no air entrainment produced, due to over dosage of a superplasticizer; air bubbles may considerably reduce the penetrability of the grout.
- Modified bleeding and fluidity curves versus water-to-solids ratios should be made available, preferably before the final mix-design.

## 3.6 A Case Study of Practical Use of the Fluidity Factor

The East and West ranges of cells of the internal yard of Daphni Monastery, Athens, Greece (World Heritage List of UNESCO) were heavily damaged, due to 1999 Athens earthquake; injection grouting was implemented for the consolidation of stone masonry walls. The design of high injectability grouts was carried out as presented in Miltiadou-Fezans et al. (2006b). After a series of laboratory tests on several mixes, a ternary grout (low alkali white cement, lime, pozzolan) was selected and applied in situ. Fluidity factor measurements were realized, both during the design and during the in-situ application.

The mix proportions of some of the grout compositions tested, along with penetrability, fluidity and stability characteristics, are summarized in Table 3.6, for various W/S ratios. The standardized sand column test method (EN 1771) was applied to check the penetrability and fluidity, along with the standard test for stability (NF P18-359). Flow time was measured using a Marsh cone (ASTM D-6910, NF P18-358) with a 4.75 mm nozzle-diameter. Fluidity factors were also measured by means of the Fluidity Factor Test, using a Marsh cone with a 3 mm nozzle-diameter. The following criteria were set for the acceptance of grouts:

- A time limit of 50 s for the sand column penetrability test ($T_{36} < 50$ s).
- A flow time of 500 ml of grout (out of 1000 ml inserted in the cone) higher than 20 s and less than 45 s, using a Marsh cone with a nozzle diameter d = 4.75 mm ($20 < t_{500\ \text{ml},\ d=4.75\ \text{mm}} < 45$ s).
- A Fluidity Factor $F_l \geq 0.70 \cdot 10^3$ mm/s corresponding to $W_{nom} \sim 205\ \mu$m.
- A maximum acceptable limit of 5% for bleeding.

**Table 3.6** Injectability characteristics of tested grout compositions (reworked from Miltiadou-Fezans et al. 2006b)

| Grout composition weight % | $I_{67}$ | $I_{70}$ | $I_{72.5}$ | $I_{75}$ | $II_{72.5}$ |
|---|---|---|---|---|---|
| White cement | 30 | 30 | 30 | 30 | 30 |
| Lime powder | 25 | 25 | 25 | 25 | 20 |
| Pozzolan | 45 | 45 | 45 | 45 | 50 |
| Superplasticizer | 1 | 1 | 1 | 1 | 1 |
| Water/Solids | 0.67 | 0.70 | 0.725 | 0.75 | 0. 725 |
| *Injectability performance* | | | | | |
| Flow time (20 s < $t_{500\ \text{ml},d=4.75\ \text{mm}}$ < 45 s) | 50 | 29 | 25 | 24 | 27 |
| Fluidity factor [$F_l \times 10^{-3} > 0.7$ (mm/s)] | 0.39 | 0.60 | 0.81 | 0.99 | 0.62 |
| Sand column [$T_{36} < 50$ (s)] | ∞ | ∞ | 43 | 26 | 94 |
| Bleeding [<5 (%)] | 1.5 | 2.5 | 3.0 | 3.5 | 5.0 |

Adapted from Miltiadou-Fezans and Tassios (2012)

**Table 3.7** Injectability characteristics of $I_{72.5}$ grout, measured during the execution of grouting works (reworked from Miltiadou-Fezans et al. 2006b)

| In situ control (serial n°) | 1st | 2nd | 3rd | 4th | 5th |
|---|---|---|---|---|---|
| $t_{500\ ml, d=4.75\ mm}$ (s) | 23 | 22 | 25 | 23 | 27 |
| $F_l \times 10^{-3}$ (mm/s) | 0.89 | 1.03 | 0.79 | 1.06 | 0.70 |
| Bleeding (%) | 3.0 | 3.5 | 1.5 | 3.2 | 3.1 |
| Apparent density (g/cm$^3$) | 1.44 | 1.40 | 1.47 | 1.41 | 1.43 |

Adapted from Miltiadou-Fezans and Tassios (2012)

The grout composition $I_{72.5}$ was selected, since it fulfilled simultaneously all the injectability criteria and had the lowest water content (W/S = 0.725), favourable for strength and durability characteristics.

Table 3.6 shows that, although the grouts $I_{70}$, $I_{72.5}$, $I_{75}$ and $II_{72.5}$ have different W/S values, the flow time of 500 ml of grout through a Marsh cone with a nozzle diameter equal to 4.75 mm is practically the same, fulfilling the criterion set ($20 < t_{500\ ml,\ d=4.75\ mm} < 45$ s).

However, only the grouts $I_{72.5}$, and $I_{75}$ are injectable, according to the sand column test.

On the other hand, the fluidity factor of the above-mentioned grouts ($I_{70}$, $I_{72.5}$, $I_{75}$ and $II_{72.5}$) varies from $0.39 \times 10^3$ to $0.99 \times 10^3$ mm/s. It has to be noted that only the fluidity factor of the two injectable grouts ($I_{72.5}$ and $I_{75}$) fulfil the criterion: $F_l \geq [1.2 - 45 \cdot (W_{nom} - 0.1)^2] \times 10^3$ mm/s ($W_{nom}$ in mm), established in Sect. 3.3 above.

This last observation highlights the practical importance of the FFT-method, as it can be used, both in Laboratory and in the worksite, where the use of sand column tests is not possible. In Table 3.7 the injectability characteristics of grout $I_{72.5}$ (measured in situ for the quality control of the grout during the works) are presented.

Grout characteristics satisfied all the injectability design criteria. The observed (rather high) scattering in measured values was expected, due to variations of weather and worksite conditions. The grout strength measured on specimens taken in situ, was also satisfactory, as reported in Miltiadou-Fezans et al. (2006b).

## 3.7 Conclusions

In conclusion, fluidity factor measurements seem to be of basic importance in designing grout mixes, along with the determination of the absolute minimum and the threshold W/S ratios. Similarly, thanks to fluidity factor measurements, the evaluation of a Cohesiveness Index is possible, useful for the estimation of the segregability of the mix.

The proposed empirical formula "Eq. 3.4", estimating the lower fluidity factor values for grouts to be injectable through sand-columns for a given $W_{nom}$-value (the lower nominal value of discontinuities widths to be injected), is very useful; thanks to this expression, the necessary fluidity ensuring penetrability in a given milieu can be estimated, before any testing of trial mixes.

On the other hand, the FFT-method can be used (along with stability tests) not only in the Laboratory, but also at the worksite, where the use of sand column test is not possible.

## References

Agulló L, Toralles-Carbonari B, Gettu R, Aguado A (1999) Fluidity of cement pastes with mineral admixtures and superplasticizer. A study based on the Marsh Cone Test. Mater Struct 32:479–485

Aïtcin PC, Ballivy G, Parizeau R (1984) The use of condensed silica fume in grouts. Innovative cement grouting ACI Special Publication SP-83, pp 1–18

Andersen PJ, Roy DM (1987) The effects of adsorption of superplasticizers on the surface of cement. Cem Concr Res 17:805–813

ASTM C939-02. Standard test method for flow of grout for preplaced-aggregate concrete

ASTM D6910-04. Standard test method for marsh funnel viscosity of clay construction slurries

Baltazar LG, Henriques FMA (2013) The use of rheology in the study of the composition effects on the fresh behavior of hydraulic lime grouts for injection of masonry walls. Rheol Acta 52:127–138. https://doi.org/10.1007/s00397-013-0674-x

Baltazar LG, Henriques FMA, Jorne F (2012) Optimization of flow behavior and stability of superplasticized fresh hydraulic lime grouts through design of experiments. Constr Build Mater 35:838–845

Baltazar LG, Henriques FMA, Jorne F, Cidade MT (2014) Combined effect of superplasticizer, silica fume and temperature in the performance of natural hydraulic lime grouts. Constr Build Mater 50:584–597

Banfill PFG (1980) Workability of flowing concrete. Mag Concr Res 32(110):17–27

Bombled JP (1974) Rhéologie des mortiers et des bétons frais, étude de la pâte interstitielle de ciment. Revue des matériaux de construction, n°688, Mai-Juin pp 137–155

Bras A, Henriques FMA (2012) Natural hydraulic lime-based grouts—the selection of grout injection parameters for masonry consolidation. Constr Build Mater 26:135–144

Bras A, Henriques FMA, Cidade MT (2010) Effect of environmental temperature and fly ash addition in hydraulic lime grout behavior. Constr Build Mater 24:1511–1517

Bras A, Henriques FMA, Cidade MT (2013) Rheological behaviour of hydraulic lime-based grouts. Shear-time and temperature dependence. Mech Time-Depend Mater 17:223–242. https://doi.org/10.1007/s11043-012-9202-0

Buil M, Paillere AM, Musicas AM (1989) Effect of superplasticizers on the pozzolanic reactivity of silica fumes. In: Proceedings of the third international conference on fly ash, silica fume and natural pozzolans in concrete, Trondheim, Norway, supplementary papers

Daimon M, Roy DM (1978) Rheological properties of cement mixes: I. Methods, preliminary experiments, and adsorption studies. Cem Concr Res 8:753–764

Daimon M, Roy DM (1979) Rheological properties of cement mixes: II. Zeta potential and preliminary viscosity studies. Cem Concr Res 9:103–110

Domone PL, Tank SB (1986) Use of condensed silica fume in Portland cement grouts. In: Proceedings of the second international conference on fly ash, silica fume, slag and natural pozzolans in concrete, vol 2, Madrid, Spain, pp 1231–1260

EN 1771 (2004) Products and systems for the protection and repair of concrete structures. Test methods. Determination of injectability and splitting test

EN 445:2007 Grout for prestressing tendons. Test methods
EN 447:2007 Grout for prestressing tendons. Basic requirements
EN 12715:2001 Execution of special geotechnical work. Grouting
Eriksson M, Friedrich M, Vorschulze C (2004) Variations in rheology and penetrability of cement-based grouts—an experimental study. Cem Concr Res 34:1111–1119
Fascicule N° 65, Décret no 85-404 du 3 avril 1985. Exécution des ouvrages de génie civil en béton armé ou précontraint, Cahier des clauses techniques générales, Marchés Publics de Travaux, Ministère de l'Urbanisme, du Logement et des Transports, Ministère de l'Economie, des Finances et du Budget, France
Fernàndez-Altable V, Casanova I (2006) Influence of mixing sequence and superplasticizer dosage on the rheological response of cement pastes at different temperatures. Cem Concr Res 36:1222–1230
Hu C (1995) Rhéologie des bétons fluides. Etudes et Recherches des Laboratoires des Ponts et Chaussées. Série ouvrages d'art OA16, LCPC, pp 1–202
Huang WH (1974) An investigation of the fluid characteristics of the cement grouts. J Test Eval JTEVA 2:516–521
Jorne F, Henriques FMA, Baltazar LG (2015) Injection capacity of hydraulic lime grouts in different porous media. Mater Struct 48:2211–2233. https://doi.org/10.1617/s11527-014-0304-9
Kalagri A, Miltiadou-Fezans A, Vintzileou E (2010) Design and evaluation of hydraulic lime grouts for the strengthening of stone masonry historic structures. Mater Struct 43:1135–1146
Khayat KH, Yahia A (1998) Simple field tests to characterize fluidity and washout resistance of structural cement grout. Cem Concr Aggregates CCAGDP 20(1):145–156
Lapasin R, Longo V, Rajgelj S (1980) The effect of the water reducer addition on the rheological properties of cement pastes. In: Proceedings of 7th international congress on the chemistry of cement, vol III, pp VI-135–VI-140
Le Roy R, Roussel N (2005) The Marsh Cone as a viscometer: theoretical analysis and practical limits. Mater Struct 38:25–30
Legrand C (1982) La structure des suspensions de ciments. Le béton hydraulique, Presses de l'École Nationale des Ponts et Chaussées, Paris, pp 99–113
Littlejohn GS (1982) Design of cement-based grouts. In: Proceedings of the conference on grouting in geotechnical engineering. ASCE, New Orleans, Louisiana, 10–12 Feb 1982, pp 1–7
Lombardi G (1985) The role of cohesion in cement grouting of rock. In: Proceedings of 15th congress on large dams, vol 3. Com. Int. des Grands Barrages, pp 235–261
Malhotra VM (1979) Performance of superplasticized concretes that have high water to cement ratios. Superplasticizers in concrete. TRF 720, National Academy of Sciences, Washington D.C., pp 28–34
Miltiadou AE (1990) Étude des coulis hydrauliques pour la réparation et le renforcement des structures et des monuments historiques en maçonnerie. Thèse de Doctorat de l'Ecole Nationale des Ponts et Chaussées. Pub. LCPC in Collection Etudes et Recherches des Laboratoires des Ponts et Chaussées, série Ouvrages d'art, OA8 ISSN 1161-028X, LCPC, Décembre 1991, Paris, France, p 278
Miltiadou-Fezans A (2000) Design of the hydraulic grouts compositions for the structural restoration of Milos catacombs. Research report, Laboratory of Restoration Materials and Techniques, Directorate for the Restoration of Byzantine and Post-Byzantine Monuments, Hellenic Ministry of Culture and Sports (in Greek)
Miltiadou-Fezans A, Tassios TP (2012) Fluidity of hydraulic grouts for masonry strengthening. RILEM Mater Struct. https://doi.org/10.1617/s11527-012-9872-8
Miltiadou-Fezans A, Tassios TP (2013a) Penetrability of hydraulic grouts. RILEM Mater Struct. https://doi.org/10.1617/s11527-012-0005-1
Miltiadou-Fezans A, Tassios TP (2013b) Stability of hydraulic grouts for masonry strengthening. RILEM Mater Struct. https://doi.org/10.1617/s11527-012-0003-3
Miltiadou-Fezans A, Papakonstantinou E, Zambas K, Panou A, Frantzikinaki K (1998) Design of hydraulic grouts of high injectability for the structural restoration of the column drums of the

Parthenon Opisthodomos. Research report, Acropolis Restoration Service, Hellenic Ministry of Culture and Sports

Miltiadou-Fezans A, Anagnostopoulou S, Gatzionis K (2001a) Investigations for the design of grouts for the structural restoration of Eptapyrgion Fortress in Thessaloniki. Research report, Laboratory of Restoration Materials and Techniques, Directorate for the Restoration of Byzantine and Post-Byzantine Monuments, Hellenic Ministry of Culture and Sports, Dec 2001 (in Greek)

Miltiadou-Fezans A, Anagnostopoulou S, Gatzionis K, Kordoulas A (2001b) Laboratory and in situ tests for the design of mortars and grouts for the structural restoration of the NE range of Cells of Hosios Loukas Monastery in Boeotia. Research report, Laboratory of Restoration Materials and Techniques, Directorate for the Restoration of Byzantine and Post-Byzantine Monuments, Hellenic Ministry of Culture and Sports, Nov 2001 (in Greek)

Miltiadou-Fezans A, Kalagri A, Kordoulas A (2003) Technical report of grouting interventions in the building SPAP II in Olympia and in the Municipal market of Pyrgos in Heli, Peloponnese. Research report, Laboratory of Restoration Materials and Techniques, Directorate for the Restoration of Byzantine and Post-Byzantine Monuments, Hellenic Ministry of Culture and Sports, Oct 2003 (in Greek)

Miltiadou-Fezans A, Kalagri A, Anagnostopoulou S (2004a) Research and investigations for the design of grouts, mortars and pastes for the restoration of the Ancient Theatre of Megalopolis. Research report, Directorate for Technical Research on Restoration of the Hellenic Ministry of Culture and Sports, Oct 2004 (in Greek)

Miltiadou-Fezans A, Kalagri A, Anagnostopoulou S (2004b). Research and investigations for the design of grouts for the restoration of the Ancient Theatre of Epidaure. Research report, Directorate for Technical Research on Restoration of the Hellenic Ministry of Culture and Sports, Oct 2004 (in Greek)

Miltiadou-Fezans A, Kalagri A, Anagnostopoulou S (2004c) Research and investigations for the design of grouts for the restoration of ranges of cells and the East gate of Daphni Monastery. Research report, Directorate of Technical Research on Restoration for the Hellenic Ministry of Culture and Sports, 2004 (in Greek)

Miltiadou-Fezans A, Kalagri A, Triantafyllou M (2006a) Research report for the design of hydraulic grouts for the structural restoration of the Katholikon of Dafni Monastery. Comparative study of hydraulic lime-based grouts. Research report, Directorate for Technical Research on Restoration, Hellenic Ministry of Culture, p 68 (in Greek)

Miltiadou-Fezans A, Anagnostopoulou S, Kalagri A (2006b) Restoration mortars and grouts for consolidation of the Cells of internal yard of Dafni Monastery. In: Proceedings of the 1st conference on restoration of the Society for research and promotion of scientific restoration of monuments (ETEPAM), Thessaloniki (in Greek)

Miltiadou-Fezans A, Kalagri A, Delinikolas N (2007a) Design of hydraulic grout and application methodology for stone masonry structures bearing mosaics and mural paintings: the case of the Katholikon of Dafni Monastery. In: Arun G (ed) Proceedings of international symposium studies on historical heritage, Antalya, Turkey, 17–21 Sept 2007, pp 649–656

Miltiadou-Fezans A, Kalagri A, Savvidou M (2007b) Investigations for the design of grouts for the structural restoration of the Church of the Assumption of the Virgin Mary in Episkopi Tegeas, Arkadia, Peloponnese. Research report, Directorate for Technical Research on Restoration of the Hellenic Ministry of Culture and Sports, 2004 (in Greek)

Miltiadou-Fezans A, Kalagri A, Kakkinou S, Ziagou A, Delinikolas N, Zarogianni E, Chorafa E (2008) Methodology for in situ application of hydraulic grouts on historic masonry structures. The case of the Katholikon of Dafni Monastery. In: D'Ayala, D., Fodde, C. R. C. (eds.) Proceedings of the 6th international conference on structural analysis of historic construction, vol II, 2–4 July, Bath, UK, Press. Balkema, Taylor and Francis Group, pp 1025–1033

Mirza J, Mirza MS, Roy V, Saleh K (2002) Basic rheological and mechanical properties of High-volume fly ash grouts. Constr Build Mater 16:353–363

Mirza J, Saleh K, Langevin MA, Mirza S, Bhutta MAR, Md TM (2013) Properties of microfine grouts at 4 °C, 10 °C and 20 °C. Constr Build Mater 47:1145–1153
Mohammed HM, Pusch R, Knutsson S, Hellström G (2014) Rheological properties of cement-based grouts determined by different techniques. Engineering 6:217–229
NF P18-358 (1985) Adjuvants pour bétons, mortiers et coulis. Coulis courant d'injection pour précontrainte. Mesure de la fluidité et de la réduction d'eau
NF P18-359 (1985) Adjuvants pour bétons, mortiers et coulis. Coulis courants d'injection pour précontrainte. Mesure de l'exsudation (stabilité)
NF P18-507 (1992) Additions pour béton hydraulique- Besoin d'eau, contrôle de la régularité. Méthode par mesure de la fluidité par écoulement au Cône de Marsh
Nguyen VH, Remond S, Gallias JL, Bigas JB, Muller P (2006) Flow of Herschel-Bulkley fluids through the Marsh cone. J Nonnewton Fluid Mech 139:128–134
Nguyen VH, Remond S, Gallias JL (2011) Influence of cement grouts composition on rheological behavior. Cem Concr Res 41:292–300
Paillère A-M, Rizoulières Y (1978) Réparation des structures en béton par injection de polymères. Bull liaison Lab Ponts Chaussées (96):17–23
Paillère AM, Buil M, Miltiadou A, Guinez R, Serrano JJ (1989) Use of silica fume and superplasticizers in cement grouts for injection of fine cracks. In: Proceedings of the third international conference on use of fly ash, silica fume, slag and natural pozzolans in concrete, vol 2, Trondheim, Norway, SP-ACI, pp 1131–1157
Paillère AM, Serrano JJ, Grimaldi G (1990) Influence du dosage et du mode d'introduction des superplastifiants sur le maintien de la maniabilité optimale des bétons à hautes performances avec et sans fumée de silice. Bull liaison Lab Ponts Chaussées 170:37–45
Papadakis M (1957) Recherches sur le malaxage "a haute turbulence" des suspensions de ciment. Publication Technique No. 82-83, Extrait de la Revue des Matériaux de Construction, CERILH, pp 1–25
Papadakis M (1959) L'injectabilité des coulis et mortiers de ciments. Revue des matériaux de construction, 531, publication technique No. 11 du CERILH, 48 p
Ranisch E-H, Rostasy FS, Herschelman F (1989) Properties of cement grouts with silica fume addition for the injection of post-tensioning ducts. In: Proceedings of the third international conference on the use of fly ash, silica fume, slag and natural pozzolans in concrete, vol 2, 18–23 June 1989, Trondheim, Norway, SP 114-56, pp 1159–117
Ritchie AGB (1965) The rheology of cement grout. In: Cement and lime manufacture, pp 9–1
Rixom M (1978) Chemical admixtures for concrete. London E. and F.N. Spon
Roussel N, Le Roy R (2005) The Marsh Cone: a test or a rheological apparatus? Cem Concr Res 35:823–830
Roy DM, Asaga K (1979) Rheological properties of cement mixes: III. The effects of mixing procedures on viscometric properties of mixes containing superplasticizers. CemConcr Res 9(6):731–739
Şahmaran M, Özkan N, Keskin SB, Uzal B, Yaman IÖ, Erdem TK (2008) Evaluation of natural zeolite as viscosity-modifying agent for cement-based grouts. Cem Concr Res 38:930–937
Saric-Coric M, Khayat KH, Tagnit-Hamou A (2003) Performance characteristics of cement grouts made with various combinations of high-range water reducer and cellulose-based viscosity modifier. Cem Concr Res 33:1999–2008
Svermova L, Sonebi M, Bartos JM (2003) Influence of mix proportions on rheology pf cement grouts containing limestone powder. Cem Concr Compos 25:737–749
Tattersall GH, Baker PH (1988) The effect of vibration on the rheological properties of fresh concrete. Mag Concr Res 40(143):79–89
Toumbakari E-E (2002) Lime-pozzolan-cement grouts and their structural effects on composite masonry walls. Ph.D Thesis, Departement of Civil Engineering, Katholieke Universiteit Leuven, Sept 2002
Toumbakari EE, Van Gemert D, Tassios TP, Tenoutasse N (1999) Effect of mixing procedure on injectability of cementitious grouts. Cem Concr Res 29:867–872

Valluzzi MR, da Porto F, Modena C (2003) Grout requirements for the injection of stone masonry walls. In: International conference on performance of construction materials—a new era of building, 18–20 Feb 2003, Cairo, Egypt

Van Rickstal F (2000) Grout injection of masonry, scientific approach and modeling. Ph.D Thesis, Departement of Civil Engineering, Katholieke Universiteit Leuven, May 2000, p 194

Viseur V, Barrioulet M (1998) Critères d'injectabilité de coulis de ciments ultrafins. Mater Struct/Matér Constr 31:393–399

Yahia A, Khayat KH (2003) Applicability of rheological models to high-performance grouts containing supplementary cementitious materials and viscosity enhancing admixture. Mater Struct/Matér Constr 36:402–412

# Chapter 4
# Stability

**Abstract** This Chapter deals with the third basic component of grout's injectability, i.e., the *stability* of the suspension against bleeding and segregation. The grout should keep its homogeneity up to its setting, as much as possible. When the bleeding remains lower than a certain limit, the grout can be used; otherwise, the grout may be unable to flow in and adhere to the internal surfaces of the discontinuities. Similarly, because of inadequate internal cohesion, larger solid particles of the gout may quickly settle, resulting in non-homogeneous layered structure of the mix. Excessive bleeding and segregation may produce blockage of the flow and sudden dangerous increase of pressure. The most prevailing parameters shaping stability characteristics, are water content and percentage of ultrafine materials. After a brief literature survey, an over-simplified predictive model of bleeding is firstly proposed and then its validity is confirmed using the results of an experimental study. The role of superplasticizers is also discussed. In both cases, with and without superplasticizer, semi-empirical formulae are proposed, that are useful for the design of a grout composition. Finally, the chapter presents experimental results demonstrating the role of water and super-plasticizer content in the appearance of segregation; some empirical formulae are also proposed for the estimation of the critical water content initiating segregation.

## 4.1 Introduction

Besides its sufficient penetrability and fluidity, a hydraulic grout in order to be injectable should also exhibit an appropriate *stability*; the grout has to be stable after mixing, and should maintain its stability, i.e., to be able to keep its *homogeneity* during the injection process, up to its setting. Thus, stability of the suspension against segregation or excessive bleeding should also be ensured in order to achieve a satisfactory injectability. Otherwise, blockage may soon appear and the quality of the intervention could be severely affected.

In this connection, it has to be reminded that the need for high fluidity may occasionally result in very diluted suspensions, being in a moderately flocculent state, potentially leading to instability.

 
A. Miltiadou-Fezans and T. P. Tassios, *Mix-Design and Application of Hydraulic Grouts for Masonry Strengthening*, Springer Tracts in Civil Engineering,
https://doi.org/10.1007/978-3-030-85965-7_4

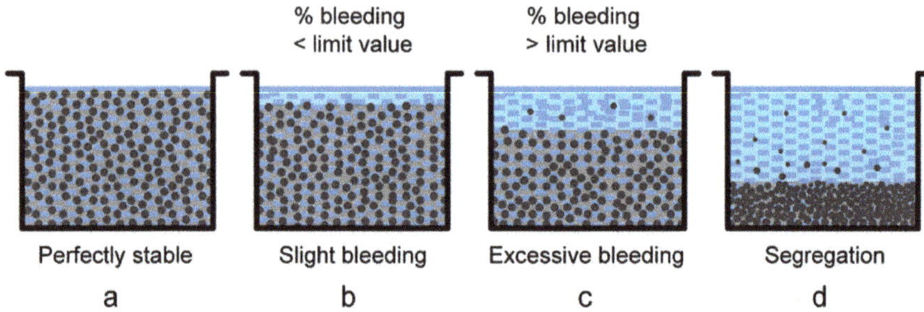


**Fig. 4.1** Schematic presentation of four main situations of suspension equilibrium (dashes represent water, circles represent grains)

By its nature, the suspension consists of grains dispersed in a liquid; it incorporates a number of interconnected channels of various dimensions, giving to the water the possibility to circulate. After mixing, the spatial distribution of the solid particles is determined by gravity and friction, as well as by repulsion interparticle forces.

Thus, as it is schematically presented in Fig. 4.1, the following four main situations of equilibrium can be distinguished (Miltiadou-Fezans and Tassios 2013b):

- The grout seems to keep its homogeneity up to its setting, and is considered as perfectly stable (Fig. 4.1a).
- A uniform slow settlement of the flocculated structure (or seepage of the water) is taking place, leaving a thin layer of clean water over the surface, called bleeding water; the interparticle forces however are sufficiently strong to hold the structure together without disrupting it. Thus, the composition of the mix is practically unchanged. That is why when the bleeding remains lower than a certain limit (depending on the grouted structure), the grout can be used despite its slight bleeding (Fig. 4.1b).
- In some cases, however excessive bleeding may result in a markedly heterogeneous grout, unable to penetrate and adhere appropriately (Fig. 4.1c).
- Occasionally, a relatively quick settlement of the individual solid particles or floccules is observed, while the solid particles are arranged in different layers according to their weight and size. This leads to the creation of zones of higher density close to the bottom of the recipient–a phenomenon called segregation. In this case, the interparticle forces were not sufficiently strong to hold the structure together; and the structure is now partially or completely disrupted. The grout cannot be used (Fig. 4.1d).

An appropriate grout mix-design should ensure that such instabilities will not occur. Segregation or bleeding could jeopardise the injectability of a grout, potentially leading to a blocking of voids, followed by a sudden dangerous increase of the pressure. Such unstable grouts may severely affect the quality of the grouted medium: lack of filling of the voids, heterogeneous adherence and strength characteristics, lower durability and resistance to environmental actions (since the solid grout will possess a less dense structure).

In the framework of a broader attempt to establish a holistic rational methodology for the design of hydraulic grouts (see Miltiadou-Fezans and Tassios 2016 and Chap. 9 of this book), the relationships between the most predominant parameters shaping stability characteristics (i.e., water content and specific surface of the solid phase, expressed by the percentage of ultrafine materials replacing cement or hydraulic lime) are examined in this chapter.

After a brief literature survey, an oversimplified predictive model of bleeding is first presented and then its validity is confirmed, using the results of an experimental study (from Miltiadou-Fezans and Tassios 2013b). The role of superplasticizers is also briefly discussed. In both cases, with and without addition of superplasticizer, semi empirical formulae are proposed that may be useful for the design of a grout composition in terms of stability. Finally, experimental results are presented in this chapter, demonstrating the role of water and superplasticizer content in the appearance of segregation: some empirical formulae are also proposed for the estimation of critical water content initiating segregation.

## 4.2 Literature Survey

### *4.2.1 Main Parameters Influencing Stability*

Instability phenomena of hydraulic grouts are similar to those present in fresh cement pastes, and depend on a multitude of parameters intervening in their properties and behaviour. The controlling factors for the stability of a suspension are the *specific surface* and the *water-to-solid ratio*; these two factors together determine the average water-film thickness and interparticle spacing (Legrand 1972; 1982; Littlejohn 1982; Vom Berg 1979; Bombled 1974; Powers 1968; Ish-Shalom and Greenberg 1962; Papadakis 1959). The mixing method also has a direct impact on stability, as it was demonstrated as early as in 1957 by Papadakis (1957), followed by many other researchers. The chemical and physical properties of the solid phase and the possible presence and quantity of any admixtures are as well important factors, as they intervene in the equilibrium of the interparticle forces.

Papadakis (1959) reported that bleeding diminishes when the *specific surface* increases and it is almost linearly increased with the water content; bleeding starts after a certain water content, depending on the specific surface of the solid phase. The author has also noted the importance of the granular composition of the grout.

According to Powers (1968), because of flocculation forces, each composition of a cement paste takes a "base" structure, and the pastes are able to bleed because of the *presence of water in excess* of what is required for the "base" structure. Moreover, according to Powers, «whatever the amount of "dilation" of the base volume by excess water may be, the maximum settlement is limited to about one-third of the excess water, unless particles can shift laterally as they settle or unless some of the particles nest more compactly into "pockets" presumably created by the excess water. The forces of flocculation are expected to prevent lateral shifting during settlement. The very small difference between the initial volume and the "base volume" of a given sample is considered to nullify the nesting effect». On this basis, Powers proposes an empirical equation to estimate the bleeding capacity of cement pastes, as a function of the water content used in the grout, and a base water content after which the specific composition starts bleeding. The empirical equation proposed by Powers (1968) is given below "Eq. 4.1".

$$\Delta\mathrm{H} = \frac{(1/3)\cdot\left[\left(WV_w/CV_c^o\right) - \left(WV_w/CV_c^o\right)_B\right]}{1+\left(WV_w/CV_c^o\right)} \tag{4.1}$$

where:

$\Delta$H: bleeding capacity (settlement per unit original height).
$WV_w/CV_c^o$: water/cement ratio by volume.
$\left(WV_w/CV_c^o\right)_B$: base water/cement ratio after which bleeding starts.

For a grout containing a solid phase composed not only of cement but of a combination of cement, lime, pozzolan, etc., it can be presumed according to Powers, that the following relation could give an estimation of bleeding capacity.

$$b = \frac{(1/3)\cdot\left[\left(\frac{W_v}{S_v}\right) - \left(\frac{W_v}{S_v}\right)_0\right]}{1+\left(\frac{W_v}{S_v}\right)} \tag{4.2}$$

where:

$b$: bleeding capacity or normalized bleeding value,
$\left(\frac{W_v}{S_v}\right)$: water to solids ratio by volume,
$\left(\frac{W_v}{S_v}\right)_0$: water to solids ratio initiating bleeding by volume.

Bombled (1974) also notes the importance of *specific surface*, demonstrating that a cement with a double specific surface has produced four times lower bleeding than the reference cement. The author reports also the improvement of stability, thanks to the ultrafine materials contained in the cement or fillers combined with it; thus, a higher specific surface and a higher water retention capacity is offered to the system, provided an adequate mixing was used to achieve the wetting of all grains. He also noted that fluidifiers can decrease bleeding since their use results in higher interstitial water retention.

Aïtcin et al. (1984) and Domone and Tank (1986) reported the improvement of stability when the solid phase of the grout contains cement and silica fume (noting at the same time that the use of superplasticizers is indispensable to obtain a satisfactory fluidity).

Miltiadou (1990) has investigated bleeding and segregation of hydraulic grouts composed by various combinations of cement and ultrafine materials (namely densified silica fume, lime and Santorini earth). She has undertaken an experimental study using a great variety of hydraulic grouts, both with and without superplasticizer and she has proposed some empirical formulae for the selection of water and superplasticizer content, as a function of the percentage of ultrafines. To demonstrate the segregation phenomenon in unstable grouts, she has used laser granulometry for the measurement of the evolution of grout particle size distribution in samples taken in a depth of 2/5 from the upper surface of the grout volume, in various time intervals after mixing. She has also proposed a practical test for characterising segregation, by measuring the thickness of a distinct denser layer sedimented at the bottom of the recipient; some of her results are presented in more details in what follows (Sect. 4.5).

The beneficial effect of the combination of cement with natural Pozzolan and silica fume on bleeding was also reported by Shannag (2002).

Shimoda and Ohmori (1982), Zebovitz et al. (1989) and Benhamou (1994) reported great improvement of stability if, instead of an ordinary cement grout, a microfine cement grout was used, thanks to the higher fineness of the latter.

Benhamou (1994) and Viseur and Barrioulet (1998), reiterated the need to ensure compatibility of the superplasticizer with the binder. The different role played by high range water reducers and viscosity-enhancing admixtures and their impact to the behaviour of grouts, has been reported by Yahia and Khayat (2003) and Saric-Coric et al. (2003). The need for a careful selection of admixtures is also demonstrated by the results on stability and fluidity obtained by Valluzzi et al. (2003).

The importance of the *mixing* procedure on stability was already studied by Papadakis (1957) and Bombled (1974). The authors have shown the beneficial effect of a high turbulence mixing on the stability of grouts. For a given water content, increasing of the mixing intensity, results in a considerable reduction of bleeding (followed by an increase of the shear stress limit). However, during continuous mixing, in-time viscosity may increase due to the abrasion of the particles (resulting in increase of specific surface), whereas hydration may be accelerated due to temperature elevation. Reduced bleeding has also been noted by Markestad (1976), by using high turbulence mixing with external cooling. On the same subject have also worked the authors mentioned in Sect. 2.2.4.2 on penetrability, and in Sect. 3.4 on fluidity. More specifically, regarding the influence of mixing method on stability, Miltiadou (1990) and Toumbakari et al. (1999), have came to the conclusion that ultrasound mixing produces more stable suspensions than high turbulence, in most cases.

### 4.2.2 Main Tests to Measure Stability

The standardized stability tests used in literature concern the measurement of bleeding water, for prestressing tendons grouts and grouts for preplaced aggregate concrete. There is not any specific standard for masonry grouting. In this section are presented the bleeding measurement tests proposed by the following standards:

- NF P 18-359 (1985): Adjuvants pour bétons, mortiers et coulis. Coulis courants d'injection pour précontrainte—Mesure de l'exsudation,
- EN 445: 2007: Grout for prestressing tendons—Test methods,
- ASTM C 940–16: Expansion and bleeding of freshly mixed grouts for preplaced-aggregate concrete in the laboratory.

To our knowledge there are not standards regarding the measurement of segregation. Some testing methods used by some researchers in the literature are presented below.

#### 4.2.2.1 Bleeding Measurements

According to NF P 18-359

Freshly mixed grout is poured into a transparent clean and dry volumetric cylinder of 100 ml, graduated in cubic centimeters, having diameter and height equal to 25 mm and 250 mm respectively, and placed on a stable horizontal surface situated in the field away from adverse weather, shocks and any vibration, and in the Lab protected from jolting at 20 ± 2 °C and 65% relative humidity (Fig. 4.2). The graduated cylinder is filled with 95 to 100 ml of grout and it remains covered with a plate or a rubber stopper at a shadowed place, protected from any evaporation. After a certain period of time, the volume of the water that has appeared at the top surface of the grout is measured directly by reading the number of cubic centimeters of water covering the grout paste (Fig. 5.2).

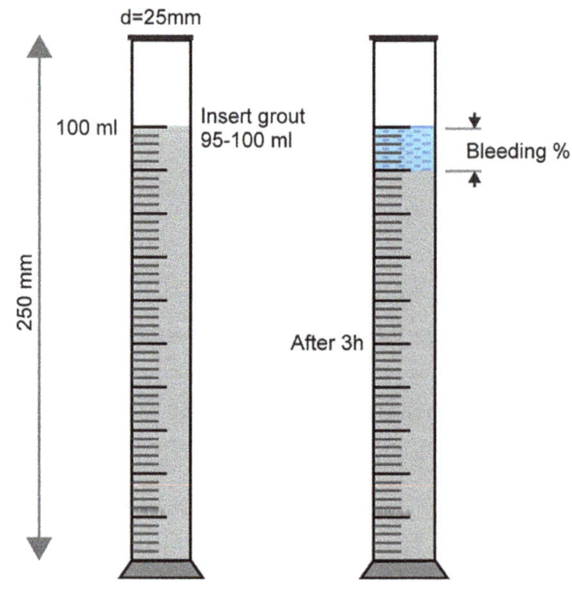


**Fig. 4.2** Measurement of bleeding according to NF P 18–359

Bleeding is expressed in volume percentage, as the ratio of the volume of bleeding water to the volume of the grout poured into the cylinder (normally 100 ml). After a period of 3 h the total bleeding water is measured, by reading directly the volume in $cm^3$ of water on the top of the grout. In case the initial volume is not equal to 100 ml the final bleeding is calculated according to the following expression:

$$Final\ Bleeding(\%) = \frac{V_w}{V_{g,i}} \times 100$$

where:

$V_w$: volume of bleed water (ml).

$V_{g,i}$: initial volume of the grout at the beginning of test (ml).

A second measurement is carried out after 24 h in order to evaluate the total percentage of water reabsorbed or bleed.

Each measurement should be repeated three times. The bleeding of the grout is given by the average value of the three tests after a period of 3 h. Temperature is also noted when the test is carried out in the worksite.

According to ASTM C 940

Into a 1000 ml glass graduated cylinder, 800 ml ± 10 ml of freshly mixed grout is poured; the cylinder is placed on a level surface free of vibration at 23 ± 2 °C and covered to prevent evaporation of the bleed water (Fig. 4.3). Take and record the readings, estimated to the nearest 1 ml, of the upper surfaces of the grout and bleed

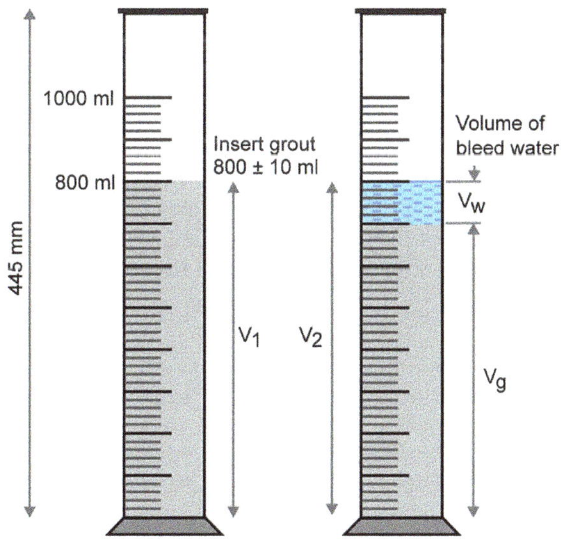


**Fig. 4.3** Measurement of bleeding according to ASTM C940

water of the specimen. The volume of the grout and the bleed water, is measured at 15-min intervals for the first 60 min, and thereafter at hourly intervals, until two equal successive readings are obtained, showing no further bleeding. In case the test is used in connection with "C937 Specification for grout Fluidifier for Preplaced Aggregate Concrete", the test shall be discontinued 3 h after the initial reading.

At the conclusion of the test, decant the bleed water into a 25 ml graduate cylinder by tilting the specimen and drawing water off with a pipet. Record the final volume of bleed water to the nearest 0.5 ml.

The bleeding is calculated according to the following expression:

$$Bleeding\,at\,prescribed\,intervals,\,(\%) = \frac{V_2 - V_g}{V_1} \times 100$$

$$Final\,Bleeding,\,(\%) = \frac{V_w}{V_1} \times 100$$

where:

$V_1$ volume of the specimen at beginning of the test, (ml).

$V_2$ volume of the specimen at prescribed intervals, measured at the upper surface of water layer (ml).

$V_g$ volume of grout portion of specimen at prescribed intervals, at upper surface of grout (ml).

$V_w$ volume of decanted bleed water (ml).

According to EN 445 (Wick-Induced Test)

A tube with an internal diameter of approximately 60 mm to 80 mm and a height of approximately 1 m is used. The tube is posed in a vertical position with its open end at the top (Fig. 4.4). Inside the tube a 7-wire strand approximately 900 mm long is inserted to fit. Install the strand in the tube, ensuring that it is firmly supported on the base, and held centered. The grout is poured into the tube at a steady flow rate to ensure there is no trapped air, and the tube is filled up to a height, $h_0$, about 10 mm above the top of the steel. Record the starting time $t_0$ and the height $h_0$ of the grout. Record the height of the grout, $h_g$, at 15 min intervals during the first hour, and subsequently at 2 h, 3 h and 24 h (noted as $t_n$ in Fig. 4.4). Record the height of bleed water, $h_w$, at the same intervals as for the grout. Bleeding is expressed as,

$$b = \left({}^{h_w}/_{h_0}\right) \times 100\%$$

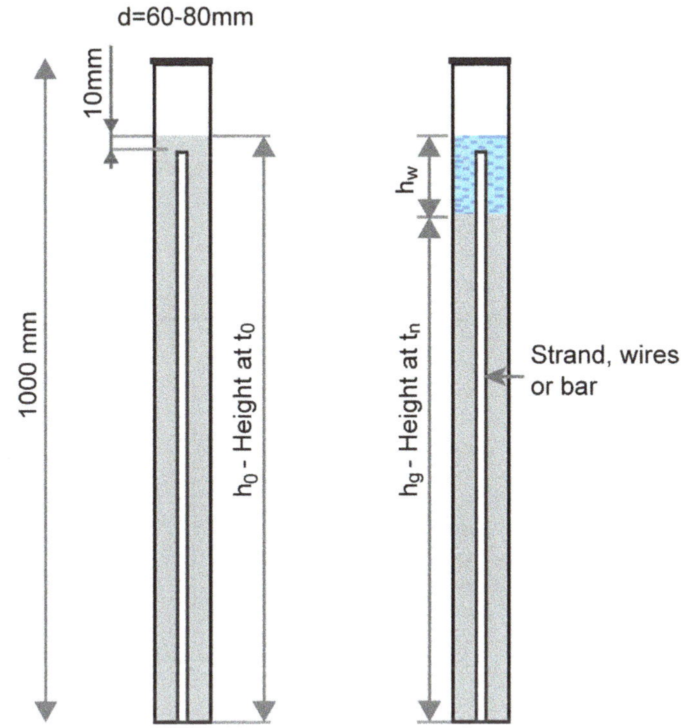


**Fig. 4.4** Measurement of bleeding according to EN 445

where, $h_w$ = height of bleed water at prescribed intervals and $h_0$ = height of the initial specimen of the grout at starting time $t_0$.

### Comparison of Standards for Bleeding Testing

In all the three standards presented above, the procedure followed is similar, but there exist some differences, which have to be mentioned, allowing for the selection of the most appropriate test for masonry grouting.

- In the French standard NF P 18-359, a substantially smaller quantity of grout is necessary, in comparison with the EN 445 and ASTM C 940 (Fig. 4.5).
- In the NF P 18-359 standard test, the measurement of total bleeding after a time interval of three hours (3 h) is proposed. A second measurement of bleeding in 24 h is as well proposed, in order to determine the total percentage of reabsorbed or bleed water. The other two standards, ASTM C940 and EN445, propose bleeding to be measured at various time intervals to study the evolution of bleeding as a function of time. Thus, when comparing the results in the relative literature, comparison is not easy, since different time intervals are used for different

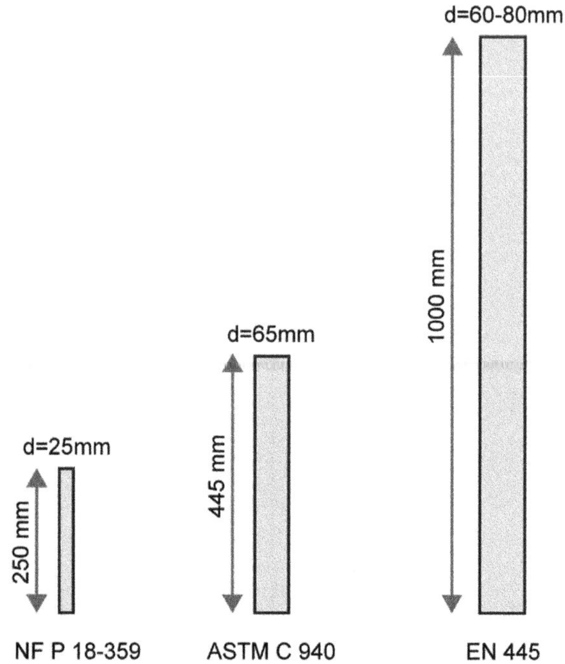


**Fig. 4.5** Schematic comparison of the size of the graduate cylinders used in the three standardized bleeding measurement tests

compositions, or from researcher to researcher. It has to be noted, however, that in EN447:2007 (Grout for prestressing tendons-Basic requirements) the value of bleeding in three hours is proposed as a criterion for design purposes.

- It may be said also, that the ASTM C940 test presents another inconveniency. The bleed water at the end of the test has to be collected in another recipient to measure its volume. This procedure may introduce some uncertainties about the reproducibility of the test. Measuring bleeding by direct reading on the graduated cylinder proposed in other standards, is without any doubt a more convenient procedure, especially in worksite conditions.

Taking into account that in the case of grouts of high injectability for masonry strengthening, the size of voids and discontinuities to be filled are relatively small, the use of the NF P 18–359 seems preferable.

Thus, bleeding measurement in the Laboratory and in the field are facilitated, as smaller grout quantity is required for testing, and consequently smaller size of graduate cylinders (Fig. 4.5) and worksite laboratory facilities.

Moreover, it is worth to be noted that in the particular case of masonry grouting, measuring of bleeding at 3 h both in Lab and in-situ (as proposed by NF P 18–359 and EN 445) is considered the most adequate. After 3 h, a considerable part of the entire bleeding is manifested, thus allowing for an early judgement of the suitability of a mix to be used, even in the worksite.

Obviously, the procedure proposed by ASTM C940 is very interesting, since more intermediate measurements give the possibility to produce a curve, presenting the evolution of bleeding as a function of time; such a procedure may be very useful for laboratory research. However, when testing bleeding in the field, for grout quality control, it is not possible to take a series of measurements. Thus, it is preferable to have a fixed time after mixing (e.g., 3 h) to measure bleeding both in the Lab and in-situ. Comparison with the acceptable limit, imposed as criterion by the mix-design is feasible and quality control is facilitated (see Chap. 10).

On the basis of this reasoning, the NF P 18–359 is preferable for testing bleeding of high injectability grouts for masonry strengthening.

#### Criteria for Maximum Acceptable Bleeding

As it has been presented above, there is not any standard for measuring bleeding for masonry grouts. Subsequently, there is not any standard providing criteria for the approval of masonry grouts for mix-design purposes. In what follows the criteria proposed in other fields and in the literature for masonry grouting are presented.

For the prestressing tendons, bleeding has to be lower than 2% in 3 h, while the bleed water has to be completely reabsorbed after 24 h (Fascicule N° 65: Décret no 85–404 du 3 avril 1985, France). According to EN 447: 2007, which also concerns prestressing tendons, bleeding should not exceed 0.3% of the initial volume of the grout after 3 h kept at rest. The volume change of the grout at rest for 24 h should be within the range of −1% and +5%.

Lombardi (1990, 1985) working on grouting in fissured rock masses states that a stable mix is one that presents less than 5% bleeding in 2 h or 3 h.

Miltiadou (1990) proposed as well the limit of 5% in 3 h for masonry grouting (measured according to NF P 18–359). This limit was also used by Kalagri et al. (2010), Badogiannis et al. (2012), Jorne et al. (2015), Papayianni and Pachta (2015), Nikolopoulou et al. (2018), while the limit of 2% was used by Miltiadou-Fezans et al. (2005) for the grouts designed to fill the fine cracks of the marble drums of the columns of the Parthenon of the Acropolis of Athens.

The limit of 5% in 2 h was considered by Toumbakari (2002) studying also masonry grouting, by Mirza et al. (2002) studying grouts for fine fissures of soil, by Mirza et al. (2013) studying grouts for concrete, by Ozcan et al. (2005) studying grouts for geotechnical works and Bras and Henriques (2012) studying natural hydraulic lime grout compositions for masonry.

#### 4.2.2.2 Estimation of Segregation

Regarding tests to measure segregation, it has to be noted that, to our knowledge, do not exist standards to measure this instability characteristic. Instead, it is clearly recognized that a grout is considered stable only when segregation is not present.

Thus, some tests have been proposed in the literature in order to identify the appearance of such an instability; moreover, some qualitative segregation estimators were also developed, as presented below.

Miltiadou (1990) used laser granulometry for the measurement of the grout particle grain size distribution at specific depths from the upper surface of the grout volume after mixing and in 15 min time intervals until 60 min after mixing (see Miltiadou-Fezans and Tassios 2013b and Sect. 4.5 of this book). Miltiadou (1990) proposed as well another estimator of segregation, as presented in Sect. 4.5.

Toumbakari (2002) has also used laser granulometry for the measurement of the grout particle grain size distribution in various time intervals after mixing. The Author has demonstrated that in case of a stable composition, the particle size distribution remained practically unaltered until 120 min after mixing.

Van Rickstal (2000) proposed a test for the evaluation of stability, based on the changes of density of the upper zone of a grout volume in rest in a receptacle. The density variations are estimated on the basis of the measurement of the differences between the weight of an object located at a constant and known depth inside the grout and the buoyancy force exerted on it. The same author proposed to accept a fluctuation of $\pm 5\%$ for the measured density as a stability criterion. This same principle, as proposed by Van Rickstal (2000) for density measurement to control grout stability, was also used by Bras and Henriques (2012).

For measuring grout density, a procedure similar to the Andersen pipette analysis for soil testing was adopted by Toumbakari (2002), using as stability criterion the one proposed by Van Rickstal (2000).

Miltiadou-Fezans et al. (2005, 2008) have also used the measurement of the apparent density of the grout after mixing and in various time intervals (using samples taken from the upper third of the height of a transparent volumetric tube, where the grout remains stagnant), as an indicator of grout segregation tendency. The same authors proposed as well to check the apparent density of grout samples taken from outflows during the application of the injections. A fluctuation of $\pm 1\%$ of the apparent density values was considered as acceptable after mixing, while for the grout taken from outflows a fluctuation of $\pm 5\%$ was adopted as acceptable.

Rosquoët et al. (2003) studied the instability of cement grouts as a function of the water content; they followed the in-time evolution of the specific density by means of a $\gamma$-densitometer bench. They demonstrated that for W/C ratios ranging from 0.40 to 0.45, the density does not vary much with time and the pastes remain homogeneous, while for higher ratios (0.5–1.0), density increases sharply with time at the bottom of the column, and the paste is heterogeneous when it exhibits higher bleeding indications. The authors have also linked these results to the degree of hydration of the solidified grout.

## 4.3 An Oversimplified Predictive Model of Bleeding

For the limited scope of this paragraph, the suspension after complete bleeding will be represented as in Fig. 4.6a: A mono-granular material will be considered, with cubic grains of total number «n» and width « d». A layer of inter-granular free water (of thickness equal to «$2t_0$») will be considered; bonded water will not be taken into account in this oversimplified model at this stage.

Total volumes of materials (see Fig. 4.6a):

Neglecting several important mechanisms of water retention during the entire process, the volume of "total" water would be estimated as follows.

"Total" water:

$$W_v = (d + 2t_0)^2 \cdot B + \mathrm{n} \cdot \left[(d + 2t_0)^3 - d^3\right] \quad (4.3)$$

Solids [cement (C) or natural hydraulic lime (NHL) and fines "F"]:

$$S_v = C_v[or\ (NHL)_v] + F_v = \mathrm{n} \cdot d^3 \quad (4.4)$$

where:

"n": denotes the "equivalent" number of solid-cubes, with an average width "d".

"$S_A$": Average Specific Surface (cm$^2$/g) by definition:

$$S_A = \frac{1}{\lambda} \cdot \frac{6d^2}{\rho_{ms} \cdot d^3} = \frac{6}{\lambda \cdot d \cdot \rho_{ms}} \quad (4.5)$$

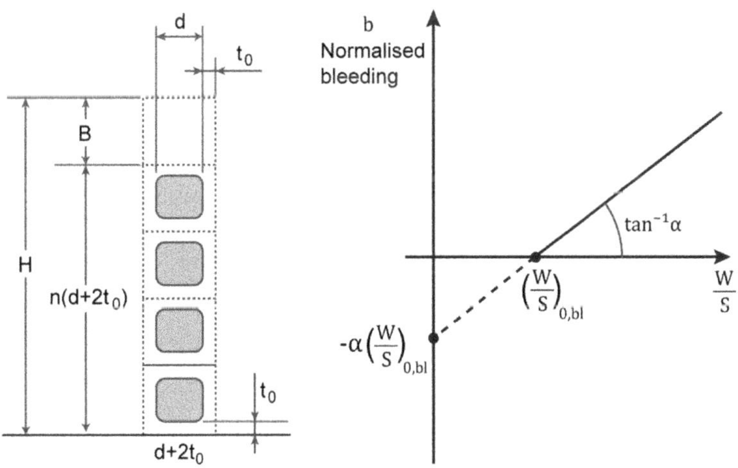


**Fig. 4.6** **a** An oversimplified model for bleeding: a column of grains with their adsorbed water ($t_o$) and its free bleeding (B). **b** Normalised bleeding (b = B/H) as a function of water to solids ratio, described by "Eqs. 4.17, 4.18 and 4.19". Adapted from Miltiadou-Fezans and Tassios (2013b)

where, λ is a form correction factor and $\rho_{ms}$ (g/cm³) is the average density of the solids.

Height of bleed water, from "Eq. 4.3", when bonded water is not yet taken into account:

$$B = \frac{W_v}{(d + 2t_0)^2} - n\frac{(d + 2t_0)^3 - d^3}{(d + 2t_0)^2} \tag{4.6}$$

Putting now H $\approx n \cdot (d + 2t_o)$, the normalised height of the initially calculated bleeding "$b_0$" is found:

$$\frac{B}{H} = b_0 \cong \frac{W_v}{n \cdot (d + 2t_0)^3} - \frac{(d + 2t_0)^3 - d^3}{(d + 2t_0)^3} \tag{4.7}$$

Neglecting higher power values of the ratio $\frac{t_0}{d}$ (whenever this is acceptable) and taking into account "Eq. 4.4", the following approximate expression is derived:

$$\mathrm{n} \cdot (d + 2t_0)^3 \cong \mathrm{n} \cdot d^3\left(1 + 6\frac{t_0}{d}\right) = S_v + S_v \cdot \frac{6t_0}{d} \tag{4.8}$$

where, "Sv" = n · d³ the volume of all grains of the solids. Thus, "Eq. 4.7" is now written as follows:

$$b_0 \cong \left(\frac{1}{1 + \frac{6t_0}{d}}\right) \cdot \frac{W_v}{S_v} - \left(1 - \frac{1}{1 + \frac{6t_0}{d}}\right) \cong \left(\frac{1}{1 + \frac{6t_0}{d}}\right) \cdot \frac{W_v}{S_v} + \frac{1 - 1 - \frac{6t_0}{d}}{1 + \frac{6t_0}{d}}$$

Or:

$$b_0 \cong \left(\frac{1}{1 + \frac{6t_0}{d}}\right) \cdot \left(\frac{W_v}{S_v} - \frac{6t_0}{d}\right) \tag{4.9}$$

The ratio $\frac{6t_0}{d} = \frac{6t_0 d^2}{d^3}$ equals the inter-granular water-to-solids ratio by volume $\left(\frac{W_v}{S_v}\right)_{0,bl}$ for zero bleeding, neglecting bonded water.

Consequently, "Eq. 4.9" may now be written as follows:

$$b_0 \cong \alpha_0\left[\frac{W_v}{S_v} - \left(\frac{W_v}{S_v}\right)_{0,bl}\right], \text{ with } \alpha_0 = \frac{1}{\left(1 + \frac{6t_0}{d}\right)}, \tag{4.10}$$

From "Eq. 4.5",

$$\left(\frac{W_v}{S_v}\right)_{0,bl} = \frac{6 \cdot t_0}{d} = \lambda \cdot \rho_{ms} \cdot t_0 \cdot S_A, \text{ not yet accounting for bonded water.} \tag{4.11}$$

$$\text{Thus, } a_0 = 1/(1 + \lambda \cdot \rho_{ms} \cdot t_0 \cdot S_A) \tag{4.12}$$

$$\text{and } b_0 = \left(\frac{W_v}{S_v} - \lambda \cdot \rho_{ms} \cdot t_0 \cdot S_A\right) / (1 + \lambda \cdot \rho_{ms} \cdot t_0 \cdot S_A) \tag{4.13}$$

It has to be reminded, however, that "Eq. 4.6" and further on, have not taken into account bonded water. Therefore, the $b_0$-values should be considerably reduced accordingly. For the actual state-of-the Art, it is proposed to apply an empirical reduction factor "$\gamma$" equal to 1/2.5 ($\gamma \approx \frac{1}{2.5}$), instead of the value $\gamma \approx \frac{1}{3}$ used by Powers (1968) under analogous conditions.

Thus, from "Eq. 4.13", the actual normalized height of bleeding "b" can be approximately predicted as follows:

$$b = \frac{1}{2.5} \cdot \frac{1}{1 + \lambda \cdot \rho_{ms} \cdot t_0 \cdot S_A} \cdot \left(\frac{W_v}{S_v} - \lambda \cdot \rho_{ms} \cdot t_0 \cdot S_A\right) \tag{4.14}$$

In order now to convert the water-to-solids ratio $\frac{W_v}{S_v}$ (by volume), into a water-to-solids ratio $\frac{W}{S}$ (by mass), the respective densities will be introduced $W_v = W/\rho_w$, $S_v = S/\rho_{ms}$.

Thus, "Eq. 4.14" is now written as follows

$$b = \frac{1}{2.5}\left(\frac{\rho_{ms}}{\rho_w} \cdot \frac{W}{S} - \lambda \cdot \rho_{ms} \cdot t_0 \cdot S_A\right) / (1 + \lambda \cdot \rho_{ms} \cdot t_0 \cdot S_A) \tag{4.15a}$$

or, if $\rho_w = 1\ \text{g/cm}^3$ (and $S_A$ in $\text{cm}^2/\text{g}$),

$$b = \frac{1}{2.5}\rho_{ms}\left(\frac{W}{S} - \lambda \cdot t_0 \cdot S_A\right) / (1 + \lambda \cdot \rho_{ms} \cdot t_0 \cdot S_A),\ [in\ cm, g] \tag{4.15b}$$

$$b = \left(\frac{1/2.5}{\frac{1}{\rho_{ms}} + \lambda \cdot t_0 \cdot S_A}\right) \cdot \left(\frac{W}{S} - \lambda \cdot t_0 \cdot S_A\right), [\text{in cm, g}] \tag{4.15c}$$

It is noted that because of the omission of the water density $\rho_w = 1\text{g/cm}^3$, "Eq. 4.15b" seems to be dimensionally incompatible; however, if $S_A$ and $\rho_{ms}$ are introduced in [cm, gr], this equation is numerically correct. "Equation 4.15c", may be written as follows:

$$b = \alpha \cdot \left(\frac{W}{S} - \lambda \cdot t_0 \cdot S_A\right), [\text{in cm, g}] \tag{4.16}$$

where:

$$\alpha = 1/[2.5\left(\frac{1}{\rho_{ms}} + \lambda \cdot t_0 \cdot S_A\right)], [\text{in cm, g}] \tag{4.17}$$

Taking into account "Eq. 4.11" and because it was taken that $\rho_w = 1$ g/cm$^3$

$$\left(\frac{W_v}{S_v}\right)_{0,bl} = \left(\frac{W}{S}\right)_{0,bl} \cdot \rho_{ms} = \lambda \cdot \rho_{ms} \cdot t_0 \cdot S_A \Rightarrow \left(\frac{W}{S}\right)_{0,bl} = \lambda \cdot t_0 \cdot S_A\,, \quad [\text{in cm, g}] \tag{4.18}$$

Finally, "Eqs. 4.16 and 4.17" can be written as follows

$$b \cong \alpha \cdot \left(\frac{W}{S} - \left(\frac{W}{S}\right)_{0,bl}\right) \tag{4.19}$$

with "$\alpha$" as in "Eq. 4.17", and $\left(\frac{W}{S}\right)_{0,bl}$ as in "Eq. 4.18".

It should however be noted that this quasi-linear dependence of the constants of "Eqs. 4.16 and 4.17" on «$S_A$», should be accepted only for a limited interval of specific surface values (possibly for $S_A$ values lower than, say, 9000 cm$^2$/g), for which $t_0/d$ may be assumed as considerably lower than unity, as hypothesised.

For a given specific surface $S_A$, "Eq. 4.19" describes the well-known linear relationship between normalised bleeding and water-to-solids ratio (Fig. 4.6b).

In other words, this model (despite its oversimplification) suggests that:

- The water-to-solids ratio "Eq. 4.18" which starts producing bleeding, increases «linearly» with the specific surface of the solids,
- The rate of dependence of bleeding, "Eq. 4.19", on the water to solids ratio, decreases when the specific surface of the solids increases, "Eq. 4.17".

Both conclusions were qualitatively expected, and they are also confirmed by means of the experimental investigation which follows.

Moreover, it is worth noting that the well-known concept of "water retention capacity" as a parameter against bleeding, is also recognised in the preceding analysis: the anti-bleeding constant $(W/S)_{0,bl} = \lambda\, t_0\, S_A$ is a function of the form of grains ($\lambda$), of the inter-granular water ($t_0$), and of the fineness ($S_A$) of the grains–namely of water retention related factors.

## 4.4 Experimental Investigation on Bleeding

### *4.4.1 Bleeding of Grouts Without Added Superplasticizer*

This experimental study was carried out following the French Standard NF P18-359; the water volume left on top of the grout (protected against any evaporation) is measured after three hours. The grouts were prepared using ultrasounds (US) as mixing method.

The US mixing method consisted of an ultrasound dispersion at 20 kHz and a simultaneous mechanical stirring at 300 rpm.

The mixing time was 2 minutes for each ultrafine material, followed by 2 more minutes after the addition of the cement. Figures 4.7, 4.8 and 4.9 show the bleeding values obtained, as a function of the water-to-solids ratio ($W/S$) by mass, for grouts made of cement (C10) and densified dilica fume (DSF) or Santorini earth (SE) or lime (L), respectively (Miltiadou 1990).

Figures 4.10 and 4.11 show those obtained for binary grouts made by cement C23 and Santorin earth SE1, and for ternary grouts made of cement C24, Milos pozzolan (MP4) and lime (L3) respectively (Miltiadou-Fezans et al. 2007). These results clearly confirm the well-known reduction of bleeding thanks to ultrafine materials.

On the other hand, the $(W/S)_{0,bl}$ value for which bleeding starts, is increased from 0.43 (plain C10 cement grout) to 0.65, 0.74 or 0.87, if 10% or 25% or 40% of cement is replaced by Santorini earth (SE), or from 0.59 (plain C24 cement grout)

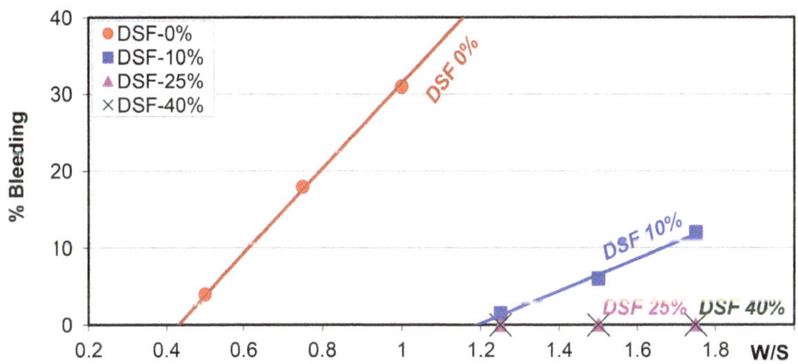


**Fig. 4.7** Bleeding values as a function of water-to-solids ratio for grouts made of cement C10 and densified silica fume (DSF). Adapted from Miltiadou-Fezans and Tassios (2013b)

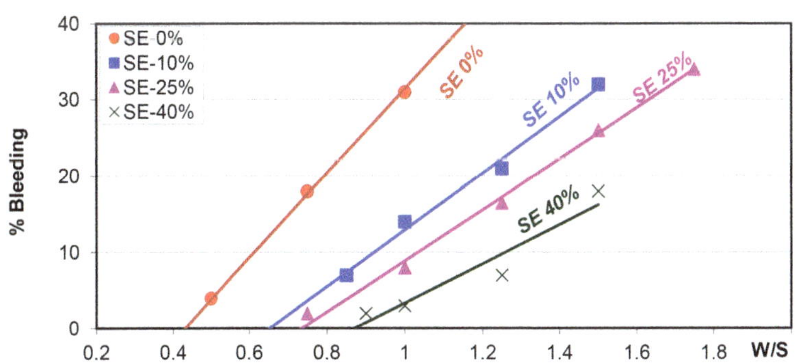


**Fig. 4.8** Bleeding values as a function of water-to-solids ratio for grouts made of cement C10 and Santorini earth (SE). Adapted from Miltiadou-Fezans and Tassios (2013b)

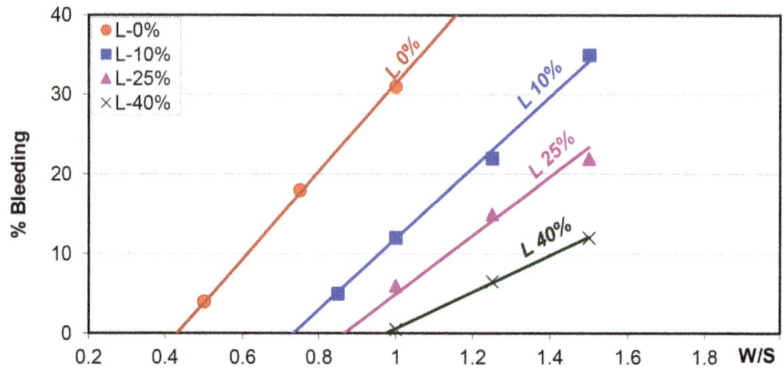


**Fig. 4.9** Bleeding values as a function of water-to-solids ratio for grouts made of cement C10 and lime (L). Adapted from Miltiadou-Fezans and Tassios (2013b)

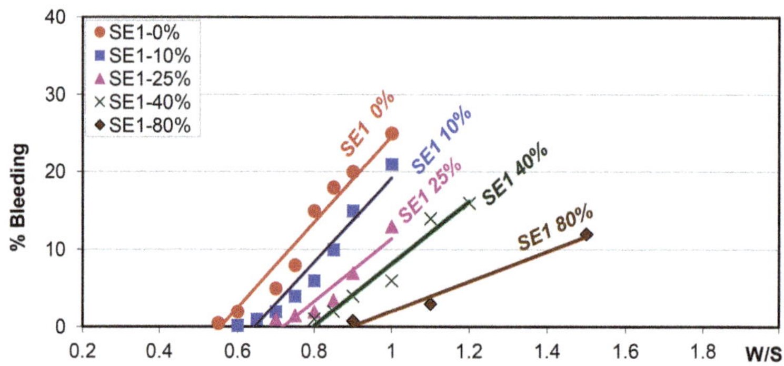


**Fig. 4.10** Bleeding values as a function of water-to-solids ratio for grouts made of cement C23 and Santorini earth (SE1). Adapted from Miltiadou-Fezans and Tassios (2013b)

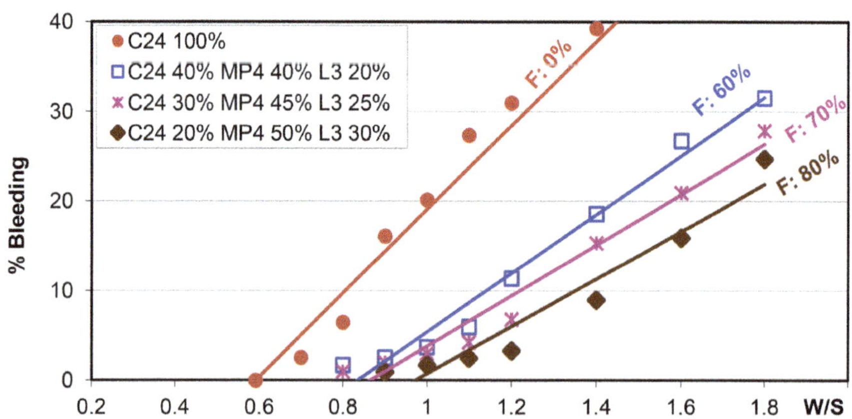


**Fig. 4.11** Bleeding values as a function of water-to-solids ratio for ternary grouts made of cement C24, Milos Pozzolan MP4 and lime L3, (F = MP4 + L3). Adapted from Miltiadou-Fezans and Tassios (2013b)

to 0.84, 0.87 or 0.97 when 60% or 70% or 80% of cement C24 is replaced by a combination of Milos pozzolan (MP4) and lime (L3), as presented in Figs. 4.8 and 4.11, respectively.

The influence of silica fume is more pronounced (see Fig. 4.7); a 10% DSF grout starts bleeding only when $(W/S)$ ratio is as high as 1.20. For 25–40% DSF blended grouts, bleeding remains nil for $(W/S)$ values lower than 1.75.

Besides, as predicted by "Eq. 4.17", the slope of each "line of bleeding" is considerably reduced if more fine solid materials are used (higher $S_A$ value).

On the other hand, these five diagrams reconfirm that bleeding is a linear function of the water-to-solids ratio, as it was predicted by "Eq. 4.19" (see below "Eq. 4.19a",

$$b \cong \alpha \cdot \left[ \frac{W}{S} - \left( \frac{W}{S} \right)_{0,bl} \right] \tag{4.19a}$$

for, however, low $t_0/d$ values (and practically for $S_A \leq 9000 cm^2/gr$).

The value $(W/S)_{0,bl}$, after which bleeding starts, is of a high practical importance, since it is an indication of an upper limit of water-to-solids ratio for non-bleeding grouts. "Equation 4.18" has shown that $(W/S)_{0,bl} = \lambda \cdot t_0 \cdot S_A$, [in cm, g]; this dependence on the "average fineness", (i.e., average specific surface $S_A$), is approximately confirmed by the experimental results shown in Table 4.1, where the value "$1/(\lambda \cdot t_0)$" was found to be roughly constant (±15%).

One may now conclude that Fig. 4.6b is a valid representation of the phenomenon of bleeding.

Regarding the $S_A$- limit values (noted as well in Sect. 4.3), it seems that a more practical value $S_{A,max} = 8{,}000 cm^2/g$ can be adopted in all cases—since for grouts with solid phase of relatively high fineness, the simple analytical models presented along the previous paragraphs may be numerically less accurate.

In Table 4.1, the experimental values of Blaine specific surface ($S_b$) for cement, natural hydraulic lime and ultrafines are shown. A series of tests were also carried out using the laser granulometer for various ultrasound mixing periods, in order to determine their grading characteristics.

It has to be noted however that the assessment of grading characteristics of ultrafines is a very delicate operation, and the results obtained depend (very often significantly) on the test conditions and on the type of equipment used.

The values of fineness of Santorini earth (SE and SE1), of Milos Pozzolan (MP4) and Lime (L and L3) retained for the purposes of the present experimental investigation were the highest values obtained in a series of measurements. The values of average fineness ($S_A$) of the solid phases of the grouts were calculated on the basis of the percentages of the constituent materials.

Figure 4.12 shows graphically that critical water -to-solids ratio is a linear function of the average fineness of the solids of the grouts, as it was approximately predicted by "Eq. 4.18".

Furthermore, Table 4.1 does confirm that the rate factor « α» of "Eq. 4.19" is inversely proportional to the global specific surface of the solids included in the mix,

**Table 4.1** Experimentally obtained values for $(W/S)_{0,bl}$ and slope "$\alpha$", as a function of the fineness of the solid phase of the grout

| Grout | % Fines | Fineness $S_b$: Specific Surface Blaine ($cm^2/g$) | | | Calculated average fineness of the solid phase of the grout ($S_A$) ($cm^2/g$) | Experimental values | | Calculated values |
|---|---|---|---|---|---|---|---|---|
| | | Cement or Hydraulic Lime | Pozzolan | Lime | | $(W/S)_{0,bl}$ (Figs. 4.8, 4.9, 4.10 and 4.11) | Slope «$\alpha$» | $[1/(\lambda t_0)]$ ($cm^2/g$) |
| C10 | 0 | 3489 | | | 3489 | 0.43 | 0.55 | 8092 |
| 90C10/10SE | 10 | 3489 | 13500 | | 4490 | 0.65 | 0.37 | 6887 |
| 75C10/25SE | 25 | 3489 | 13500 | | 5992 | 0.74 | 0.33 | 8149 |
| 60C10/40SE | 40 | 3489 | 13500 | | 7493 | 0.87 | 0.26 | 8614 |
| 90C10/10L | 10 | 3489 | | 18000 | 4940 | 0.73 | 0.45 | 6737 |
| 75C10/25L | 25 | 3489 | | 18000 | 7117 | 0.87 | 0.37 | 8209 |
| 60C10/40L | 40 | 3489 | | 18000 | 9293 | 0.97 | 0.23 | 9535 |
| C23 | 0 | 4100 | | | 4100 | 0.55 | 0.55 | 7388 |
| 90C23/10SE1 | 10 | 4100 | 8596 | | 4550 | 0.64 | 0.52 | 7150 |
| 75C23/25SE1 | 25 | 4100 | 8596 | | 5225 | 0.72 | 0.40 | 7274 |
| 60C23/40SE1 | 40 | 4100 | 8596 | | 5898 | 0.80 | 0.40 | 7402 |
| 20C23/80SE1 | 80 | 4100 | 8596 | | 7697 | 0.89 | 0.19 | 8605 |
| C24 | 0 | 4550 | | | 4550 | 0.59 | 0.47 | 7691 |
| 40C24/40MP4/20L3 | 60 | 4550 | 7280 | 10500 | 6832 | 0.84 | 0.33 | 8181 |
| 30C24/45MP4/25L3 | 70 | 4550 | 7280 | 10500 | 7266 | 0.87 | 0.28 | 8398 |
| 20C24/50MP4/30L3 | 80 | 4550 | 7280 | 10500 | 7700 | 0.97 | 0.26 | 7922 |

(continued)

**Table 4.1** (continued)

| Grout | % Fines | Fineness $S_b$: Specific Surface Blaine ($cm^2/g$) | | | Calculated average fineness of the solid phase of the grout ($S_A$) ($cm^2/g$) | Experimental values | | Calculated values |
|---|---|---|---|---|---|---|---|---|
| | | Cement or Hydraulic Lime | Pozzolan | Lime | | $(W/S)_{0,\,bl}$ (Figs. 4.8, 4.9, 4.10 and 4.11) | Slope «$\alpha$» | $[1/(\lambda t_0)]$ ($cm^2/g$) |
| NHL5 | 0 | 8000(*) | | | 8000 | 0.82 | 0.28 | 9782 |
| NHL3.5 Z | 0 | 8568 | | | 8568 | 0.93 | 0.16 | 9215 |
| | | | | | | | **Average** | **8068≃8000** |

In the last column, values for $1/\lambda t_0$ are also calculated. Adapted from Miltiadou-Fezans and Tassios (2013b)
(*)Value indicated by the manufacturer

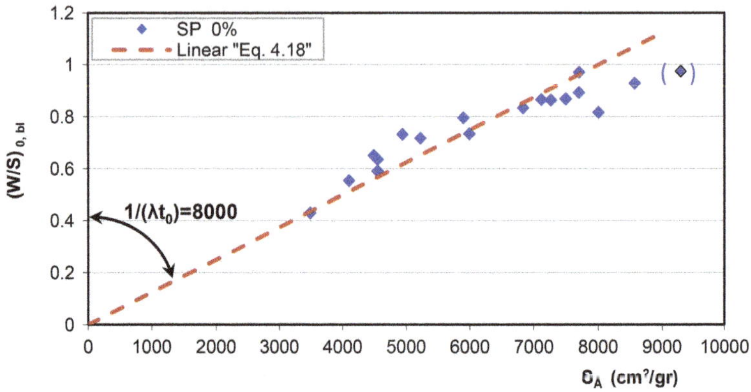


**Fig. 4.12** $(W/S)_{0,bl}$ versus the specific surface of the solid phase of the grout, for "average fineness" values lower than 8,000 $cm^2/g$; the dotted line corresponds to the approximate expression proposed by "Eq. 4.18". Adapted from Miltiadou-Fezans and Tassios (2013b)

as it was predicted by "Eq. 4.17". Its numerical verification (see Fig. 4.13) seems to be rather satisfactory.

In conclusion, within the limited interval of parameters studied in the present experimental investigation, it can be taken that:

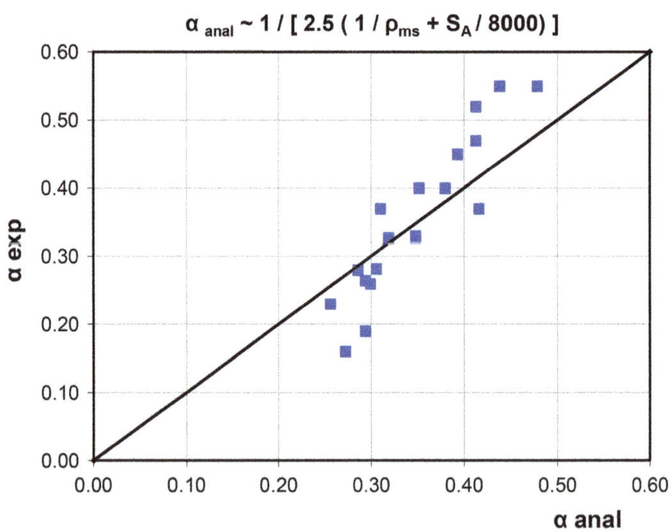


**Fig. 4.13** Analytical values versus experimental values of the rate of "normalised bleeding", as a function of water-to-solids ratio. The value of $\rho_{ms}$ was taken approximately equal to 2.5 $g/cm^3$. Adapted from Miltiadou-Fezans and Tassios (2013b)

$$\alpha = 1 / [2.5\left(\frac{1}{\rho_{ms}} + \frac{S_A}{8000}\right)], \; [\text{in cm, g}] \tag{4.20}$$

$$\left(\frac{W}{S}\right)_{0,bl} \cong \frac{S_A}{8000}, \; [\text{in cm, g}] \tag{4.21}$$

Thus, "Eq. 4.19" could finally be written as follows:

$$b \cong \left(\frac{1/2.5}{\frac{1}{\rho_{ms}} + \frac{S_A}{8000}}\right)\left[\left(\frac{W}{S}\right) - \left(\frac{S_A}{8000}\right)\right], \; [\text{in cm, g}] \tag{4.22}$$

where $S_A$ is the average Blaine fineness of the solids (in $cm^2/g$) and $\rho_{ms}$ (in $g/cm^3$) is the average density of the grains of the solid phase.

It is reminded that because of the omission of the water density $\rho_w = 1 g/cm^3$, "Eq. 4.22" seems to be dimensionally incompatible; however, if $S_A$ and $\rho_{ms}$ are introduced in [cm, g], this equation is numerically correct.

Although a broader validity of this expression cannot be claimed, "Eq. 4.22" can nevertheless be very helpful in the preliminary grout mix design. Putting b = 5%, an approximate *critical* $(W/S)_{bl,crit.}$ value is found; the actual $(W/S)$ value of the mix should be lower than $(W/S)_{bl,crit.}$, which can be calculated by the following expression:

$$\left(\frac{W}{S}\right)_{bl,cr} = 1.1 \cdot \frac{S_A}{8000} + 0.04, \; [\text{in cm, g}] \tag{4.23}$$

Based on the data of Figs. 4.7, 4.8, 4.9, 4.10 and 4.11, the comparison of experimental to analytical values for this critical 5% bleeding seems satisfactory (see Fig. 4.14) for $S_A \leq 8000$ $cm^2/g$. However, the more specific role of the mixing method could be further taken into account.

### 4.4.2 The Role of Superplasticizers

As it is well known, superplasticizers are also added in hydraulic grouts containing ultrafine elements, in order to help deflocculation; thus, fluidity is further enhanced.

As far as the bleeding of these grouts is concerned, the same oversimplified model developed in Sect. 4.3 may also be used, as modified in Sect. 4.4.1. In fact, it is expected that the average thickness of inter-granular water « $t_0$» will be reduced thanks to the presence of the superplasticizer (SP). This reduction could be roughly estimated as follows:

$$(t_0)_{SP} = t_0 - \zeta \cdot (SP) \text{ and } \lambda \cdot (t_0)_{SP} = \lambda \cdot t_0 - \zeta \cdot \lambda \cdot (SP) \tag{4.24}$$

where $\zeta$ would be a constant, and (SP) the percentage of superplasticizer.

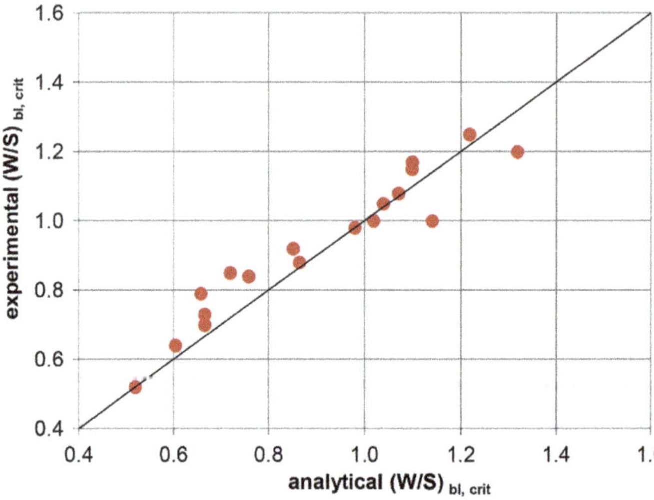


**Fig. 4.14** Comparison of analytical and experimental values of critical water-to-solids ratio leading to 5% bleeding. Adapted from Miltiadou-Fezans and Tassios (2013b)

Putting $\lambda \cdot t_0 = k_0$ in (g/cm$^2$) according to the units' simplification introduced in Sect. 4.4.1, it is possible to write

$$\lambda \cdot (t_0)_{SP} = k_{SP} = k_0 - \delta \cdot (SP), \ [g/cm^2] \tag{4.25}$$

where $\delta$ is another constant.

With reference to "Eq. 4.19", it is possible to predict the bleeding of grouts containing a percentage (SP) of superplasticizer, as follows:

$$b = \alpha_{SP}\left[\frac{W}{S} - \left(\frac{W}{S}\right)_{0,bl,SP}\right] \tag{4.26}$$

where $\alpha_{SP} = 1 / [2.5\left(\frac{1}{\rho_{ms}} + k_{SP} \cdot S_A\right)]$ and $\left(\frac{W}{S}\right)_{0,bl,sp} = k_{SP} \cdot S_A$, [in cm, g]

In what follows, the results of an experimental investigation on the subject are presented. Figures 4.15 and 4.16 (Miltiadou 1990) show the development of bleeding as a function of the water-to-solids ratio for several grouts made of cement C10 and Santorini earth (SE) or Lime (L), respectively, in which a melamine formaldehyde resin was used as superplasticizer in a percentage of dry resin of 0.33 of the weight of the solids of the grout (SP = 0.33%).

Furthermore in Fig. 4.17 are presented the results obtained for ternary grouts made by cement C24, Milos pozzolan (MP4) and lime (L3), in which the same type and SP percentage (0.33%) was also used (Miltiadou-Fezans et al. 2007).

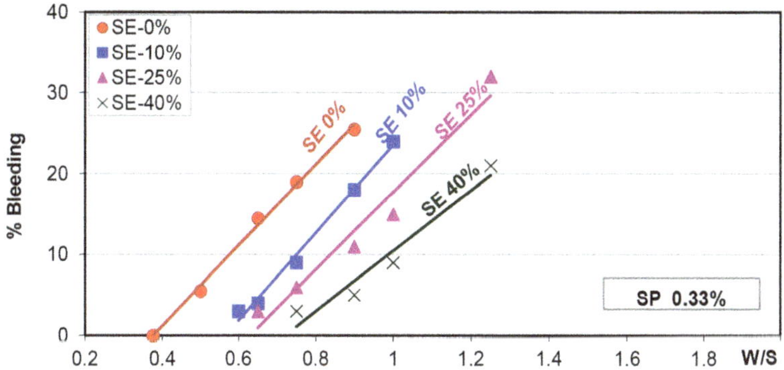


**Fig. 4.15** Bleeding versus (W/S), for cement C10 and Santorini earth (SE) grouts containing SP = 0.33%. Adapted from Miltiadou-Fezans and Tassios (2013b)

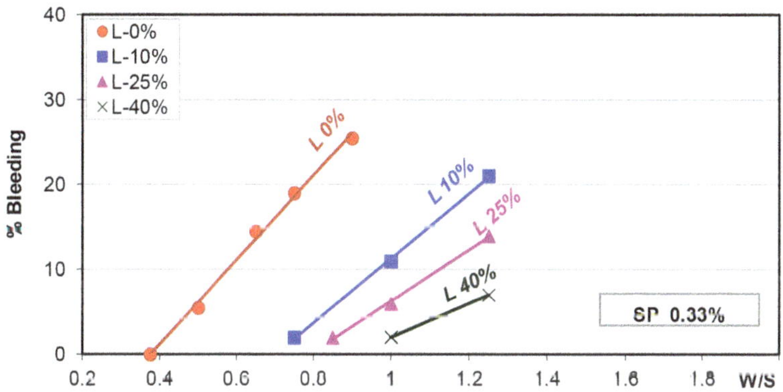


**Fig. 4.16** Bleeding versus $(W/S)$, for cement C10 and lime (L) grouts containing SP = 0.33%

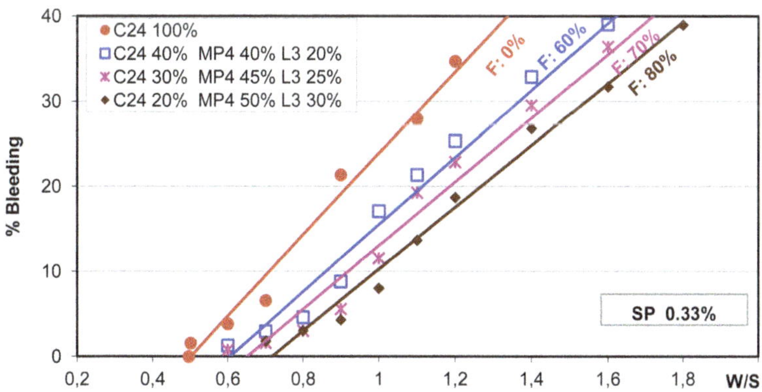


**Fig. 4.17** Bleeding values as a function of water-to-solids ratio for ternary grouts made of cement C24, Milos pozzolan MP4, lime L3 and SP = 0.33% (F = MP4 + L3). Adapted from Miltiadou-Fezans and Tassios (2013b)

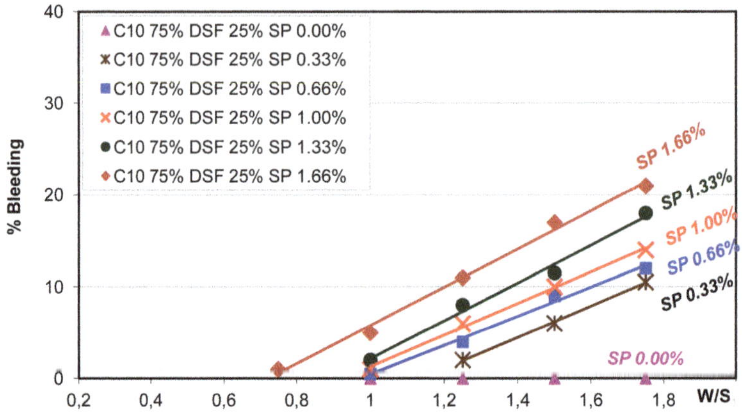


**Fig. 4.18** Bleeding versus $(W/S)$ for a 75% cement C10 and 25% densified silica fume grout (DSF), containing several percentages of superplasticizer. Adapted from Miltiadou-Fezans and Tassios (2013b)

In Table 4.2 the results of two more compositions of natural hydraulic lime grouts are also presented. Based on these experimental data, a calibration of "Eq. 4.24" is attempted in what follows.

Similar results were found in the case of superplasticizers used in grouts containing densified dilica fume (DSF) (Fig. 4.18, for SP-content equal or higher than 0.33%).

As expected, for each group of grout-materials, the slope of the straight-lines is moderately affected by the percentage of the superplasticizer (compare Fig. 4.15 to Fig. 4.8, as well as Fig. 4.17 to Fig. 4.11).

However, it is worth noting that in the case of binary "Cement + Lime" grouts (compare Fig. 4.16 with 4.9), the addition of a superplasticizer does not seem to substantially affect the bleeding; this phenomenon needs more specific investigation. On the other hand, it is very important to observe that, as it is known, in the specific case of DSF the obligatory use of SP produces a *very important increase* of bleeding rate as it is clearly demonstrated in Fig. 4.18.

Table 4.2 shows experimentally found $(W/S)_{0,bl,SP}$-values, as a function of the calculated average fineness of solids of each grout, for various percentages of superplasticizer (percentages of dry melamine formaldehyde resin, referred to the weight of the solids of the grout).

Based on the data of Table 4.2, the constant of "Eq. 4.25" may be estimated as follows:

$$(k)_{SP} = k_0 - 0.5 \cdot 10^{-4} \cdot (SP)^{\%} \text{ with } k_0 \approx 1.25 \cdot 10^{-4} = 1/8000\ [g/cm^2] \tag{4.27}$$

Thus, the other constant of "Eq. 4.25" is found to be roughly equal to $\delta \cong 0.5 \cdot 10^{-4}$. Consequently,

**Table 4.2** Values of $(W/S)_{0,bl,SP}$ as a function of the Blaine fineness of the solid phase of the grout, for various percentages of superplasticizer

| Grout | % Fines | Calculated average fineness ($S_A$) ($cm^2/g$) | 0% SP | | 0.16% SP | | 0.33% SP | |
|---|---|---|---|---|---|---|---|---|
| | | | $(W/S)_{0,bl.sp}$ | Calculated $10^4$ $k_0$ ($g/cm^2$) | $(W/S)_{0,bl.sp}$ | Calculated $10^4$ $k_{SP}$ ($g/cm^2$) | $(W/S)_{0,bl.sp}$ | Calculated $10^4$ $k_{SP}$ ($g/cm^2$) |
| C10 | 0 | 3489 | 0.43 | 1.24 | 0.41 | 1.17 | 0.38 | 1.09 |
| 90C10/10SE | 10 | 4490 | 0.65 | 1.45 | 0.62 | 1.38 | 0.57 | 1.27 |
| 75C10 /25SE | 25 | 5992 | 0.74 | 1.23 | 0.70 | 1.17 | 0.63 | 1.05 |
| 60C10/40SE | 40 | 7493 | 0.87 | 1.16 | 0.80 | 1.07 | 0.72 | 0.96 |
| 90C10/10L | 10 | 4940 | 0.73 | 1.48 | 0.72 | 1.46 | 0.70 | 1.42 |
| 75C10/25L | 25 | 7117 | 0.87 | 1.22 | 0.84 | 1.18 | 0.80 | 1.12 |
| 60C10/40L | 40 | 9293 | 0.97 | 1.05 | 0.94 | 1.01 | 0.90 | 0.97 |
| C24 | 0 | 4550 | 0.59 | 1.30 | | | 0.49 | 1.09 |
| 40C24/ 40MP4/20L3 | 60 | 6832 | 0.84 | 1.22 | | | 0.61 | 0.89 |
| 30C24/ 45MP4/25L3 | 70 | 7266 | 0.87 | 1.19 | | | 0.65 | 0.90 |
| 20C24/ 50MP4/30L3 | 80 | 7700 | 0.97 | 1.26 | | | 0.74 | 0.97 |
| NHL5 | 0 | 8000 | 0.82 | 1.02 | | | 0.66 | 0.82 |
| NHL3.5 Z | 0 | 8568 | 0.86 | 1.09 | | | 0.77 | 0.90 |
| **Average** | | | | **1.22 ~ 1.25** | | **1.21 ~ 1.20** | | **1.03 ~ 1.05** |

Adapted from Miltiadou-Fezans and Tassios (2013b)

$$b = \frac{1}{2.5\left[\frac{1}{\rho_{ms}} + \left(1.25 - 0.5 \cdot SP^{\%}\right) \cdot 10^{-4} \cdot S_A\right]} \times$$
$$\times \left[\left(\frac{W}{S}\right) - \left(1.25 - 0.5 \cdot SP^{\%}\right) \cdot 10^{-4} \cdot S_A\right], \;[\text{in cm, g}] \qquad (4.28)$$

Given that from "Eq. 4.26" $\left(\frac{W}{S}\right)_{0,bl,SP} = k_{SP} \cdot S_A$, the following expression can be written

$$\left(\frac{W}{S}\right)_{0,bl,SP} \cong \left(1.25 - 0.5 \cdot SP^{\%}\right) \cdot 10^{-4} \cdot S_A \;(\pm 0.10), \;[\text{in cm, g}] \qquad (4.29)$$

"Equation 4.29" is valid for this specific superplasticizer and for various compositions containing cement in combination with ultrafine materials or hydraulic lime. A correction term "± 0.10" was added here, in recognition of the simplicity of the analytical expression proposed.

In Fig. 4.19a and b the experimental values of $(W/S)_{0,bl,SP}$ are plotted as a function of the mean fineness (expressed in terms of calculated $S_A$) of the solid phase of the grouts, for 0.16% and 0.33% of superplasticizer respectively. In the same figure, the dotted lines predicted by "Eq. 4.27" (for 0.16% and 0.33% of SP respectively) are also presented; their validity is limited up to $S_A$-values lower than 8000 $cm^2/g$.

In conclusion, it is reconfirmed that superplasticizer, together with the enhancement of fluidity, produce a higher sensitivity to bleeding. For the particular condition of this investigation, it is of practical importance to predict, roughly though, the water-to-solids critical ratio resulting in a 5% bleeding as follows:

$$\left(\frac{W}{S}\right)_{bl,SP,crit} \cong 1.1 \cdot \left(1.25 - 0.5 \cdot SP^{\%}\right) \cdot 10^{-4} \cdot S_A + 0.04, \;[\text{in cm, g}] \qquad (4.30)$$

valid for a superplasticizer of melamine formaldehyde resin.

Based on the preceding methodology, similar expressions can be made ready, regarding other superplasticizers and/or other ultrafines. However, densified silica fume mixes are not easily amenable to this simplified methodology, because of the disproportionately high specific surface of their grains.

In a more practical fashion, however, the case shown in Fig. 4.18 could be empirically treated, e.g., by a best fitting curve:

$$\left(\frac{W}{S}\right)_{bl,SP,crit.} = 1.55 \cdot \left(1 - \frac{SP^{\%}}{4.30}\right) \quad \text{for SP} \geq 0.20\% \qquad (4.31)$$

an expression valid only for a binary grout "75% cement C10 + 25% densified silica fume" and the above-mentioned superplasticizer.

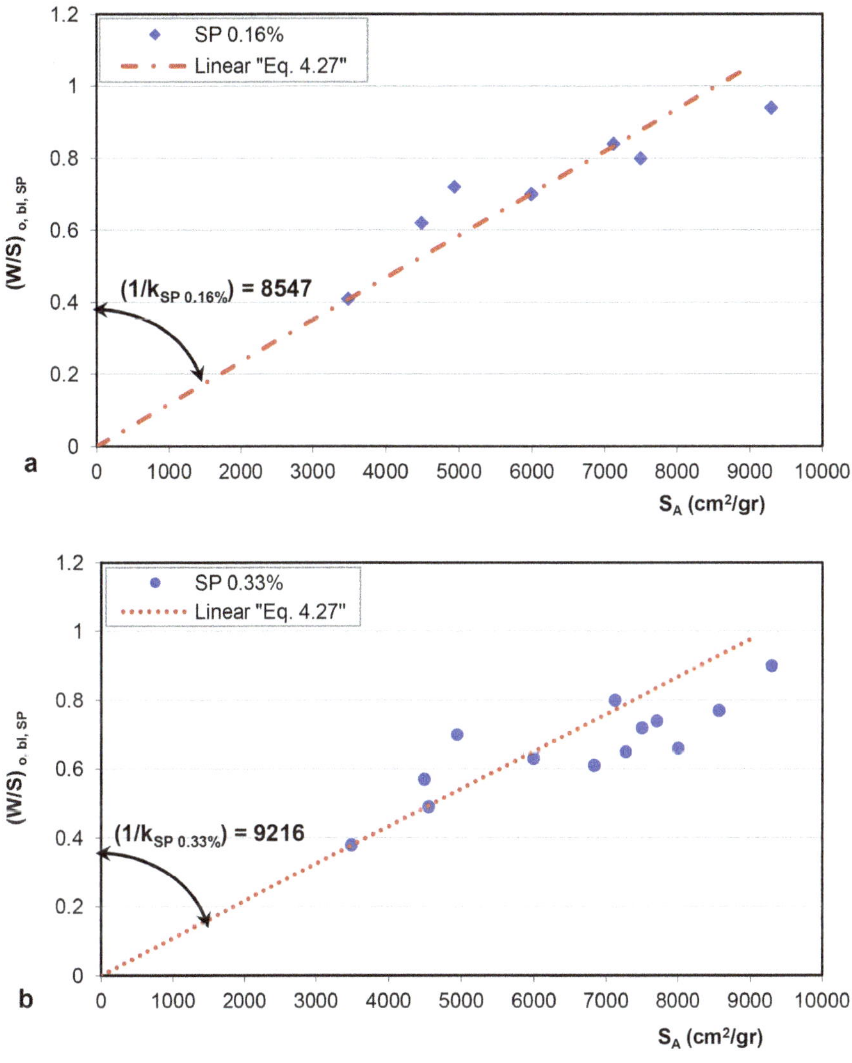


**Fig. 4.19** Experimental values of $(W/S)_{0,bl,SP}$ versus the mean fineness of the solid phase of the grouts, **a** for 0.16% and **b** for 0.33% of superplasticizer, against analytical predictions (dotted lines) made by means of the approximate expression proposed in "Eq. 4.27". Adapted from Miltiadou-Fezans and Tassios (2013b)

Interestingly enough, even in this case, linearity of relationships continues to be the rule.

## 4.5 Segregation

Hydraulic grouts may contain relatively high water and/or superplasticizers' percentages in order to satisfy fluidity requirements. Consequently, when at rest, such grouts tend to segregate; in other words, larger grains tend to move downwards.

Thus, grading within a vertically stored suspension mass is modified in time; upper masses get "thinner", and lower masses get "thicker". This "internal" differentiation of the mix is an additional instability phenomenon, as opposed to bleeding which does not necessarily affect the composition of the settled mass underneath the bleeded water. A qualitative estimator of this segregation is sought in what follows.

### *4.5.1 In-Time Modification of Grading of an Unstable Grout*

Suppose that small samples are taken from an upper layer of such a "grout column" at different times "$t_1$" and "$t_2$", after mixing. Samples are analysed using a laser granulometer; their gradings are found, and the respective $d_{50}$ values are read: $d_{50,1}$ at $t_1$ and $d_{50,2}$ at $t_2$ (see Fig. 4.20). One could formulate an empirical "Instability Estimator" (IE) as follows:

$$(IE) \cong \left(1 - \frac{d_{50,2}}{d_{50,1}}\right) \tag{4.32}$$

Its values may range between almost unity ("complete instability") and zero ("perfect stability").

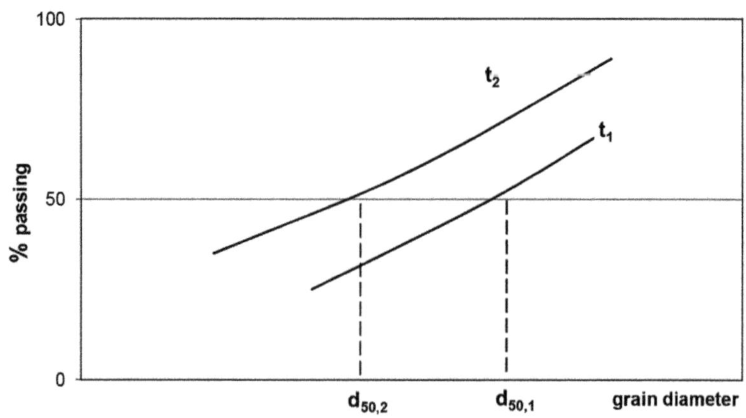


**Fig. 4.20** Grading curves of grout samples taken from an upper layer of a segregation-sensitive "grout column", at times $t_1$ and $t_2$. Adapted from Miltiadou-Fezans and Tassios (2013b)

Miltiadou (1990) has explored this approach for the assessment of segregation. Analysing samples taken from several suspensions (as indicated in Figs. 4.21 and 4.22), at several time-intervals, she was able to produce Fig. 4.23, where *shifting of* $d_{50}$*-values in time* is apparent (as observed in a purposely tall grout-container, so that the segregation tendency becomes more apparent).

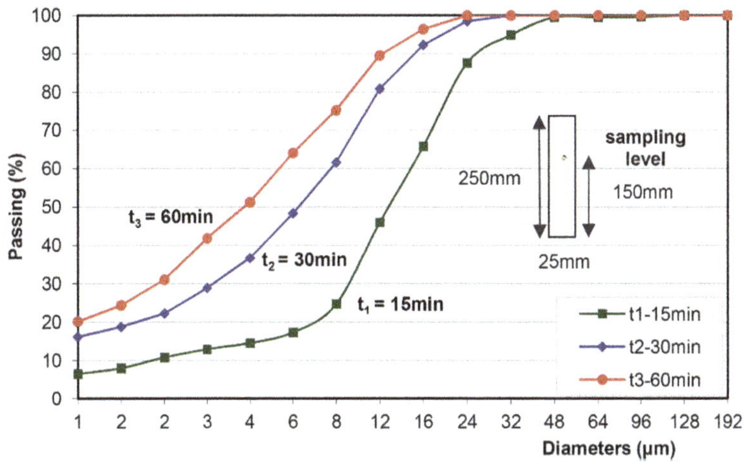


**Fig. 4.21** In-time modification of grading of an unstable grout containing cement and lime, exhibiting segregation (cement C10 75%, lime (L) 25%, (W/S) = 0.75, SP = 0.83% of the solid phase). Adapted from Miltiadou-Fezans and Tassios (2013b)

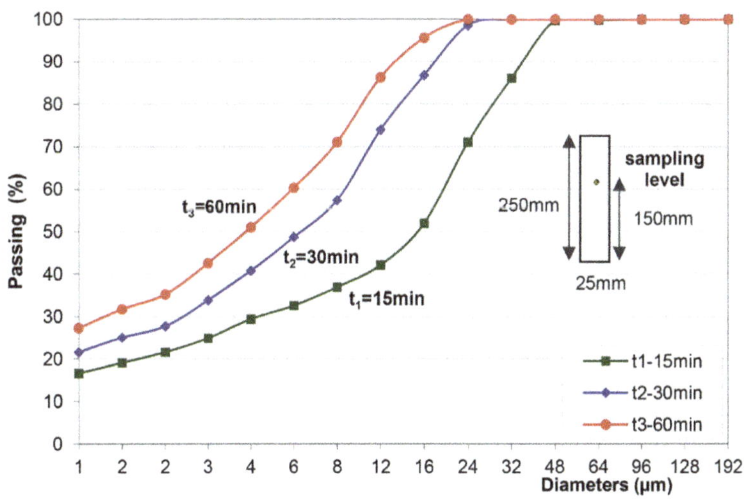


**Fig. 4.22** In-time modification of grading of an unstable grout containing cement and DSF, exhibiting segregation (cement C10 75%, DSF 25%, (W/S) = 0.75, SP = 1.33% of the solid phase). Adapted from Miltiadou-Fezans and Tassios (2013b)

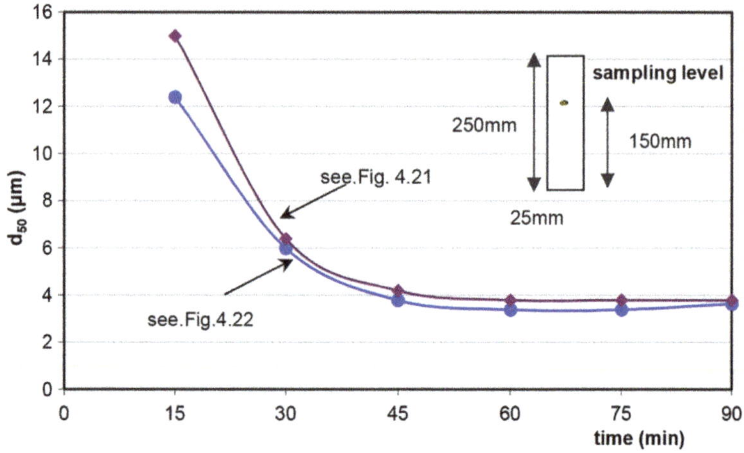


**Fig. 4.23** In-time "thinning" of suspension material, sampled in the upper part of the container (at a level 60% of its height). In unstable mixes (of Figs. 4.21 and 4.22), the mean diameter $d_{50}$ tends to drastically decrease in time. Adapted from Miltiadou-Fezans and Tassios (2013b)

For the same grout mixes, the value of the "IE" "Eq. 4.32" was calculated in the time-interval $t_1 = 15$ min and $t_2 = 30$ min, and a value $IE = 1 - \frac{6}{14} = 0.55$ was found.

Such an in-time differentiation in grading affects the apparent density of grout, consecutively sampled from the same area: Unstable mixes tend to exhibit lower apparent in-time densities of inevitably small samples taken.

### 4.5.2 In-Space Differentiation of Grading

Besides this "in-time" modification, "in-space" drastic differentiation of grading may also be observed in unstable mixes.

Such a differentiation, 15 min after mixing, is shown in Fig. 4.24, regarding samples of an unstable cement grout taken from near the top, the middle and near the bottom of a 250 mm tall container.

### 4.5.3 A Simple Analytical Model

A simple analytical model following the (IE) "Eq. 4.32" concept may be helpful in estimating the critical water-to-solids ratio, leading to an expression of "acceptable" degree of segregation.

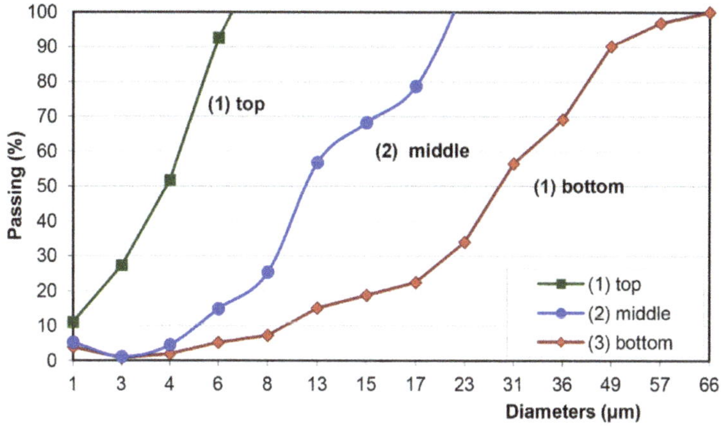


**Fig. 4.24** In-space modification of grading curves of samples of an unstable cement-grout (cement C10, W/C = 0.75 and SP = 0.66%), taken from the top, middle and bottom of the recipient, 15 min after mixing. Adapted from Miltiadou-Fezans and Tassios (2013b)

Roughly speaking, the ratio $\frac{d_{50,2}}{d_{50,1}}$ depends mainly on the internal cohesiveness "C" of the suspension. This cohesiveness depends on (i) the specific surface "$S_A$" and (ii) the water and "equivalent" superplasticizers contents.

An "equivalent W/S" value $\omega_{equ.}$ is introduced:

$$\omega_{equ.} = \frac{W}{S} + \mu(SP) \tag{4.33}$$

where "μ" denotes an equivalence-constant.

Thus, one could write:

$$\frac{d_{50,2}}{d_{50,1}} = \beta_1 S_A - \beta_2 \omega_{equ.} \tag{4.34}$$

for normal $S_A$-values (since otherwise, this linear model would be inappropriate).

Now, in order to formulate an expression for an "acceptable" (highest) value of instability estimator, $(IE)_{crit}$, one could possibly seek two critical parameters leading to an undesirable segregation, jeopardising injectability: First the equivalent water-to-solids ratio (accounting for the role of superplasticizers as well), and second, an index of *non-uniformity* of the grading of solid grains (such as the ratio $d_{95}/d_5$) enhancing the sedimentation of larger grains. Consequently, the "Eq. 4.34" should now be modified as follows:

$$\frac{d_{50,2}}{d_{50,1}} = \beta_1 S_A - \beta_1^{'} \left(\frac{d_{95}}{d_5}\right) - \beta_2 \omega_{equ.} \tag{4.35}$$

Finally, the instability estimator could be expressed as follows:

$$(IE) \cong 1 - \beta_1 S_A + \beta_1^{'} \left( \frac{d_{95}}{d_5} \right) + \beta_2 \omega_{equ.} \tag{4.36}$$

However, since the role of the non-uniformity index of the grading of the solids is not clearly known, "Eq. 4.36" may be more practically modified as follows:

$$(IE) \cong z - \beta_1 S_A + \beta_2 \omega_{equ.} \tag{4.37}$$

With $z > 1$, an unknown parameter expressing the role of the non-uniformity of grading of the solids.

Now, "Eq. 4.37" could be written as follows:

$$\beta_2 \omega_{crit} = [(IE)_{crit} - z] + \beta_1 S_A \tag{4.38}$$

or, via "Eq. 4.33",

$$\left( \frac{W}{S} \right)_{segr} = \left[ \frac{(IE)_{crit} - z}{\beta_2} \right] + \frac{\beta_1}{\beta_2} S_A - \mu(SP) \tag{4.39}$$

or

$$\left( \frac{W}{S} \right)_{segr} = \lambda_1 + \lambda_2 S_A - \lambda_3 (SP) \tag{4.40}$$

where:

$\lambda_1$: a constant, reflecting the lowest acceptable level of instability $(IE)_{crit}$, as well as the non-uniformity of the grading of the solids.
$\lambda_2$, $\lambda_3$: numerical factors.
$S_A$: global specific surface of solids of the mix, and.
*SP*: percentage of the superplasticizer.

"Equation 4.40" seems to be promising for further research, towards a further rationalisation of the segregability criteria.

However, instead of this analytical approach, a more practical one is described in the following paragraph.

### 4.5.4 Practical Criterion of Segregability

Another more practical criterion of segregability is introduced: The trial mix is poured into a 10 cm diameter container, 15 cm high. After approximatively 15 min, the bottom of the container is observed and the possible appearance of a distinct denser layer sedimented at the bottom is examined. The thickness of such a layer is measured, and the following empirically found criterion is applied: If this thickness is greater than approximately 1 mm, the mix is considered as unstable; this segregation makes it inappropriate for use, and the mix has to be rejected.

Such a criterion is applied in what follows.

For each solid phase (combination of cement C10 and various percentages of SE, L and DSF) and for a given $(W/S)$ a series of grouts was tested for different SP (melamine formaldehyde resin) contents (Miltiadou 1990). In Table 4.3 are reported the values of $(W/S)$ and SP%, for which a "distinct denser layer" (thickness greater than 1 mm) appears at the bottom of the container. These experimental results are also presented in Fig. 4.25.

For Santorini earth and lime ultrafines, the straight lines of Fig. 4.25 may be roughly expressed as follows (where average values of constants were selected):

$$\left(\frac{W}{S}\right)_{segr} \cong 0.80 + \left(\frac{S_A}{6000}\right) - 1.7 \cdot (SP)^{(\%)}, \text{ where } \mathrm{S_A}\left[\mathrm{cm^2/g}\right] \tag{4.41}$$

**Table 4.3** Values of the SP%, in combination with various water contents, for which a "distinct denser layer" appeared on the bottom of the recipient, considered as an initiation of unacceptable segregation

| Solid phase of grout | $S_A$ cm²/g | SP% upper values for which segregation is taking place corresponding to various water to solid ratios | | | | | | | | |
|---|---|---|---|---|---|---|---|---|---|---|
| | | W/S | | | | | | | | |
| | | 0.50 | 0.65 | 0.75 | 0.90 | 1.00 | 1.15 | 1.25 | 1.50 | 1.55 |
| 90C10 10SE | 4490 | 0.50 | 0.50 | 0.416 | 0.33 | 0.25 | | 0.166 | | |
| 75C10/ 25SE | 5992 | | 0.58 | 0.50 | 0.417 | 0.33 | | 0.25 | | |
| 60C10/ 40SE | 7493 | | 0.75 | 0.66 | 0.58 | | | 0.40 | | |
| 90C10/ 10L | 4940 | | 0.66 | | | 0.50 | | 0.33 | | |
| 75C10/ 25L | 7117 | | 0.83 | 0.83 | | 0.66 | | 0.50 | | |
| 60C10/ 40L | 9293 | | | 1.00 | | 0.83 | | 0.75 | 0.58 | |
| 90C10/ 10DSF | 24780* | | 1.00 | | | 0.75 | | 0.66 | 0.50 | |
| 75C10/ 25DSF | 54100* | | | 1.00 | | 0.83 | | 0.75 | 0.58 | |
| 60C10/ 40DSF | 86560* | | | 2.33 | | | 2.00 | 2.00 | | 1.83 |

(*) The $S_A$ values of the grout solid phase were calculated using the specific surface value of DSF measured by nitrogen adsorption method (BET), which was found to be equal to 21.64 m²/g.
Adapted from Miltiadou-Fezans and Tassios 2013b

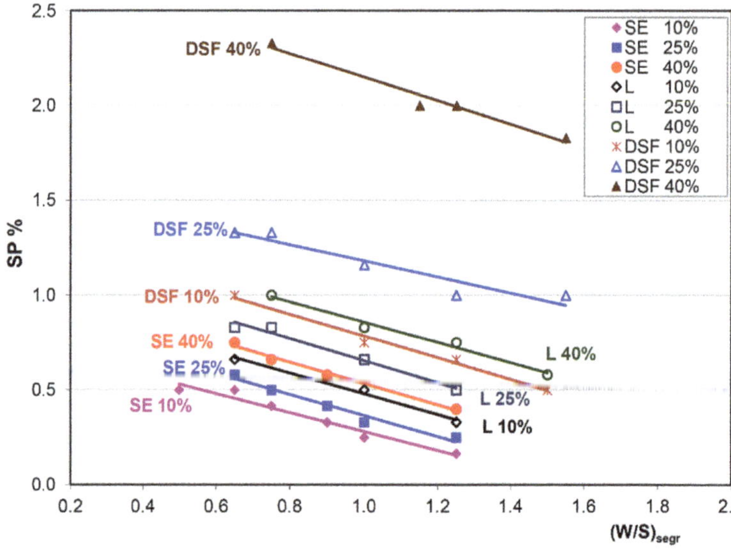


**Fig. 4.25** Combinations of $(W/S)_{segr}$ and SP% values, for which a "bottom distinct denser layer" appears in various grouts, respecting penetrability and fluidity criteria. Their solid phase was cement C10 and various percentages of SE, L and DSF. Adapted from Miltiadou-Fezans and Tassios (2013b)

It is very *interesting* to observe that "Eq. 4.41" is similar to "Eq. 4.40", although it was based on a different and much more empirical experimental methodology.

Although in the case of densified silica fume the $S_A$-values mentioned in Table 4.3 have only a nominal significance, one could also write for DSF-containing grouts an expression similar to that of "Eq. 4.41":

$$\left(\frac{W}{S}\right)_{segr} \cong 1.60 + \left(\frac{S_A}{30000}\right) - 1.7 \cdot (SP)^{(\%)}, \text{ where } S_A\left[\text{cm}^2/\text{g}\right] \qquad (4.42)$$

It is also interesting to observe that the "sensitivity" of all grouts versus potential segregation caused by excessive percentage of superplasticizers, is independent of the nature of grouts (observe the common constant equal to 1.7 in all cases).

### 4.5.5 Cohesiveness Index

In this paragraph a reference will be made to the practical Cohesiveness Index (C.I.), based on the Fluidity Factor Test (FFT) method (see Fig. 3.6 and Table 3.3). In Table 4.4 (essentially, a repetition of Table 3.3) are presented the rheological characteristics of cement C23 [Blaine Specific surface $S_b = 4100$ cm$^2$/g, satisfying the penetrability grading rules proposed by Miltiadou-Fezans and Tassios (2013a) for a $W_{nom} =$ 0.108 mm, corresponding to the 0.63–1.25 mm sand column], plus Santorini earth

**Table 4.4** Rheological characteristics of cement C23 and Santorini earth SE1 grout compositions (SP% = 0), measured by means of the FFT method, and compared to limit W/S-values observed in the sand-column test and in stability tests. Adapted from Miltiadou-Fezans and Tassios (2012)

| SE1 % | $S_A$ cm$^2$/g | Fluidity test (FFT) | | Penetration in the sand column 0.63–1.25 mm | Fluidity factor $10^{-3}$ max $F_l$ (mm/s) | C.I (s/mm) (Fig. 3,4) | Initiation of bleeding | Initiation of segregation "Eq. 4.41" |
|---|---|---|---|---|---|---|---|---|
| | | $(W/S)_{o.f}$ | $(W/S)_{u.f}$ | $(W/S)_{sand\text{-}col}$ | | | $(W/S)_{0\text{-}bl}$ | $(W/S)_{segr}$ |
| 0 | 4100 | 0.47 | 0.75 | 0.70 | 2.13 | 131 | 0.55 | 1.48 |
| 10 | 4550 | 0.50 | 0.77 | 0.75 | 2.00 | 135 | 0.64 | 1.56 |
| 25 | 5225 | 0.53 | 0.82 | 0.75 | 1.78 | 163 | 0.72 | 1.67 |
| 40 | 5898 | 0.58 | 0.92 | 0.80 | 1.73 | 196 | 0.80 | 1.78 |
| 80 | 7697 | 0.63 | 1.04 | 1.10 | 1.56 | 264 | 0.89 | 2.08 |

$S_A$: Calculated average fineness of the solid phase of the grout, on the basis of the measured fineness of its constituents

$(W/S)_{o,f}$: Absolute minimum possible water-to-solids ratio for fluidity

$(W/S)_{u,f}$: Ultimate value of water-to-solids ratio, which practically results in the maximum possible fluidity factor

$(W/S)_{segr}$: Limit water to solid phase ratio resulting in segregation, calculated using the "Eq. 4.41"

SE1 (with max d < 20 μm and $S_b$ = 8596 cm$^2$/g) grout compositions (with SP% = 0), measured by means of the FFT method, and compared to limit W/S-values observed in the sand-column test and in stability tests (bleeding and segregation) (Miltiadou-Fezans and Tassios 2012).

So, it was proven that such a cohesiveness index is proportional to the critical W/S ratio initiating segregation, calculated using the "Eq. 4.41". Consequently, it seems that it is possible to use this practical cohesiveness index as another criterion of the stability of a grout. Despite the inadequacy of actually available data, the comparison of columns 7 and 9 of Table 4.4, seems to suggest that, very roughly though,

$$\left(\frac{W}{S}\right)_{segr} \approx \frac{C.I.^{[s/mm]}}{90} \quad (\pm 10\%) \tag{4.43}$$

for $S_A$-values lower than 7000 cm$^2$/g.

Further experience is needed in order to customise such a relationship, in terms of more specific data. For the time being, however, a comparison between results of "Eqs. 4.43 and 4.41" (see Table 4.5), seems to be practically satisfactory, although the two equations under consideration, result from different series of tests.

Finally, it is worth noting that, as it is known the so called "water retention capacity" of the grains is an important parameter against segregation, recognised in the preceding analysis as well, via the Cohesiveness Index. In fact, referring to Fig. 3.6, where the concept of Cohesiveness Index is introduced, the following clarifications are needed: The $(W/S)_{0,f}$ -value corresponds to the initiation of "separation" of solid grains, due to a surplus of water content.

**Table 4.5** Comparison between results of "Eqs. 4.43 and 4.41" for the estimation of the critical W/S ratio initiating segregation $(W/S)_{segr}$

| | | | $(W/S)_{segr}$ | |
|---|---|---|---|---|
| SE % | C.I | $S_A$ $cm^2/g$ | "Equation 4.43" | "Equation 4.41" |
| 0 | 131 | 4100 | 1.45 | 1.48 |
| 10 | 135 | 4550 | 1.5 | 1.56 |
| 25 | 163 | 5225 | 1.81 | 1.67 |
| 40 | 196 | 5898 | 2.18 | 1.78 |

For the examined grouts, see Table 4.4. Adapted from Miltiadou-Fezans and Tassios (2012)

Additional water put then into the mix may be largely retained by the grains. But, later on, at $(W/S)_{u.f}$, when the fluidity factor cannot be further increased, the retention capacity is practically overcome, and abundant free water circulates in-between the grains. That is why, larger values of the difference $\{(W/S)_{u.f}\text{-}(W/S)_{0,f}\}$ correspond to higher water retention capacities (i.e. to higher content of ultrafine materials). Thus, the water retention capacity of the grains may be expressed via a "Cohesiveness" Index, which was rightly recognized to constitute an anti-segregation criterion.

## 4.6 Conclusions

Grout mixes exhibiting appropriate penetrability and fluidity, should also be checked against their possible instability. Otherwise, blockage may soon appear, and the quality of the intervention could be severely affected. In order to achieve a satisfactory injectability, stability of the suspension against excessive bleeding or segregation should be ensured.

- In the framework of a broader attempt to establish a holistic rational methodology for the design of hydraulic grouts, the critical relationships between the most predominant parameters shaping stability characteristics (i.e. water content and specific surface of the solid phase expressed by the percentage of ultrafine materials replacing cement or hydraulic lime) were examined in this chapter.
- To this end, an oversimplified predictive model of bleeding was proposed and its validity was confirmed using the results of an extensive experimental study.
- Semi-empirical formulae were proposed for the estimation of bleeding for various grout compositions (based on cement or hydraulic lime associated to ultrafine materials, with or without superplasticizer), that may be useful for the design of a grout composition.
- Thus, acceptable normalised bleeding equal to 5% may be secured by means of the predictive "Eq. 4.23" or "Eq. 4.30", as a function of the water-to-solids ratio, the average specific surface of the solids and the percentage of the superplasticizer used.

- On the other hand, it was proposed that segregation of relatively thin grouts could theoretically be checked by means of a general expression like "Eq. 4.40". Thus, the critical water-to-solids ratio resulting in segregation is found to depend on the acceptable degree of instability, the average specific surface of solids and the percentage of superplasticizer used. However, more data are necessary for further exploitation of this analytical approach.
- Moreover, based on experimental results, the empirical formulae "Eqs. 4.41 and 4.42" were proposed for the estimation of the critical water content initiating segregation, that may also be useful for the design of a grout composition.

It has to be noted, however, that those formulae may be practically employed for specific cases, similar to those examined in the experimental studies carried out in the framework of this book; further research is necessary for their broader validation. Nevertheless, even under different circumstances, it is believed that the quantative guidance given in this chapter, may be of some practical utility for a pre-design stage of grouts.

## References


Aïtcin PC, Ballivy G, Parizeau R (1984) The use of condensed silica fume in grouts. Innovative cement grouting ACI Special Publication SP-83, pp 1–18

ASTM C 940–16 Standard test method for expansion and bleeding of freshly mixed grouts for preplaced aggregate concrete in the laboratory. ASTM International, 2016

Badogiannis E, Miltiadou-Fezans A, Kalagri A, Ipsilanti E, Dedeloudis C, Vintzileou E (2012). Mix design and performance evaluation of grouts with superfine natural pozzolan. In: Jasieńko J (ed) Structural analysis of historical constructions. DWE Wrocław Poland, pp 773–780

Benhamou O (1994) Comportement rhéologique des coulis de liants hydrauliques ultrafins destinés à l'injection. Ph.D. Thesis, ENS des Mines de Paris, pp 1–318

Bombled JP (1974) Rhéologie des mortiers et des bétons frais, étude de la pâte interstitielle de ciment, Revue des matériaux de construction, n°688, Mai-Juin pp.137–155

Bras A, Henriques FMA (2009) The influence of the mixing procedures on the optimization of fresh grout properties. Mater Struct 42:1423–1432

Bras A, Henriques FMA (2012) Natural hydraulic lime-based grouts—The selection of grout injection parameters for masonry consolidation. Constr Build Mater 26(2012):135–144

Domone PL, Tank SB (1986) Use of condensed silica fume in portland cement grouts In: Proceedings of the second international conference on fly ash, silica fume, slag and natural Pozzolans in concrete, Madrid, Spain, Vol. 2, pp 1231–1260

EN445:2007—Grout for prestressing tendons—Test methods

EN447:2007—Grout for prestressing tendons—Basic requirements

Fascicule No 65 : Exécution des ouvrages de génie civil en béton armé ou précontraint (1985). In : Cahier des clauses techniques générales- Décret no 85–404 du 3 avril 1985. Marchés Publics de Travaux. Ministère de l'Urbanisme, du Logement et des Transports - Ministère de l'Économie, des Finances et du Budget, France.

Ish-Shalom M, Greenberg SA (1962) The rheology of fresh portland cement paste. In: Proceedings IV international on symposium on the chemistry of cement, Washington 1960, Washington D.C., National Bureau of Standards, 1962, Monograph 43, Vol. II, pp. 731–748

Jorne F, Henriques FMA, Baltazar LG (2015) Injection capacity of hydraulic lime grouts in different porous media. Mater Struct 48:2211–2233. https://doi.org/10.1617/s11527-014-0304-9

Kalagri A, Miltiadou-Fezans A, Vintzileou E (2010) Design and evaluation of hydraulic lime grouts for the strengthening of stone masonry historic structures. Mater Struct 43:1135–1146, doi 10.1617/s11527-009-9572-1
Legrand C (1972) Contribution à l'étude de la rhéologie du béton frais. Matériaux et Constructions, Vol. 5, No. 29, pp 275–295, and No. 30, pp 379–393
Legrand C (1982) La structure des suspensions de ciments. Le béton hydraulique, Presses de l'École Nationale des Ponts et Chaussées, Paris, pp 99–11
Littlejohn GS (1982) Design of cement based grouts. In: Proceedings of the conference on grouting in geotechnical engineering. ASCE, New Orleans, Louisiana, February 10–12, 1982, pp 1–7
Lombardi G (1985) The role of cohesion in cement grouting of rock. In: Procedings of 15th congress on large dams. Com Int Des Grands Barrages 3:235–261
Lombardi G (1990) La perméabilité et injectabilité des massifs rocheux fissures. Rev. Français Géotechnique 51:5–29
Markestad SA (1976) The Gukild-Carlsen method for making stabilized pastes of cement. Mater Struct 9(50):115–117
Miltiadou AE (1990) Étude des coulis hydrauliques pour la réparation et le renforcement des structures et des monuments historiques en maçonnerie. Thèse de Doctorat de l'Ecole Nationale des Ponts et Chaussées. Pub. by LCPC in Collection Etudes et Recherches des Laboratoires des Ponts et Chaussées, série Ouvrages d'art, OA8 ISSN 1161–028X, LCPC, Décembre 1991, Paris, France, p 278
Miltiadou-Fezans A, Papakonstantinou E, Zambas K, Panou A, Frantzikinaki K (2005) Design and application of hydraulic grouts of high injectability for the structural restoration of the column drums of the Parthenon Opisthodomos. In: Brebbia CA, Torpiano A (eds) Structural studies, repairs and maintenance of heritage architecture IX, WIT Press, Southampton, pp 461–472
Miltiadou-Fezans A, Tassios TP (2012) Fluidity of hydraulic grouts for masonry strengthening. RILEM Mater Struct https://doi.org/10.1617/s11527-012-9872-8
Miltiadou-Fezans A, Tassios TP (2013a) Penetrability of hydraulic grouts. RILEM Mater Struct https://doi.org/10.1617/s11527-012-0005-1
Miltiadou-Fezans A, Tassios TP (2013b) Stability of hydraulic grouts for masonry strengthening. RILEM Mater Struct https://doi.org/10.1617/s11527-012-0003-3
Miltiadou-Fezans A, Tassios TP (2016) Holistic methodology for the mix design of hydraulic grouts in strengthening historic masonry structures. In: Papagianni I, Stefanidou M, Pachta V (eds) Proceedings of 4th historic mortars conference (HMC2016), 10th-12th Ocrtober 2016, Santorini, Greece, pp 580–587
Miltiadou-Fezans A., Kalagri A, Savvidou M (2007) Investigations for the design of grouts for the structural restoration of the Church of the Assumption of Virgin Mary in Tegea Arkadias. Research report, Directorate for Technical Research on Restoration of the Hellenic Ministry of Culture and Sports (in Greek)
Miltiadou-Fezans A, Kalagri A, Kakkinou S, Ziagou A, Delinikolas N, Zarogianni E, Chorafa E (2008) Methodology for in situ application of hydraulic grouts on historic masonry structures. The case of the Katholikon of Dafni Monastery. In: D' Ayala D and Fodde E (eds), Structural Analysis of Historic Construction, Preserving Safety and Significance, Proceedings of 6th International Conference on SAHC, 2–4 July, Bath, CRC Press/Balkema, Taylor and Francis Group, London UK, Vol. II, pp 1025–1033
Mirza J, Mirza MS, Roy V, Saleh K (2002) Basic rheological and mechanical properties of High-volume fly ash grouts. Constr Build Mater 16:353–363
Mirza J, Saleh K, Langevin MA, Mirza S, Bhutta MAR, Tahir MMD (2013) Properties of microfine grouts at 4° C, 10° C and 20° C. Constr Build Mater 47(2013):1145–1153
NF P18–359 (1985) Adjuvants pour bétons, mortiers et coulis. Coulis courants d'injection pour précontrainte—Mesure de l'exsudation (stabilité)
Nikolopoulou V, Adami CE, Karagiannaki D, Vintzileou E, Miltiadou-Fezans A (2018) Grouts for strengthening two- and three-leaf stone masonry, made with earthen mortars. Int J Arch Heritage Conserv Anal Restorat 13(5):663–678. https://doi.org/10.1080/15583058.2018.1463414
Ozcan T, Zaimoglou AS, Hinislioglou S, Altun S (2005) Taguchi approach for optimization of the bleeding on cement-based grouts. Tunn Undergr Space Technol 20(2005):167–173

Papadakis M (1957) Recherches sur le malaxage “a haute turbulence” des suspensions de ciment. Publication Technique No. 82–83, Extrait de la Revue des Matériaux de Construction, CERILH, pp. 1–25

Papadakis M (1959) L’injectabilité des coulis et mortiers de ciments. Revue des matériaux de construction, 531, publication technique No. 11 du CERILH, 48 pages

Papayianni I, Pachta V (2015) Experimental study on the performance of lime-based grouts used in consolidating historic masonries. Mater Struct 48(7):2111–2121

Powers T-C (1968) The properties of fresh concrete. Wiley, New York, pp 533–599

Rosquoët F, Alexis A, Khelidj A, Phelipot A (2003) Experimental study of cement grout: Rheological behavior and sedimentation. Cem Concr Res 33(2003):713–722

Saric-Coric M, Khayat KH, Tagnit-Hamou A (2003) Performance characteristics of cement grouts made with various combinations of high-range water reducer and cellulose-based viscosity modifier. Cem Concr Res 33:1999–2008

Shannag MJ (2002) High-performance cementitious grouts for structural repair. Cem Concr Res 32:803–808

Shimoda M, Ohmori H (1982) Ultra fine grouting material. Grouting in Geotechnical Engineering, New Orleans, 10–12 February, pp 77–91

Toumbakari E-E (2002) Lime-pozzolan-cement grouts and their structural effects on composite masonry walls. Ph.D. Thesis, Departement of Civil Engineering, Katholieke Universiteit Leuven, September 2002

Valluzzi MR, da Porto F, Modena C (2003) Grout requirements for the injection of stone masonry walls. International Conference on Performance of Construction Materials—A New Era of Building, February 18–20, 2003, Cairo, Egypt

Van Rickstal F (2000) (2000) Grout injection of masonry, scientific approach and modeling. Ph.D Thesis, Departement of Civil Enginering, Katholieke Universiteit Leuven, May 2000

Viseur V, Barrioulet M (1998) Critères d’injectabilité de coulis de ciments ultrafins. Mater Struct 31:393–399

Vom Berg W (1979) Influence of specific surface and concentration of solids upon the flow behaviour of cement pastes. Magazine of Concrete Research, Vol. 31. No. 109:211–216

Yahia A, Khayat KH (2003) Applicability of rheological models to high-performance grouts containing supplementary cementitious materials and viscosity enhancing admixture. Materials and Structures, Vol. 36, July 2003, pp 402–412

Zebovitz S, Krizek RJ, Atmatzidis DK (1989) Injection of sands with very fine cement grout. Journal of Geotechnical Engineering, Vol. 115, no 12, December 1989, ASCE, ISSM 0733–9410/89/0012–1717, paper 24121

# Chapter 5
# Guidelines for the Estimation of $W_{nom}$

**Abstract** This Chapter describes the categories of possible internal discontinuities of masonry; it is because of such discontinuities (pores, local interface detachments, local sliding's, cracks) that masonry strength may be reduced. The filling of these discontinuities by means of an appropriate grout may increase masonry strength, provided that the grout was able to penetrate the body of masonry, to reach most of these discontinuities and flow along each of them. In order to decide the necessary "penetrability" capacity of the grout, a rough evaluation of a critical value "$W_{nom}$" of the opening of these discontinuities of masonry, is needed. This chapter examines several possibilities of quantification of such a representative opening value for several categories of masonry. Finally, an easy to apply practical approach of "opening classes" is proposed, and relevant examples are given. Thus, for each specific case, the selection of grout solids' grading is facilitated, in order to satisfy penetrability requirements established in Chap. 2.

## 5.1 Introduction

Scope of this chapter is first to describe the categories of possible internal discontinuities of masonry, since it is because of such discontinuities (pores, local interface detachments, local slidings, cracks) that masonry-strength is reduced. The filling of these discontinuities by means of an appropriate grout may increase masonry strength, provided that the grout was able to penetrate the body of masonry, to reach most of these discontinuities and flow along each of them. In order to decide the necessary "penetrability" capacity of the grout, a rough evaluation of a *critical*value "$W_{nom}$" of the opening of these discontinuities is needed; the concept of this nominal discontinuity width was introduced in Sect. 2.2.2.

In what follows, a qualitative description of several categories of possible internal discontinuities of masonry is presented.

**Generalized tensile or shear cracks through the body of masonry.** A possible repair-strategy aiming at the mere "filling" of these cracks (Fig. 5.1a), without any strengthening of the surrounding masonry regions, could hardly be considered as an appropriate engineering solution: Grout should penetrate through masonry body, long

 
A. Miltiadou-Fezans and T. P. Tassios, *Mix-Design and Application of Hydraulic Grouts for Masonry Strengthening*, Springer Tracts in Civil Engineering,
https://doi.org/10.1007/978-3-030-85965-7_5

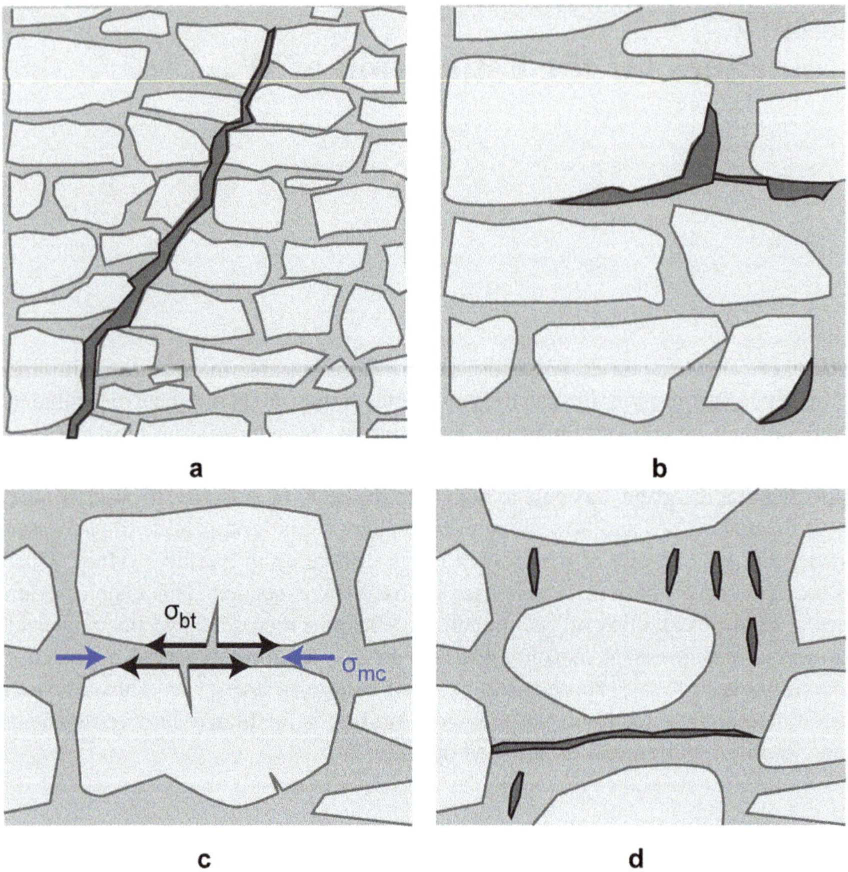


**Fig. 5.1** Schematic presentation of several categories of possible internal discontinuities of masonry

enough as to reach all the crack depth. Besides, most probably, other areas around the crack may be submitted to higher stresses in the future; thus, they necessitate an early strengthening before being potentially cracked.

In conclusion, even in the case of generalized cracks, decision making regarding the representative $W_{nom}$-value should be based on discontinuity categories other than these cracks.

**Mortar-to block interfaces.** Such interface-discontinuities (Fig. 5.1b) may be produced because of:

- Building gross-errors
- High shrinkage values of thick mortar layers
- Hidden sliding shear cracks along masonry joints.

Their filling with grout is of major structural importance. However, to this end, appropriate flow-routes to reach them should be available, thanks to the selection of an adequately small value of $W_{nom}$, possibly smaller than the opening of the aforementioned discontinuities.

**Cracks on blocks** may appear (Fig. 5.1c), due to:

- The incompatibility of Poisson's ratio between blocks and mortar
- Local shear stresses
- Structural deficiencies of blocks.

Here again, appropriate flow-routes of grouts are of primary importance.

**Internal mortar discontinuities** (Fig. 5.1d), such as:

- Construction large voids
- Tensile mortar cracks due to internal thermal stresses
- Sliding shear cracks through thick mortar layers.

Obviously closed pores of constitutive materials are not considered as "discontinuities".

All the aforementioned categories are often coexisting in masonry structures. Figure 5.2 offers an impression of some of these voids in three-leaf masonry. Furthermore in Figs. 5.3 and 5.4 some sections of two and three-leaf masonries are presented. As it can be clearly seen, large and fine cracks are present, both in case of low quality and good quality masonries, built with lime mortar or lime and earth mortar or earth mortar.

An attempt is made in what follows, to formulate some rules assisting the designer of the grout to evaluate the average level of hindrance ("dis-penetrability") exhibited by a masonry to be penetrated by an appropriate grout. In doing so, however, it is important to note that our scope is not to fill "all" voids of masonry:

- First, because of the need of masonry-wall to "breath".

**Fig. 5.2** A three-leaf masonry. **a** The entire section and **b** a detail of the section

**Fig. 5.3** Sections of two-leaf masonries, where, large and fine cracks and other discontinuities are present

**Fig. 5.4** Sections of three-leaf masonries, where, large and fine cracks and other discontinuities are present. 2nd and 4th photo courtesy of AB Raptis

- Second, because a considerable increase of the strength of masonry elements may be achieved after filling voids of a nominal minimum width of the order of 100–300 μm, as proved by relevant experimental research works (Miltiadou 1990; Vintzileou and Tassios 1995; Valuzzi 2000; Toumbakari et al. 2005; Vintzileou and Miltiadou-Fezans 2008; Kalargi et al. 2010; Mazzon 2010; Silva 2012; Vintzileou et al. 2015; Nicolopoulou et al. 2018; Mouzakis et al. 2017; Luso and Lourenço 2019; see also Chap. 6).
- Third, because the grouting-costs are disproportionally higher when the targeted opening $W_{nom}$ of voids would be considerably decreased.

Thus, an optimum selection of the targeted $W_{nom}$-value should be sought. Experience shows that on account of the aforementioned reasons, masonries with $W_{nom}$-values lower than 0.1mm do not need to be grouted. Nevertheless, fissured large blocks incorporated in masonry, may need special grouting even for lower $W_{nom}$-values, as it will be judjed by the structural Engineer. In such cases, commun hydraulic grouts are not adequate; special ultrafine materials have to be used.

## 5.2 Information on Existing Internal Discontinuities

For the design of a grout, the advantages of the $W_{nom}$ approach are quite clear: a quantification of the penetrability problem is feasible, and a more general quasi-theoretical handling of it is facilitated. However, the difficulties in estimating such a "representative effective minimum opening" of discontinuities should not be underestimated. The following methods could possibly be envisaged for such an estimation:

**Statistical evaluation in case of a network of visible cracks** (see Figs. 5.5 and 5.6).

**Examination of endoscopical observations,** may provide very useful information. In Fig. 5.7 endoscopy images from the interior of the masonries of the Katholikon of Dafni Monastery (Vintzileou 2003) are presented. Various types of discontinuities are present.

Figure 5.8 shows endoscopy images from the masonry of the Athens's Cathedral (Geoereuna Laboratory-O.T.M, EPE 1995).

More precise information one may have when using endoscopical observations in the form of printed developments of cylindrical surfaces. A more sophisticated technique (Borehole Image Processing System) allows a colour picture of the entire surface of the hole (360°) developed on a plane to be taken. The use of such a technique in the case of the structural restoration of the Pisa Tower gave the possibility to measure the width of cracks and cavities and to detect the detachment between the marble facing and the infill (See Fig. 10.33 in Chap. 10, reproduced from Macchi and Ghelfi 2005).

**Examination of large diameter cores** of substantial length, extracted from the masonry, whenever possible: a listing of widths of discontinuities observed along

**Fig. 5.5** Crack pattern of the internal face of the North wall of Aghia Eirini church in Aiolou Street, in Athens. Adapted from Miltiadou-Fezans (2004)

**Fig. 5.6** Coexistence of cracks of various widths in a masonry face

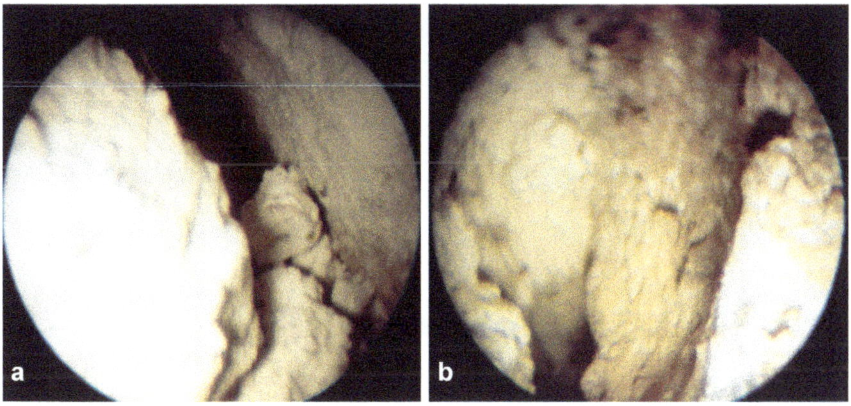


**Fig. 5.7** Endoscopies from the masonry of the Katholikon of Daphni Monastery, Attica, Greece. **a** Empty joint and cracked mortar. **b** Mortar detached from the block. Adapted from Vintzileou (2003). Courtesy of E. Vintzileou and V. Pallieraki

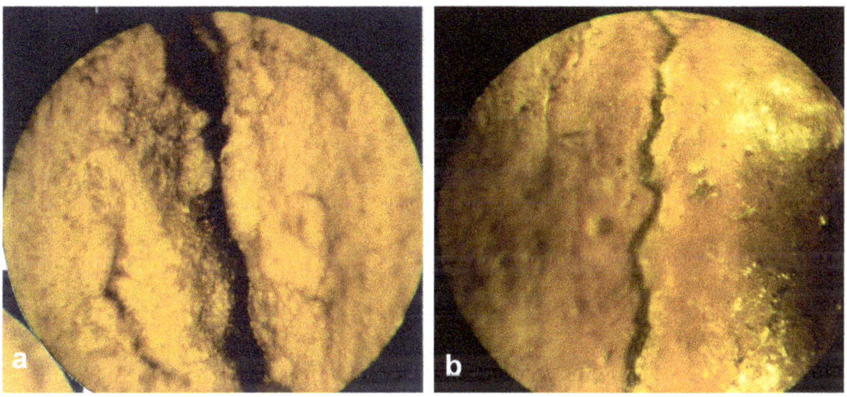


**Fig. 5.8** Endoscopy images from the masonry of the Athens's Cathedral. **a** Insufficient filling of joint between blocks or crack through joint. **b** Old crack of a block, filled with soot. Adapted from Geoereuna Laboratory-O.T.M. EPE (1995)

several generatrices and along transversal circles may be helpful. Furthermore, the observation of the internal face of the hole after taking out of the core, offers additional information on internal discontinuities (Fig. 5.9).

**Visual observation of the masonry** near doors or windows or after a local removement of the rendering and of a couple of blocks (Figs. 5.10a, b, c and 5.11a) or just of the rendering (Fig. 5.11b).

Finally, the role of structural damage on grout penetrability has to be discussed here: As soon as some structural damage of masonry occurs, hydraulic frictions along the flow-route of grout are reduced since "overall permeability" is increased,

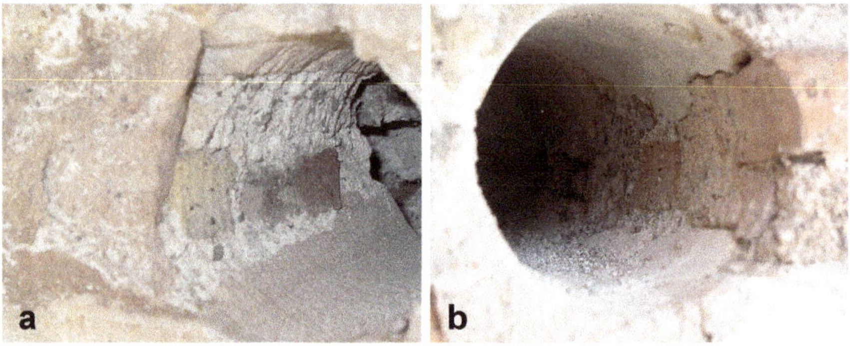

**Fig. 5.9** Discontinuities and cracks are detected along the internal faces of the masonry, after taking out a core

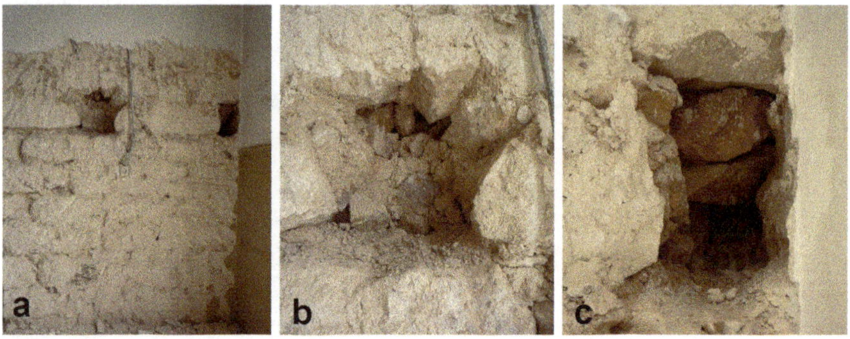

**Fig. 5.10** **a** Local removement of renderings and of a couple of blocks in two areas. **b** Observation of the masonry interior in the area to the left and **c** in the area to the right

whereas local "fine voids penetrability" remains to be secured by the penetrability rules of Chap. 2 (Sect. 2.6).

As soon as a $W_{nom,\,m}$ of a masonry ("m") is estimated, a sand column ("s. c.") is selected with its corresponding $W_{nom,\,s.\,c.}$ value, as close as possible to the $W_{nom,\,m}$ of the masonry.

It is however noted that the scope of the entire procedure does not necessitate a high precision–it is merely a starting stage; after all, the proposed design methodology will be subsequently submitted to several performance checks.

More generally, based on broad experience, a conventional classification of "dis-penetrability" may be agreed for various typologies of masonry, various types of mortars, various degrees of cracking, etc.; each of those cases may be represented by one nominal W-value.

The alternative approach of trying to build a simulacrum of masonry within a transparent mould, despite its representativeness' problems, may also profitably be used

**Fig. 5.11** **a** Uncovered area of the masonry's interior in the building of Athens's Cathedral. Adapted from Geoereuna Laboratory-O.T.M. EPE (1995). **b** Masonry observation after local removement of renderings. Courtesy of AB Raptis

in parallcl, whcncvcr thc grout dcsigncr avails of sufficicntly dctailcd information regarding the composition and density of the milieu under consideration.

## 5.3 Quantification Attempts

In order to serve a quantification attempt, the information gathered may be further elaborated, as follows:

1. Large stone blocks of masonry if it is significantly cracked because of several causes, such as mineralogical imperfections, hindered thermal deformations, fire, etc., may be strengthened by means of grouting of such cracks. A representative effective minimum opening "$W_{nom, m}$" may be estimated based on an appropriate scanning and digitization of these cracks. A practical lower percentile value of the crack openings may be selected as a "$W_{nom, m}$" value.
2. In the case of available sufficiently long and large diameter cores, there is a possibility of recording the widths of discontinuities observed along the lateral surface of the cylinder. To this end, a sufficiently dense network of generatrices and circles may be drawn on this surface, and the encountered width values

"$W_i$" be listed. Here again, an appropriate lower percentile (see Sect. 5.4) may be selected as a first estimate of "$W_{nom, m}$".

3. Additional information may be taken by visual observation of broken areas of masonry or of the transverse sections of masonry near doors or windows, after a local removement of some blocks. Quantification of discontinuities width is rather difficult in this case; however, such observations may assist experienced Engineers in their final decision making about the targeted penetrability level of a given masonry.

**Note:** In some cases of deficient building technology, mainly granular materials with inadequate percentage of silt or clay are used as masonry mortars. As a result, the scope of the grouting is to fill the *voids* of this granular material. The estimation of a minimum representative diameter of these voids, may be based on the expression

$$W_{nom, m} \sim 0.15\, D_{15}$$

estimating the diameter of the smallest path passing through grains of the size $D_{15}$ according to Dantu (1961) (Chap. 2, Sect. 2.2.2).

## 5.4 Practical Approach

The method presented in the preceding paragraph 5.3 constitutes a *descriptive* approach. Although this method is characterized by significant uncertainties, it is suggested to try its application, in order to achieve the maximum possible understanding of the milieu.

However, its results may be considered only as indicative. That is why a *more practical* approach should also be followed, in order to assist the designer to select a starting-value of $W_{nom}$.

1. Conceptually, the target $W_{nom}$-value for the mix-design of the grout should observe the following two requirements.

   – Economical requirement: A value $(W_{nom})_e$ should be selected such that any lower value would result in disproportionately higher costs of grouting in order to further increase the final strength[1] of masonry.
   – Hydraulic-communication requirement: An adequate percentage[2] of discontinuities with openings larger than $(W_{nom})_h$ should allow the grout to flow through and fill these discontinuities.

---

[1] In fact, for a very low $W_{nom}$-value, (i) the cost for further finer binders is much higher, but also (ii) the contribution of filled fine voids to the strength is smaller because of the small percentage of the areas of such fine voids in the critical cross section.

[2] It does not make sense to require the hydraulic communication of *all* small discontinuities, since we do avail of other means for communication enhancement: namely the face and depth-variability of the intake-pipes.

In both cases, a reasonably *small* W-value is sought. In order to observe both requirements, a design nominal opening $(W_{nom})_d$ can be selected, satisfying the following condition

$$(W_{nom})_{\boldsymbol{d}} = \boldsymbol{max}\{(W_{nom})_{\boldsymbol{e}}, (W_{nom})_{\boldsymbol{h}}\}$$

Obviously, this is only a qualitative criterion, since the state-of-the-art does not allow for further quantitative application.

2. The following completely empirical suggestions offer a practical assistance to the designer of a grout-mix to select a preliminary value of "$W_{nom}$", if available data, as those in Sect. 5.3 are not conclusive.

$$W_{nom} \sim \lambda \cdot k_1 \cdot k_2 \cdot k_3 \cdot V_0 \ [mm]$$

where the variables and constants may take the following values:

$V_o$-*values* (expressing the conditions of the building mortar and the infill):
Almost non permeable mortar (e.g., clay mortar), $V_0 = 0.2$ mm.
Relatively permeable mortar (e.g., normal lime mortar), $V_0 = 0.3$ mm.
Permeable mortar, $V_0 = 0.4$ mm.
$k_i$-values are suggested in Table 5.1
$\lambda$-*values* may take the following values:
One leaf masonry: 1.0
Two leaf masonry: 1.1
Three leaf masonry: 1.2

Numerical examples are given in Table 5.2. Very good quality undamaged and bad quality damaged ashlar and rubble masonries (one-leaf, two-leaf and three-leaf) are presented, for two types of mortar permeability.
For the good quality masonries, presented in the Table 5.2, a $W_{nom}$-value $< 0.1$ mm is found. In such cases grouting is not recommended, as explained in Sect. 5.1.

**Table 5.1** Values of $k_i$ -variables in function to the planeity of block's sides, the filling of joints by mortar and the visible cracks and discontinuities

| | a | b | c | |
|---|---|---|---|---|
| Planeity of blocks' sides | Ashlar stones | Semi-regular stones | Rubber stones | $k_1$ |
| | 0.5 | 0.7 | 0.9 | |
| Filling of joints by mortar | Full | Usual | Deficient | $k_2$ |
| | 0.5 | 0.6 | 1.0 | |
| Visible cracks in the mortar, or block/mortar detachments | Nil | Rare | Frequent | $k_3$ |
| | 0.6 | 0.7 | 1.0 | |

**Table 5.2** Numerical examples of calculated $W_{nom}$-values as a function of construction characteristics, type of mortar and type and pathology of masonry

| Construction characteristics of masonry | Type of mortar | Calculated $W_{nom}$-values for the three types of masonry | | |
|---|---|---|---|---|
| | | One-leaf masonry | Two-leaf masonry | Three-leaf masonry |
| Very good ashlar masonry: $k_1 = 0.5$ Very good filling of joints: $k_2 = 0.5$ No cracks or block/mortar detachment: $k_3 = 0.6$ | Relatively permeable mortar, $V_0 = 0.3$ | 0.045 | 0.049 | 0.054 |
| | Almost non-permeable mortar, $V_0 = 0.2$ | 0.03 | 0.033 | 0.036 |
| Very good rubble masonry: $k_1 = 0.9$ Very good filling of joints: $k_2 = 0.5$ No cracks or block/ mortar detachment: $k_3 = 0.6$ | Relatively permeable mortar, $V_0 = 0.3$ | 0.081 | 0.089 | 0.097 |
| | Almost non-permeable mortar, $V_0 = 0.2$ | 0.054 | 0.059 | 0.065 |
| Very bad quality damaged ashlar masonry: $k_1 = 0.5$ Deficient filling of joints: $k_2 = 1.0$ Frequent cracks or block/ mortar detachment: $k_3 = 1.0$ | Permeable mortar, $V_0 = 0.4$ | 0.20 | 0.22 | 0.24 |
| | Almost non-permeable mortar, $V_0 = 0.2$ | 0.1 | 0.11 | 0.12 |
| Very bad quality damaged rubble masonry: $k_1 = 0.9$ Deficient filling of joints: $k_2 = 1.0$ Frequent cracks or block/ mortar detachment: $k_3 = 1.0$ | Permeable mortar, $V_0 = 0.4$ | 0.36 | 0.396 | 0.432 |
| | Almost non-permeable mortar, $V_0 = 0.2$ | 0.18 | 0.198 | 0.216 |

In the case of damaged bad quality ashlar masonries examined in Table 5.2, a $W_{nom}$-value of 0.2 to 0.24 mm may be retained for one-leaf masonry to three-leaf masonry, respectively. Higher values are calculated in case of bad quality damaged rubble masonries. As presented in Table 5.2, a $W_{nom}$-value of 0.36 to 0.43 mm is found for one-leaf masonry to three-leaf masonry, respectively.
It has to be noted however that, in both cases, when an almost non-permeable mortar is considered, the corresponding $W_{nom}$-values would be much smaller, namely, 0.1–0.12 mm for damaged bad quality ashlar masonries, and 0.18 to 0. 216 mm for damaged bad quality rubble masonries, expressing the penetrability difficulties in such type of structures.

3. It is reminded that on the basis of the extensive literature survey in Sect. 2.2, and our calibration (Sect. 2.4), the practical values of nominal opening of discontinuities of masonries were found to vary from 0.1 mm up to 0.5 mm. Consequently, the preceding practical suggestions are purposely tuned to yield similar numerical results. But because of the clearly rough approximations used, the designer may select for the preliminary design any round value close to the above numerical results.

## References


Dantu P (1961) Etude mécanique d'un milieu pulvérulent formé de sphères égales de compacité maxima. In: 5ème Congrès International de Mécanique des sols et des Travaux de Fondations. Paris, Dunod, Publication, 61–3, 10 pages

Geoereuna Laboratory-O.T.M, EPE (1995) Report on Investigations of construction materials of Athens's Cathedral)

Kalagri A, Miltiadou-Fezans A, Vintzileou E (2010) Design and evaluation of hydraulic lime grouts for the strengthening of stone masonry historic structures. Mater Struct 43:1135–1146, doi 10.1617/s11527-009-9572-1

Luso E, Lourenço P (2019) Mechanical behaviour of two leaf masonry wall-strengthening using different grouts. J Mater Civ Eng 31(7):04019096

Machi G, Ghelfi S (2005) Interventi definitivi di stabilizzazione strutturale. In: Fermo A, Francescone L, Lunetti D, Tursi L. Valente D (eds) La Torre restituita: Gli studi e gli interventi che hanno consentito la stabilizzazione della Torre di Pisa, Bollettino d' Arte, Ministero per i Beni e le Attività Culturali, Volume Speciale 2005, Volume III, Roma, pp 173–215

Mazzon N (2010) Influence of Grout injection on the dynamic behaviour of stone masonry buildings. Ph.D Thesis, University of Padova

Miltiadou AE (1990) Étude des coulis hydrauliques pour la réparation et le renforcement des structures et des monuments historiques en maçonnerie. Thèse de Doctorat de l'Ecole Nationale des Ponts et Chaussées. Pub. LCPC in Collection Etudes et recherches des Laboratoires des Ponts et Chaussées, série Ouvrages d'art, OA8. ISSN 1161–028X. LCPC, Décembre 1991. Paris, France, p 278

Miltiadou- Fezans (2004) Structural restoration of the West part of the church of Aghia Eirini in Athens. In: Trakossopoulou K, Doussi M, Xatzitrifon N (eds) Proceedings of the 2nd National Congress Appropriate interventions for the safeguarding of monuments and historical buildings, Vol 2, 14–16 October 2004, Thessaloniki, Greece, Hellenic Ministry of Culture and Technical Champer of Greece, pp 173–183

Mouzakis C, Adami CE, Karapitta L, Vintzileou E (2017) Seismic Behaviour of timber-laced stone masonry before and after interventions: shaking table testes on two-storey masonry model. Bull Earthquake Eng. https://doi.org/10.1007/s10518-017-0220-9

Nikolopoulou V, Adami CE, Karagiannaki D, Vintzileou E, Miltiadou-Fezans A (2018) Grouts for strengthening two- and three-leaf stone masonry, made with earthen mortars. Int J Archit Heritage Conserv Anal Restorat 13(5):663–678

Silva B (2012) Diagnosis and strengthening of historical masonry structures: numerical and experimental analyses. Ph.D Thesis, University of Brescia, p 407

Toumbakari EE, Van Gemert D, Tassios TP, Vintzileou E (2005) Experimental investigation and analytical modeling of the effect of injection grouts on the structural behaviour of three-leaf masonry walls. In: Modena c, Lourenco P, Roca P (eds) Proceedings of 4th structural analysis of historical constructions. Padova, Taylor and Francis Group, London, pp 707–717

Valluzzi, M-R (2000) Comportamento meccanico di murature storiche consolidate con materiali e tecniche a base di calce. Ph.D Thesis, University of of Trieste, p 276

Vintzileou E, Miltiadou-Fezans A (2008) Mechanical properties of three-leaf stone masonry grouted with ternary or hydraulic lime-based grouts. Eng Struct 30(8):2265–2276

Vintzileou E, Mouzakis C, Adami CE, Karapitta L (2015) Seismic Behavior of three-leaf stone masonry buildings before and after interventions: Shaking table tests on two-storey masonry model. Bull Earthquake Eng. https://doi.org/10.1007/s10518-015-9746-xs

Vintzileou E, Tassios TP (1995) Three leaf stone masonry strengthened by injecting cement grouts. J Struct Eng ASCE 121(5):848–856 15

Vintzileou E (2003) The use of radar and endoscopy to investigate the masonry of the Katholikon of Daphni Monastery. In: Research report Investigation of the mechanical behaviour of brick enclosed byzantine masonry, bearing or not mural mosaics, and of the methods of its repair: the case of the Katholikon of Daphni Monastery, Laboratory of Reinforced Concrete, National Technical University of Athens, Directorate for Technical Research on Restoration, Hellenic Ministry of Culture and Sports (in Greek)

# Chapter 6
# Strength-Related Data of Grouts

**Abstract** This Chapter deals with grout-mix design issues related to the targeted *strength* of the masonry to be grouted. Only two parameters enter the discussion: targeted $f_{wc}$-value and corresponding required $f_{gr,c}$-value. The chapter explores how a range of required $f_{gr,c}$-values suffices to be related to a targeted $f_{wc}$-value. This loose correlation is due to the fact that grout-to- stone bond properties are shaping the final structural behaviour of grouted masonry. Thus, tensile rather than compressive strength of the grout is the relevant parameter. Besides, dehydration of grout entering the masonry takes place; consequently, some additional rules regarding mix-composition are respectively derived. Finally, several empirical relationships are offered predicting masonry compressive strength before and after grouting. Obviously, among the grout compositions resulting in $f_{gr,c}$-values within the required strength range, those mixes will be retained respecting the other required performances. The Chapter ends with a long Appendix presenting detailed strength results (both tensile and compressive) for several grout compositions, described in literature.

## 6.1 Introduction

The scope of this Chapter is the following:

1. To assist the Designer to estimate the required strength of the grout to be injected, (a "target strength" range), taking into account the desired final strength of the masonry under consideration.
2. To assist the Designer to select appropriate grout compositions that will develop strengths approximately equal to the target strength of the grout estimated in § 1 above.
3. To assist the Designer to select some alternative compositions of the grout, (able to lend to it approximately equal strengths as per § 1 above), in order to better satisfy other performance requirements of the grout.

It is important to remind (see Chaps. 7 and 8) that, besides the necessary strength, other desired properties of the grout should be taken into account in order to end-up with some candidate compositions, before testing:

 
A. Miltiadou-Fezans and T. P. Tassios, *Mix-Design and Application of Hydraulic Grouts for Masonry Strengthening*, Springer Tracts in Civil Engineering,
https://doi.org/10.1007/978-3-030-85965-7_6

- Chemical compatibility with materials existing in the masonry considered or with its surface decoration elements. In Chap. 7 guidance is given on this performance requirement.
- Penetrability, Fluidity and Stability of the grout, needed in each particular case, will be considered in selecting those candidate compositions, able in principle to observe these three basic requirements.

## 6.2 In Situ Modifications of the Grout After the Injection

### *6.2.1 Dehydration of Grouting Entering the Masonry*

For a complete understanding of the mechanical properties of the grouted masonry, it is needed to examine the final condition of the grout in the voids of the masonry:

1. When the mix-design rules and the grouting application specifications are followed, bleeding and segregation during the penetration of the grout is not expected. However, dehydration is always a possibility. In the specific case of unstable mixes, dehydration is a high risk.
2. Due to the surface porosity of the blocks and of the existing mortars of the grouted masonry, the initial water content of the grout may be substantially reduced. Depending on the absorptiveness of the aforementioned in-situ building materials of the masonry, Miltiadou (1990) for stable binary grouts, has measured water-content losses, both for relatively impermeable blocks and for porous limestones: Indicatively, in the case of a very porous stone (St Maximin stone, with porosity n = 26%), it was found that the water content of grout changes in few minutes from the value of 75% to ~ 20%, while when a non-porous stone is used (Marquise stone, porosity n = 1–2%) the grout water content even after 3 h was only slightly changed from 75 to 70%.
   In this respect, the following basic phenomena should be considered:
   - First, possible increase of bond strength will be observed at the interface between the masonry blocks and the grout, due to the absorption of water, as well as due to the penetration of some fines of the grout into the pores of the block.
   - On the other hand, there is a risk of disproportionately high loss of water of the very first grout material entering the masonry. Because of these water-losses, the absorptiveness of the surrounding masonry blocks will be reduced; thus, the incoming grout masses will be submitted to lower water losses.
     Thus, the following two undesirable consequences should be discussed:
     (a) The first grout-quantity may show lower fluidity: However, the continuous pumping of grout behind this "front" material, may normally remedy such situations.
     (b) It is also possible, the initial grout to remain non-hydrated, resulting in lower strength. However, this undesirable local effect may be partly

counteracted by the broader beneficial effects regarding the grout-to-block bond.

### 6.2.2 Measures Against the Dehydration

In any case, the adverse effects of such in situ modifications of the grout due to block porosity should be minimized.

To this end, the following simple rules may be helpful:

- In order to increase the water-percentage "retained" by the grains of the grout, higher fines-content is needed (Sect. 4.3) and an appropriately drastic mixing method should be used (Sects. 2.2.4 and 3.4). Experimental evidence on this matter is offered by Miltiadou (1990), based on estimation of non-hydrated binder by means of XRD measurements and differential thermal analysis on grout samples taken from a 1 mm joint between two stone cylinders (of four types of stones with porosity varying from n = 1% to n = 26%): The increase of ultrafine material content resulted in considerable decrease of non-hydrated binder, thanks to the higher water retention capacity of the ultrafine materials, acting as a reserve of water, indispensable for binders' hydration.
- Grouting operations should preferably be avoided during very hot and dry periods, in order to reduce the absorption capacity of the materials within the masonry.

## 6.3 Grout Strength Versus Masonry Strength Required

### 6.3.1 Introduction

In structural redesign of masonry buildings, the required strength of the masonry "$f_{wc,requ.}$" is first calculated under the actions to be considered; and it has to be admitted that, compressive strength of masonry allows for strengths under other action-effects (such as shear or tensile strength) to be approximately estimated. Thus, the required $f_{wc,requ.}$-value (compressive strength) is the first information needed for the mix-design of the grout. If the actual compressive strength of masonry is lower than the aforementioned $f_{wc,requ.}$-value, then an appropriate grouting should be decided, although grouting may be needed also in the case of a sufficiently strong masonry damaged by severe cracking. In this case, however, a broader structural redesign is needed, in which the grouting may also be used, in combination with other intervention means.

The next step is to estimate (roughly though) a corresponding required range of compressive strength ($f_{gr,c}$) of this grout, although this is not a direct criterion of the overall effectiveness of grouting. But, because of its simplicity, it will be used, (mainly by means of empirical relationships), so that subsequently, more pertinent criteria will be estimated, such as the tensile $f_{gr,t}$ (see Sect. 6.4.4.3) or the bonding strength (see Sect. 6.5) of the grout.

## 6.3.2 *Estimation of the Strength of Existing Stone-Masonry Before and After Grouting*

Since the necessary $f_{gr,c}$-values will be estimated on the basis of the required $f_{wc}$-values, this section includes some guidance for the estimation of the compressive strength of existing stone masonry, before and after grouting. To this end, previous experience and available empirical formulae may be used. Distinction should be made between intact and damaged masonries. In each country, local experience and calibrated empirical formulae should be used. In what follows, however, some possible solutions are presented, that might be helpful to this end.

### 6.3.2.1 Intact Masonry

One-leaf Stone Walls

*Masonry strength before grouting*

$$f_{wc,0} = \xi\left[\left(\frac{2}{3}\sqrt{f_{bc}} - f_0\right)\right] + \lambda \cdot f_{mc} \text{ (Tassios and Chronopoulos 1986)}$$

where:

$f_{bc}$: compressive strength of blocks.
$\lambda$: mortar-to-stone bond factor, taken as.

$\lambda = 0.50$ for rough stones and
$\lambda = 0.10$ for very smooth-surface stones.

$f_0$: a reduction due to the inhomogeneity of construction, and depending on non-orthogonality of blocks, taken as (in MPa):

$f_0 = 0.00$ for ashlar (block) stones.
$f_0 = 0.50$–$1.00$ for semi-regular stones.
$f_0 = 0.50$–$2.50$ for rubble stones,
$\xi$: a factor expressing the adverse effect of thick mortar joints
$\xi = \frac{1}{1+3.5(k-k_0)} \leq 1.0$

$k = \frac{V_m}{V_w}$,

with:

$V_m$, $V_w$ = the volume of mortar and the volume of masonry respectively.
$k_0 = 0.30$.

For $k = \frac{V_m}{V_w} \leq 0.30$, a constant ξ-value will be taken, $\xi = 1.00$.

The formula is valid for the following materials' strength values: stones' strength $f_{bc} = 25$–$75$ MPa and mortar strength $f_{mc} = 0.5$–$2.5$ MPa.

Among its drawbacks, this empirical formula completely disregards interlocking of blocks along and transversally of the face of the wall.

## Two-leaf Stone Walls

### *Masonry strength before grouting*

The same empirical formula may be used in the case of two leaf masonry walls, if the two leaves are connected with "diatonoi or semidiatonoi" stones (i.e. through-the-section stones). If such connections are not present, each leaf has to be considered independent, and the compressive strength of the masonry would conservatively be taken equal to the lower of the strengths of each leaf.

## One or Two-leaf Stone Walls

### *Masonry strength after grouting*

In the absence of more specific data, the compressive strength of the above two categories of masonry after grouting may be roughly estimated by means of the following expression

$$f_{wc,s} = f_{wc,0} + \Delta f_0 + \lambda \cdot n \cdot f_{gr,c} \ (\pm 35\%) \tag{6.1}$$

where:

$f_{wc,0}$: masonry strength before grouting.
$n$: ratio of the volume of grout embodied to the masonry, normalized to the total volume of the mortar of the single leaf masonry (indicatively ~ 10–30%).
$\lambda$: mortar-to-stone bond factor, taken as.

$\lambda = 0.50$ for rough stones and.
$\lambda = 0.10$ for very smooth-surface stones.

$f_{gr,c}$: compressive strength of the grout.

$\Delta f_0 = f_0 / \left(1 + \frac{1}{10n}\right)$: reduction of the inhomogeneity factor (counteracting the severity of the initial $f_0$-value included in the expression of $f_{wc,0}$).

According to another approach,

$$f_{wc,s} = f_{wc,0}\left[1 + 0.013 \cdot (100G_{gr} : G_w)^3\right] \text{ (Vintzileou and Tassios 1995)} \quad (6.2)$$

where:

$f_{wc,0}$, $f_{wc,s}$: compressive strength of masonry before and after grouting respectively.

$G_{gr}$ and $G_w$: weight of the injected grout and initial weight of the single leaf wall.

Note that this formula disregards the strength of the grout.

Three-leaf Stone Walls

*Masonry strength before grouting*

Based on Egermann (1993), the following expression was developed by Vintzileou and Tassios (1995), assuming that the compressive strength of this type of masonry depends only on the compressive strength of the external masonry leaves, especially for walls in which the filling material represents 1/3 to 1/2 of the total cross-sectional area:

$$f_{wc,0} = \left(\frac{V_{ext}}{V_w}\right) \cdot f_{wc,e}$$

where:

$V_{ext}$: volume of the external leaves.
$V_w$: total volume of the wall.
$f_{wc,e}$: compressive strength of the external masonry leaf.

This expression was satisfactorily recalibrated by Silva et al. (2014).

A more analytical expression was proposed by Tassios (2004):

$$f_{wc,0} = \left(2 \cdot \lambda_e \cdot \delta \cdot f_{wc,e} + \lambda_i \cdot f_{wc,i}\right) / (1 + 2\delta) \text{ (Tassios 2004)}$$

Or in case the external leaves have not the same thickness

$$f_{wc,0} = \left[\lambda_e\left(\delta_{e1} \cdot f_{wc,e1} + \delta_{e2} \cdot f_{wc,e2}\right) + \lambda_i \cdot f_{wc,i}\right] / (1 + \delta_{e1} + \delta_{e2}) \text{ (Tassios 2004)}$$

where:

$f_{wc,e}$: compressive strength of the external masonry leaf.

$f_{wc,i}$: compressive strength of the infill.
$\delta = t_e/t_i$: ratio of the respective thickness of external leaf and infill (in case that the external leaves have the same thickness).
$\delta_{e1} = t_{e1}/t_i$: ratio of the respective thickness of external leaf "1" and infill.
$\delta_{e2} = t_{e2}/t_i$: ratio of the respective thickness of external leaf "2" and infill.
$\lambda_e, \lambda_i$: correction factors in order to take into account the interaction between the external leaves and the infill ($\lambda_e$ = reduction of compression strength of external leaf due to transversal horizontal deformations imposed by the infill material to the leaves and $\lambda_i$ = increase of compression strength of infill material due to its favourable confinement by the external leaf: $\lambda_e < 1$, $\lambda_i > 1$).
These factors may be roughly taken equal to 0.80 and 1.20 respectively.

*Masonry strength after grouting*

The following expression may be used for the estimation of the compression strength of three-leaf stone masonry after grouting

$$f_{wc,s} = f_{wc,0}\left(1 + \frac{V_i}{V_w}\frac{f_{wc,i,s}}{f_{wc,0}}\right) \ (\pm 35\%) \ \text{(Vintzileou and Tassios 1995)} \qquad (6.3)$$

where:

$f_{wc,s}$: compressive strength of the three-leaf masonry strengthened with grouting.
$f_{wc,0}$: compressive strength of the three leaf masonry before grouting.
$V_i$: the volume of the initial infill material within the entire masonry.
$V_w$: volume of the entire masonry wall.
$f_{wc,i,s}$: the compressive strength of the infill after grouting.

It has to be noted that the $f_{wc,i,s}$-values, along and across the wall, are expected to be very scattered, drastically depending on the skillfulness of the mason and the binding capacity of the infill material, as well as its initial voids. The following possible approximations are available in literature, based only on the quality of the grout.

$$f_{wc,i,s} \sim 1.25\sqrt{f_{gr,c}} \ \text{(Vintzileou and Tassios 1995)}$$

$$f_{wc,i,s} \sim 0.31\, f_{gr,c}^{1.18} \ \text{(Valluzzi et al. 1994)}$$

where, $f_{gr,c}$ denotes the compressive strength of the grout.

Note however that, according to Silva et al. (2014), compressive strengths of the grout higher than 5-times the initial strength of the masonry, cannot substantially contribute to the after-grouting strength of walls. This limit ratio appears to be even lower (~3) on the basis of data presented by Vintzileou (2011), Fig. 6.

Because of the numerous uncertainties accompanying the phenomena discussed in the preceding paragraphs, future research and more rational proposals will further assist the design Engineer on these particular matters.

#### 6.3.2.2 Damaged Masonry

Empirical estimation of residual compressive strength of masonry after damage is a rather risky exercise. Consequently, the selection of the required compressive strength $f_{gr,c}$ of the grout will not be based on the initial masonry strength, as in the case of undamaged walls: In that case, the problem was to find an appropriate $f_{gr,c}$-value, in order to achieve a required increase of the initial strength of the masonry, *before* damage.

In the case of damaged walls, the grout will necessarily be used in most cases together with several other structural intervention methods,[1] in order to *reinstate* the initial strength of masonry. Consequently, the appropriate $f_{gr,c}$-values should be decided by the Structural Engineer within the design-models to be used for the re-establishment of the bearing capacity of damaged regions (such as repair of tensile or shear cracks, of compressively failed corners, etc.).

If, however, the target value of masonry strength is *higher* than its initial value, then, the strength $f_{gr,c}$ of the grout to be used to this end, will be estimated as in Sect. 6.3.2.1, since the Structural Engineer (to the best of her/his knowledge) has already reinstated the initial strength of the masonry.

#### 6.3.2.3 Commentary

It is noted that in the available empirical formulae predicting the strength of grouted masonry, the strength of the grout appears rather indirectly under the form of "$\lambda . f_{gr,c}$" "Eq. 6.1" or "$\sqrt{f_{gr,c}}$" "Eq. 6.3" or even it *disappears completely, as* in the case of "Eq. 6.2". The rationale of this situation is that a grout contributes to the strength of an existing masonry mainly via its bond resistance with the blocks and with the mortar of the masonry. And this bond-strength is closer to the tensile rather than the compressive strength of the grout.

Besides, this bond-strength drastically depends on the condition of the substratum of the grout-to-block contact: Cleanness, humidity, roughness and porosity are the conditions shaping this bond behaviour.[2] Since, however, such bond strengths are not easily determined, the tensile-strength $f_{gr,t}$ of the grout is a better estimator of bond-strength of the grout (see Sect. 6.4.4).

Consequently:

[1] Such as placing of stone or metal "keys" through cracks, deep re-pointing, possible addition of "reinforcements" along leveled joints, local rebuilding, etc.

[2] See Sect. 6.5 and Miltiadou (1990, 1998), Toumbakari (2002), Toumbakari et al. (2007), Adami and Vintzileou (2008, 2010), Adami et al. (2008), Luso and Lourenço (2017b).

- In view of the indirect and variable ways $f_{gr,c}$ is contributing to $f_{wc}$, rough approximations are sufficient in estimating the required $f_{gr,c}$-values.
- For practical reasons, only the compressive strength "$f_{gr,c}$" of the grout will be used as a criterion of the "appropriateness" of the selected grout—its significance being however reduced, the way it appears in "Eqs. 6.1 and 6.3".

## 6.4 Expected Strengths of Grouts

### *6.4.1 Introduction*

1. The designer of the grout, in order to optimize the performances of the grout (see Chap. 8), should take two basic decisions:
   - the nature of the solids, i.e., the binder to be used in the grout should be selected (hydraulic lime, cement, pozzolans, hydrated lime, or others), and
   - the number of binders included in the grout should be decided (one-binder grout, binary grout or ternary grout).
   The strength of the final grout is one, among others, of its performances; in this paragraph, some guidance is offered regarding the strength expected to be achieved by various grout compositions.
2. It has also to be reminded, however, that the final mechanical properties of a grout within the masonry, are expected to be quite different than under the initial conditions before injection (Sect. 6.2). Nevertheless, if the construction rules (see Chap. 10) are followed, the initial strength of a grout is an acceptable estimator of its strength in situ.

### *6.4.2 The Main Parameters Influencing the Strength of the Grout*

As it has been shown in Chap. 2, grouts containing only cement used in the past for grouting were proven inadequate to fill the small size voids and cracks of masonries (inadequate penetrability). Because of this drawback of plain cement grouts, the addition of ultrafine materials was decided, early enough (e.g., Paillère et al. 1986; Miltiadou 1990). On the other hand, the need for a wide range of required mechanical properties of the grouts was also recognized, together with the need of enhanced durability, since grouting is a non-reversible intervention.

Thus, binary grouts (mixes of cement and hydrated lime, natural or artificial pozzolans, silica fume, etc.) and ternary grouts (cement, hydrated lime and natural or artificial pozzolans) or hydraulic-lime-based grouts were developed (Penelis et al. 1989; Miltiadou 1990, 1998; Tomaževič and Apih 1993; Binda et al. 1993, 1994,

1997, 2003; Valluzzi 2000; Toumbakari 2002; Miltiadou-Fezans et al. 2005, 2007a, 2008; Adami and Vintzileou 2008; Kalagri et al. 2010; Badogiannis et al. 2012; Bras and Henriques 2012; Uranjek et al. 2014; Baltazar et al. 2014; Papayianni and Pachta 2015; Jorne et al. 2015; Yüzer et al. 2015; Luso and Lourenco 2017a; Nikolopoulou et al. 2018).

The most common ternary grouts are composed by 50–20% of cement and combinations of lime with pozzolan in various lime/pozzolan ratios (1/2 to 1/5) (Penelis et al. 1989; Miltiadou 1990; Toumbakari 2002; Miltiadou-Fezans et al. 2007a; Papayianni and Pachta 2015).

The natural hydraulic lime (NHL)-based grouts in most of the cases offered in the literature are composed by 100% natural hydraulic lime or in some specific cases by 90–80% NHL in association with 10–20% of pozzolana. The optimum combination of NHL to natural pozzolan has to be studied in each specific case, regarding both strength and durability characteristics (Miltiadou-Fezans et al. 2021).

Additionally, various premixed hydraulic lime or lime based solid phases of grouts were also developed in the market and tested by various researchers (Binda et al. 1993, 1994, 1997, 2003; Valluzzi 2000; Corradi et al. 2002, 2008; Oliveira et al. 2006; Kalagri et al. 2010; Mazzon 2010; Artioli et al. 2011; Silva 2012; Yüzer et al. 2015; Giaretton et al. 2017; Luso and Lourenco 2016).

As it is known, the main parameters affecting the strength of grouts are:

- The nature and granularity of binders,
- The water to solids ratio,
- The mixing method,
- The superplasticizer content.

A commentary of these parameters follows.

**Nature of binders** Grouts containing only cement (depending of course on the cement type I55, II42.5, etc., according to EN 197) may exhibit high strength values, provided that all injectability requirements are satisfied, which is rarely achieved (see Chap. 2). Thus, the occasionally needed high strength results are usually obtained by means of binary grouts, with a replacement of approximately 25% of the cement with a highly reactive artificial or natural pozzolan. In this case, the pozzolanic action is in favour of strength, since the pozzolan is more or less consumed by the calcium hydroxide released during the hydration of cement, producing stable compounds similar to those produced by the hydration of Portland cement. When higher percentages of cement are replaced by a pozzolanic material, the strength obtained is usually lower, since a certain percentage of pozzolanic material will not be activated because of the lack of calcium hydroxide.

That is (among other reasons) why ternary compositions were developed, with lower cement content, in combination with lime and a natural or artificial pozzolan. Thus, the initially developed strength is due to the stable products formed by the hydration of Portland cement; this strength is further enhanced by the pozzolanic activity, thanks to the directly available lime, leading to a denser solid. Consequently, a higher strength is achieved than in the case of a low cement content binary composition, containing cement and lime or cement and pozzolan.

Furthermore, similar results may be obtained when hydraulic lime is used, alone or in combination with a pozzolanic material (of the order of 10–20%), mainly for durability purposes. In this case, the type of the hydraulic lime (NHL5 or NHL3.5 or NHL2, according to the EN 459) plays an important role on the strength of the grout, as it is directly linked to the hydraulicity index of the binder and hence to its strength characteristics. High hydraulicity index is preferable in order to achieve a reasonable setting time and strength, as well as satisfactory durability. Nevertheless, further research is necessary on this type of grouts.

**The high water to solids ratio** plays the well-known detrimental role regarding mechanical properties and stability, although it is dictated by fluidity requirements. It is therefore recommended to make the most profitable use of appropriate superplasticizers in order to reduce water-to-solids ratios, under the conditions mentioned here below.

**The importance of mixing method** has also to be reminded here (see Chaps. 2, 3, and 4). High turbulence or ultrasound mixing has to be used, in order to ensure satisfactory injectability and strength. The mixing method is very important in order to achieve a good dispersion of grains and complete wetting of their surface, thus minimizing the risk of possible non-hydration of grout material, that would result in lower strength.

**The role of superplasticizers** is mainly linked to their capacity to reduce the water content necessary to achieve the same injectability characteristics. Nevertheless, as explained in previous Chaps. (2, 3, and 4), small changes in its content may jeopardize stability and fluidity characteristics, as well as the strength of the grout.

Moreover, the chemical nature of the superplasticizer should be suitably selected, depending on the nature of the binders of the grout.

### *6.4.3 Experimental Checking*

A commentary is needed regarding the reliability of the experimental methods used and the rather large variability of the obtained results regarding strength evaluation of grouts.

**Size and form of specimens**: Since the maximum diameter of the grains in the grout is of the order of magnitude of 0.1 mm, it is clear that very small size specimens could be suitable for the determination of the mechanical properties of the grout. However, for practical reasons, due to the lack of standards for measurning masonry grouts strength, the standards for grouts of prestressing tendons (EN 445) or those for mortars of masonry (EN 1015-11) were in the most of cases used in the literature (prismatic specimens 4 × 4 × 16 cm for flexural strength, and corresponding cubic specimens for compressive strength). Thus, grout specimens have usually a cross section of 40 × 40 mm—although criticism has been expressed against such relatively large sizes, because of the possible detrimental effects due to the differential shrinkage of outer and inner layers of material.

On the other hand, it should also be noted that smaller size specimens exhibit relatively higher strengths, which may be closer to the strength of the grout inside the small-size voids of masonry. Thus, a preference could be given to cylindrical specimens of grout of 20 mm diameter and 40 mm height, tested in compression and splitting to better estimate direct tensile strength (Miltiadou 1990, Miltiadou-Fezans 2000, Miltiadou-Fezans et al. 2021), instead of the flexural strength measured on prismatic specimens 4 × 4 × 16 cm. It is also interesting to mention the testing by splitting of injected sand-column specimens, indicated by EN1771 (Ferragni et al. 1984; Miltiadou 1990; Van Rickstal 2000; Toumbakari et al.1999, Biçer-Şimşir and Rainer 2013).

The standard-sized specimens are prepared in non-porous containers (moulds); thus, it is noticed in the literature (Holmström 1981, Schuller et al. 1994; Griffin 2004) that the results are not representative of reality, since the suction of the grout water by the porous materials of masonry drastically changes the grout composition in the interior of the masonry and its mechanical characteristics. Therefore, the fabrication of specimens using moulds with an internal porous surface has been attempted by some researchers (Griffin 2004). The results may be very useful and closer to reality. Nevertheless, there is not yet a standardized test for masonry grouts; the difficulty of reproducing the absorptive properties of the milieu is the main problem to be solved.

**Curing**: Grout-specimens cured under unsatisfactory humidity conditions, are exposed to disproportionally large differential shrinkage of outer and inner layers, resulting in the development of internal tensile stresses. Thus, compressive and mainly tensile strengths are reduced. Similar (seemingly unjustified) consequences are faced when specimens are taken out of the wet-room long before their testing.

**Tensile strength**: In previous chapters, the significance of the tensile strength of the grouts was made clear; grout-to-stone bond is better represented by the tensile strength of the grout—especially by long term tensile strength (Miltiadou-Fezans et al. 1998, 2007a; Toumbakari 2002; Kalagri et al. 2010; Miltiadou-Fezans et al. 2021). In this respect, however, it has to be noted that it is not rare in literature to observe that grout specimens aged more than approximately 6 months, tend to show *reduced* tensile strengths, whereas their compressive strengths remain stable—if not increased. (Toumbakari et al.1999; Toumbakari 2002). The size of specimens and the non-porous material of the moulds are directly linked to this phenomenon. The need for developing specific standards to be used for measuring the mechanical characteristics of masonry grouts is once again highlighted.

## 6.4.4 *Indicative Strength Values of Grouts*

### 6.4.4.1 Preamble

The multiplicity of the parameters shaping the final strength of a grout, does not leave much space for a theoretical methodology in order to assist the designer of the grout. It is also reminded that the estimation of a target $f_{gr,c}$-value (assisted by the data contained in Sect. 6.3), aims only at a rather rough approximation than at a mathematically calculated result. Consequently, a similar rough approximation is allowed in the present stage too, in selecting some candidate grout-compositions, potentially able to exhibit the decided approximate level of strength.

Nonetheless, in what follows, the designer may find a first guidance of the order of magnitude of expected strengths of various grout compositions, despite the understandable large scattering.

#### 6.4.4.2 Numerical Values

In the Tables 6.11, 6.12 and 6.13, presented in the Appendix of this Chapter, several numerical results of grout strengths are listed, as reported in literature for compositions of One binder, Two binders and Three (or more) binders. It is admitted that, for the sake of simplicity, particular properties of each binder are not described—except of some qualitative brand names; this is one of the causes of the variability of the reported results. Similarly, another important parameter is missing, i.e., the mixing method. These omissions were decided to make feasible a shorter presentation, so that the influence of the most important strength parameters be more easily distinguished.

An even shorter (and inevitably less precise) presentation will be attempted in what follows, in order to further assist the designer of a grout to pre-estimate a probable value of strengths of her/his composition. To this end, the following simplified Tables 6.1, 6.2, 6.3, 6.4, 6.5, 6.6, 6.7, 6.8 and 6.9 contain ranges of experimental results for several grout compositions.

Grouts with One Binder

*Cement Grouts*

**Table 6.1** Indicative strengths of Grouts with one binder: Cement grout compositions

| Type and % of binder | W/S | 28-days $f_{gr}$ (MPa) | | 3-months $f_{gr}$ (MPa) | |
|---|---|---|---|---|---|
| | | $f_{gr,c}$ | $f_{gr,t}$ | $f_{gr,c}$ | $f_{gr,t}$ |
| 100% Cement, including some microfine cements | 0.40–0.60[a] | 40.00–25.00 | 4.00–3.50 (s) | (50.00)–30.00 | 4.50 (f) |
| | 0.60–0.70 | 30.00–25.00 | 4.00–2.00 (f) | ~ 30.00 | 4.0 0 (f) |
| | 0.80 | 25.00–10.00 | – | – | – |
| | 1.20 | 10.00–5.00 | | | |

[a]Note that some of the results reported in this Table do not correspond to compositions injectable in fine cracks of masonries. (*s* splitting, *f* flexion)

### *Hydraulic lime-based grouts*

**Table 6.2** Indicative strengths of Grouts with one binder: Hydraulic lime grout compositions

| Type and % of binder | W/S | 28-days $f_{gr}$ (MPa) | | 3-months $f_{gr}$ (MPa) | |
|---|---|---|---|---|---|
| | | $f_{gr,c}$ | $f_{gr,t}$ | $f_{gr,c}$ | $f_{gr,t}$ |
| 100% NHL5 or NHL3.5 or NHL3.5z | 0.50–0.60[a] | 7.00–3.50 | – | – | – |
| | 0.60–0.80 | (3.50–3.00) | – | – | – |
| | 0.80–0.90 | 3.00–1.50 | 1.50–0.80 (f) | 5.00–3.00 | 2.50–1.50 (f) |
| | 0.90–1.10 | 1.50–1.00 | 0.80–0.40 (f) | 4.00–1.00 | 1.50–1.00 (f) |
| 100% NHL 2 | 0.60–0.70 | 2.50–1.70 | 1.20–1.00 (f) | 3.00–2.00 | 1.30–0.90 (f) |
| | 1.30 | 0.70 | 0.60 (f) | ~1.50 | ~0.90 (f) |

[a] Note that some of the results reported in this Table do not correspond to compositions injectable in fine cracks of masonries. *NHL* Natural Hydraulic Lime. (*f* flexion)

## Grouts with Two Binders

### *Binary grouts (Cement + Densified silica fume)*

**Table 6.3** Indicative strengths of Grouts with two binders: Cement and Densified silica fume grout compositions

| Type and % of binders | W/S | 28-days $f_{gr}$ (MPa) | | 3-months $f_{gr}$ (MPa) | |
|---|---|---|---|---|---|
| | | $f_{gr,c}$ | $f_{gr,t}$ | $f_{gr,c}$ | $f_{gr,t}$ |
| C/DSF: 90/10 | 0.75 | 20.00–25.00 | ~ 2.50 (s) | – | – |
| C/DSF: 75/25 | 0.75 | 20.00–30.00 | 2.50–3.00 (s) | – | – |
| C/DSF: 60/40 | 1.00 | 15.00–20.00 | 2.00–2.50 (s) | – | – |

*C* cement, *DSF* densified silica fume. (*s* splitting)

### *Binary grouts (Cement + Pozzolan)*

**Table 6.4** Indicative strengths of Grouts with two binders: Cement and Pozzolan grout compositions

| Type and % of binders | W/S | 28-days $f_{gr}$ (MPa) | | 3-months $f_{gr}$ (MPa) | |
|---|---|---|---|---|---|
| | | $f_{gr,c}$ | $f_{gr,t}$ | $f_{gr,c}$ | $f_{gr,t}$ |
| C/P: 90/10 | 0.70–0.75 | 25.00–20.00 | 2.00 (s)–3.50 (f) | 25.00 | 4.50 (f) |
| C/P: 75/25 | 0.75–0.80 | 20.00–12.00 | 2.00 (s)–4.00 (f) | 27.00–17.00 | 5.00–3.50 (f) |
| C/P: 60/40 | 0.65–0.75 | 25.00–15.00 | 5.00–2.50 (f) | 30.00–18.00 | 6.00–3.50 (f) |
| C/P: 60/40 | 0.85–1.00 | 10.00 | 1.30 (s)–2.50 (f) | 13.00 | 3.50 (f) |
| C/FA: 40/60 | 0.50–0.65[a] | 20.00–10.00 | - | 30.00–10.00 | |

[a] Note that some of the results reported in this Table may not correspond to compositions injectable in fine cracks of masonries. *C* cement, *P* Pozzolan, *FA* fly ash. (*f* flexion)

*Binary grouts (Cement + Lime)*

**Table 6.5** Indicative strengths of Grouts with two binders: Cement and Lime grout compositions

| Type and % of binders | W/S | 28-days $f_{gr}$ (MPa) | | 3-months $f_{gr}$ (MPa) | |
|---|---|---|---|---|---|
| | | $f_{gr,c}$ | $f_{gr,t}$ | $f_{gr,c}$ | $f_{gr,t}$ |
| C/L: 90/10 | 0.75 | 16.00 | 2.00 (s) | | |
| C/L: 80/20 | 0.85 | 15.00 | 3.50 (f) | 20.00 | 4.50 (f) |
| C/L: 75/25 | ~0.80 | 10.00–15.00 | 1.50 (s) | 20.00 | 1.50 (s)–3.00 (f) |
| C/L: 60/40 | 0.85–1.00 | 3.00–3.50 | 0.80 (s) | 5.50–4.50 | 0.90 (s)–0.60 (s) |
| C/L: 20/80 | 0.85 | 0.40 | 0.25 (f) | 0.90 | 0.35 (f) |

*C* cement, *L* Lime. (*s* splitting, *f* flexion)

*Binary grouts (NHL + Pozzolan)*

**Table 6.6** Indicative strengths of Grouts with two binders: Natural Hydraulic Lime and Pozzolan grout compositions

| Type and % of binders | W/S | 28 day $f_{gr}$ (MPa)s | | 3 months $f_{gr}$ (MPa) | |
|---|---|---|---|---|---|
| | | $f_{gr,c}$ | $f_{gr,t}$ | $f_{gr,c}$ | $f_{gr,t}$ |
| NHL5/P: 90–80/10–20 | 0.80 | 2.50–1.10 | 1.30-0.90 (f) | 5.00–4.00 | 2.80–1.70 (f) |
| NHL3.5z/P: 80–60/20–40 | 0.85–1.20 | 2.00–1.00 | 1.50–0.50 (f) | 4.00–1.50 | 2.00–0.30 (f) |
| NHL5/SF: 98–90/2–10 | 0.45–0.55 | 10.00–7.00 | | | |

*NHL* Natural Hydraulic Lime, *P* Pozzolan, *SF* Silica Fume. (*f* flexion)

## Grouts with Three or More Binders

*Ternary grouts (Cement + Lime + Densified silica fume)*

**Table 6.7** Indicative strengths of Grouts with three or more binders: Cement, Lime and Densified silica fume grout compositions

| Type and % of binders | W/S | 28-days $f_{gr}$ (MPa) | | 3-months $f_{gr}$ (MPa) | |
|---|---|---|---|---|---|
| | | $f_{gr,c}$ | $f_{gr,t}$ | $f_{gr,c}$ | $f_{gr,t}$ |
| C/L/DSF: 70/13.5/16.5/SP 1% | 0.75 | 17.5 | 1.75 (s) | | |
| C/L/DSF: 50/22.5/27.5/SP 1.66% | 1.00 | 11.00 | 1.40 (s) | | |

*C* cement, *L* Lime, *DSF* densified silica fume, *SP* superplasticizer. (*s* splitting)

## *Ternary grouts (Cement + Lime + Natural + Artificial Pozzolan)*

**Table 6.8** Indicative strengths of Grouts with three or more binders: Cement, Lime Natural and Artificial Pozzolan grout compositions

| Type and % of binders | W/S | 28-days $f_{gr}$ (MPa) | | 3-months $f_{gr}$ (MPa) | |
|---|---|---|---|---|---|
| | | $f_{gr,c}$ | $f_{gr,t}$ | $f_{gr,c}$ | $f_{gr,t}$ |
| C/L/P: 60/20/20 | 0.75–1.00 | | | 11.00–6.00 | |
| C/L/P: 50/25–20/25–30 | 0.80–1.00 | | | 10.00–5.00 | |
| | 1.00–1.25 | | | 4.50–2.50 | 2.00–1.00 (f) |
| C/L/P: 40/25–30/35–30 | 0.70–0.80 | 5.50–4.00 | 1.50–2.00 (f) | 7.50–5.50 | 2.50–2.00 (f) |
| | 1.20–1.30 | 1.50–1.00 | 0.50–0.60 (f) | 3.50–2.00 | 1.00–1.20 (f) |
| C/L/P: 40/15–20/45–40 | 0.80 | 4.20 | 2.30 (f) | 7.50 | 3.30 (f) |
| C/L/P: 30/30–10/40–60 | 0.70–0.90 | 4.00–3.00 | 2.70–1.20 (f) | 9.00–5.00 | 3.50–1.50 (f) |
| | 1.00–1.40 | 2.00–0.50 | 0.90–0.30 (f) | 3.00–1.30 | 1.50–0.70 (f) |
| C/L/P/DSF: 30/15/50–45/5–10 | 1.00–1.10 | 6.00–4.00 | 1.70–1.20 (f) | 17.00–11.00 | 3.40–2.60 (f) |
| C/L/P: 20/30/50 | 1.30 | 0.60 | 0.30 (f) | 2.80 | 2.00 (f) |
| C/L/P: 10–15/50–15/40–70 | 0.85–1.10 | 1.50–2.90 | 0.50–1.00 (f) | 3.00–5.50 | 0.30–2.00 (f) |
| C/L/P/B: 10/45/25/20 | | | | | |
| C/L/P/Cl: 10–25/50–25/30–0/10–50 | 0.90–1.10 | 1.50–0.50 | 1.00–0.20 (f) | 1.00–3.50 | 0.70–0.20 (f) |

*C* cement, *L* Lime, *P* Pozzolan, *DSF* densified silica fume, *B* brick Dust, *Cl* Clay. (*f* flexion)

## *Pre-mixed grouts*

**Table 6.9** Indicative strengths of Grouts with three or more binders: pre-mixed hydraulic grouts[a]

| Type and % of binder | W/S | 28-days $f_{gr}$ (MPa) | | 3-months $f_{gr}$ (MPa) | |
|---|---|---|---|---|---|
| | | $f_{gr,c}$ | $f_{gr,t}$ | $f_{gr,c}$ | $f_{gr,t}$ |
| FEN X-B | 0.50 | 3.20 | 0.35 (f) | | |
| FEN X-B | [b]?[b] | 12.50 | 3.00 (f) | | |
| Unilit B Fluid 0 | 0.70 | 1.50 | 1.30 (f) | 2.60 | 1.50 (f) |
| Arbaria Inezione | [b]?[b] | 7.00–20.00 | ~ 3.00 (f) | 30.00 | ~ 3.50 (f) |
| Mapei Antique I | 0.35 | 21.00 | 4.00 (f) | 23.00 | 5.50 (f) |
| Calce per consolidamento | 0.60–0.65 | 1.50 | | 2.50 | |
| Lime-Injection | [b] | 12.00 | 3.50 (f) | 13.0 | 4.00 (f) |

[a] Note that some of the results reported in this Table may not correspond to compositions injectable in fine cracks of masonries

[b] The water to solids ratio were not known to the authors of this book. As reported in the respective papers (Appendix, Table 6.11), the W/S ratio "suggested by the manufacturer" was used. (*f* flexion)

In order to make these Tables to be as simple as possible (and perhaps for a broader use), the large variability of each category of binders was not specified: Thus, the general term "cement" is used, without mentioning its strength or its fineness. Similarly, for lime-binders and pozzolans, their fineness and reactivity capacity are not specified in these Tables. The same is true for the mixing method of the grout and the use or not of superplasticizer.

It is however believed that, thanks to these omissions, the basic consequences of the major parameters on the strengths of grouts are made more evident—so that the designer be assisted in evaluating a range of strength capacity of a *preliminary* grout composition.

On the other hand, it is worth to remind here that it is mainly the tensile strength $f_{gr,t}$ of the grout that matters; and because $f_{gr,t}$-values are better correlated to the square root of $f_{gr,c}$ rather that the $f_{gr,c}$-value itself, the demand in $f_{gr,c}$-terms may not be numerically as important as in other structural problems; after all, $f_{gr,c}$ does not directly contribute to masonry strength (see Sect. 6.4.4.1). For all these reasons, the designer is entitled to select a rough target $f_{gr,c}$-value of reasonable magnitude.

#### 6.4.4.3 Empirical Relationships Between $f_{gr,t}$ and $f_{gr,c}$

As it is well known, the ratio between compressive and tensile strength of grout is not constant. According to Vintzileou (2011), $f_{gr,t}/f_{gr,c}$ varies between 0.6 to 0.2 when $f_{gr,c}$ varies between 3.0 to 10.0 MPa. Based on the data of the detailed Tables 6.11, 6.12 and 6.13 presented in the Appendix of this Chapter, the following empirical relationships between *flexural* tensile $f_{gr,t}$ and compressive strengths of the grouts are formulated. Appropriate relationship between flexural and splitting tensile strengths should be used in case that splitting tests are applied.

In Fig. 6.1, based on almost 250 tests of several independent researchers, the well-known non-linear relationship between tensile and compressive strengths of grouts is apparent. More practically, the following rule of thumb may be useful:

For:

$$f_{gr,c} < 2.0\ \text{MPa},\quad f_{gr,t}\ /\ f_{gr,c} \sim \frac{1}{2}$$

$$3.0 < f_{gr,c} < 6.0\ \text{MPa},\quad f_{gr,t}\ /\ f_{gr,c} \sim \frac{1}{3}$$

$$f_{gr,c} \sim 10.0\ \text{MPa},\quad f_{gr,t}\ /\ f_{gr,c} \sim \frac{1}{4}$$

$$f_{gr,c} \sim 20.0\ \text{MPa},\quad f_{gr,t}\ /\ f_{gr,c} \sim \frac{1}{5}$$

The $f_{gr,t}/f_{gr,c}$ ratio decreases when increasing the compressive strength of the grout. This ratio is reasonably expected to be further decreased for even higher

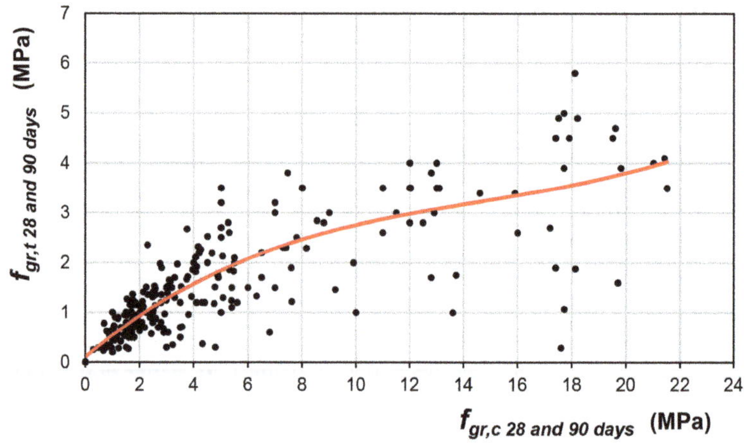


**Fig. 6.1** Flexural tensile strength versus compressive strength of grouts (at 28 and 90 days). Selected compositions having compressive strength up to 22 MPa

compressive strengths, up to the well-known value of 1/10, applicable in the case of low strength concretes.

A more direct approach is followed in Figs. 6.2 and 6.3, where $f_{gr,t}$ values are better correlated to the square root of $f_{gr,c}$ values.

Numerically, the following expressions may be used in cases where more specific information of the grout composition is missing.

For grouts of 28 days of age:

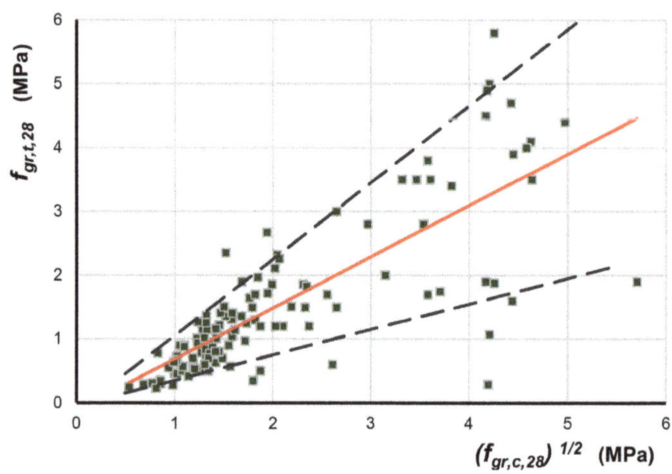


**Fig. 6.2** Flexural tensile strength of 28-days grouts, related to compressive strength (see "Eq. 6.4")

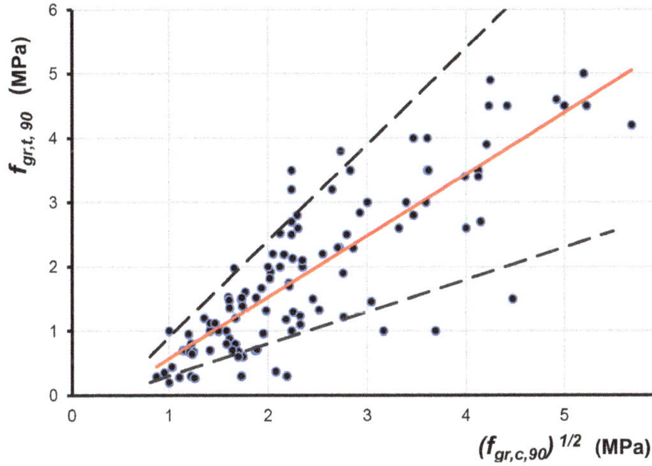


**Fig. 6.3** Flexural tensile strength of 90-days grouts, related to compressive strength (see "Eq. 6.5")

$$f_{gr,t,28} \cong \left[0.8\sqrt{f_{gr,c,28}} - 0.1\right](\pm 50\%)\ \text{MPa}\ \ [f_{gr,t},\ \text{flexural tests}] \tag{6.4}$$

For grouts of 90 days of age:

$$f_{gr,t,90} \cong \left[\sqrt{f_{gr,c,90}} - 0.4\right](\pm 50\%)\ \text{MPa}\ \ [f_{gr,t},\ \text{flexural tests}] \tag{6.5}$$

The expected large scattering is mainly due to the fact that the data bank used in this connection, contains information from more than 40 independed research works, regarding a very large variety of grout compositions (~200 compositions of one, two or three binders' grouts, using binders such as cement, hydraulic lime, pozzolans, etc.).

## 6.5 Grout-to-Stone Bond Properties

As it has been explained in other parts of this book, the bond strength of grouts to the substratum of the stone materials of masonry (Fig. 6.4) is of basic importance. The following "interface" parameters are of major significance:

- Tensile bond ($f_{gr,b,t}$), contributing to the tensile strength of the grouted masonry.
- Shear slip bond ($f_{gr,b,s}$), contributing to the compressive strength of grouted masonry (critical cracks under shear and compression).

On the other hand, the strength of the masonry itself depends also on the tensile ($f_{gr,t}$) and shear ($f_{gr,s}$) resistance of the body of the grout itself. But, the compressive strength of the grout ($f_{gr,c}$) does not seem to contribute directly to masonry strengths

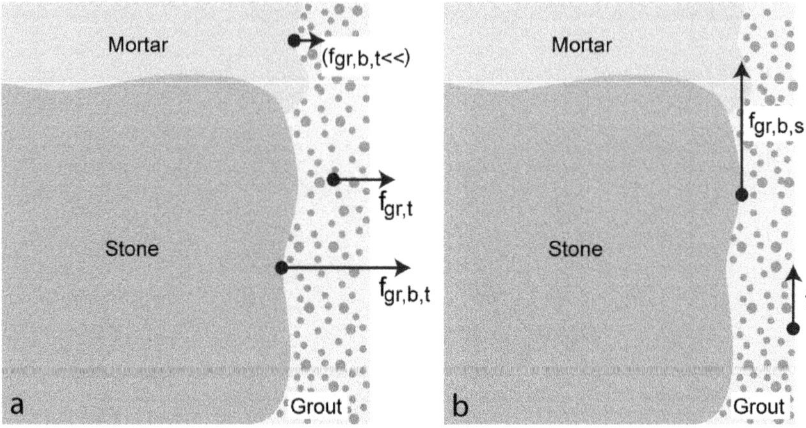


**Fig. 6.4** Schematic presentation of **a** the tensile bond strength and **b** the shear slip bond strength of grout-stone interfaces

(although the shear-strength $f_{gr,s}$ *of the grout* contributes partly to the value of the compressive strength $f_{gr,c}$ of the grout). It is therefore clear that the grout-to-stone bond strengths ($f_{gr,b,t}$, $f_{gr,b,s}$), as well as the internal tensile and shear strengths ($f_{gr,t}$, $f_{gr,s}$) of a grout, are of major importance.

The tensile strength of grouts was discussed in 6.4.4.2 and 6.4.4.3. In what follows, a brief summary of available research work on the *grout-to stone bond properties* will be given. Thus, possible particularities of some compositions regarding bond-strengths of the grouts will also be explained. This summary is mainly based on the following literature: Miltiadou (1990), Toumbakari (2002), Adami and Vintzileou (2008, 2010) and, Vintzileou and Adami (2009).

Bond properties depend on the following main parameters:

- Rugosity of the interface
- Surface porosity and pore size
- Water absorptivity
- Strength of the substratum material
- Chemical affinity of the binder to the substratum
- Thermohygrometrical conditions

Because of the multitude of these parameters, it is expected that bond properties will be subjected to a rather large *variability*. Besides, part of it, is due to the extreme sensitivity of the corresponding testing methods, which contribute to a very large scattering of experimental findings.

**Bond tensile strength of grouts**, perpendicularly unstucked to the stone substratum:

$$f_{gr,b,t} \approx 0.7\, f_{gr,t}\ (\pm 30\%) \text{ for considerable rugosities of stone surface} \qquad (6.6a)$$

$$f_{gr,b,t} \approx 0.4\, f_{gr,t}\ (\pm 30\%)\ \text{for low rugosity of stone surface} \tag{6.6b}$$

Note: More recent results by Luso and Lourenço (2017b), were not taken into account in the preceding review, since the thickness of the intermediate grout layer tested was equal to 20 mm, instead of the approximately 2 mm of the previous investigations; thus, expectedly, in the case of the Luso and Lourenço (2017b) investigation bond results were rather lower.

**Bond shear mechanical properties** As in every shear phenomenon, the *shear* resistance of grouts placed on stone substrata depends on the value of the normal stress "σ" acting on the sheared interface.

- Under zero normal stress (grout-to-stone cohesion)

$$c \approx \frac{1}{7} f_{gr,t}\ (\pm 30\%),\quad \text{for mainly cementitious grouts} \tag{6.7a}$$

$$c \approx \frac{1}{3} f_{gr,t}\ (\pm 30\%),\quad \text{for low cement tripartite grouts and hydraulic lime grouts} \tag{6.7b}$$

- Maximum friction coefficient (in $f_{gr,b,s} = \mu_{max}.\sigma$):
  For low strength grouts and rough interfaces, an oversimplified expression could be the following, for relatively low σ-values:

$$\mu_{max} \approx 1/\sigma^{[\mathrm{MPa}]}\ (\pm 20\%) \tag{6.8}$$

  A somehow sophisticated expression (similar to the one applicable along concrete-to-concrete interfaces) proposed by Adami and Vintzileou (2010) is:

$$\mu_{max} = 0.4\left(\frac{\sigma}{f_{gr,c}}\right)^{-\beta},\quad f_{gr,c} < 10\,\mathrm{MPa} \tag{6.9}$$

  where $f_{gr,c}$ denotes the compressive strength of the grout, and
  β = 0.40 for marble
  β = 0.55 for travertine
  β = 0.80 for sandstone substratum

- Residual friction coefficient (for large slips, after the falling branch of resistance):

  Only indicative values are available:

$$\mu_{res} \approx \frac{1}{3}\mu_{max},\ \text{for } \sigma = 0.3\,\mathrm{MPa} \tag{6.10a}$$

$$\mu_{res} \approx \frac{1}{4}\mu_{max},\ \text{for } \sigma = 0.6\,\mathrm{MPa} \tag{6.10b}$$

- Critical slip, leading to maximum shear bond resistance

$$s_u \approx 1.0\,(\pm 0.5)\,\text{mm} \tag{6.11}$$

It has to be noted that in the preceding analysis, the degree of saturation of stone or stone-like materials was not considered as an additional parameter, since the solid perimeters of masonry voids may only qualitatively characterized as "relatively wet", after the injection of the grout and before its setting and strengthening. Thus, it is believed that the role of the saturation degree may be covered by the scattering limits proposed.

Nevertheless, available data on bond resistance of unstucked (pulloff) interfaces, show the detrimental effect of saturation of the substratum: Such a bond reduction between the cases of "dry" and "saturated" stone materials, was found to be roughly equal to 40% in the case of Miltiadou (1990) and 50% in the case of Luso and Lourenço (2017b). Thus, it has to be reiterated that water injection to masonries just before grouting, should be avoided.

**Synopsis**

The above presented mechanical parameters may be directly useful in the case of computational models regarding local structural behaviour of masonry. Moreover, the aforementioned numerical relationships may offer some qualitative information as follows:

The tensile bond strength "Eq. 6.6a, b" is roughly equal to half the flexural tensile strength of the grout itself. However, the rugosity of the stone surface is much more important.

It is important to note that quasi vertical interfaces (under practically zero normal stress) exhibit bond-shear "c" "Eq. 6.7a, b" roughly equal to one fourth of the tensile strength of the grout. But in this case, the main parameter is the nature of the binders: low cement ternary or hydraulic lime compositions are relatively stronger in bond-shear, for the same $f_{gr,t}$-value.

The role of normal stress "Eq. 6.8" on the "$f_{gr,b,s}$" value of the maximum bond-shear-resistance cannot be overestimated:

$$f_{gr,b,s} = \mu_{max} \cdot \sigma = 0.4 \cdot f_{gr,c}^{\beta} \cdot \sigma^{1-\beta} \tag{6.12}$$

where $\beta < 1$ as in "Eq. 6.9".

**Notes**

1. The particular case of *brick* masonry should be briefly commented here. The subject of porosity and rugosity of the external surfaces of the bricks can be easily examined in each case, be it in a qualitative way. However, in the case of low strength or inadequately fired bricks, there is a possibility of detachment of a weak skin-layer of the brick, when a grout is under tension across a brick surface. Similarly, under direct shear conditions along a brick interface, failures may appear by means of a diagonal cracking of the brick substratum. In these

cases, the values of $f_{gr,b,t}$ and $f_{gr,b,s}$ are shaped only by the low tensile strengths of the external regions of the brick, independently of the composition and the mechanical properties of the grout.

2. Similar behaviour may be observed, in cases of very weak porous stones.
3. The preceding data do not cover the case of a substratum consisting of a layer of *mortar* used to build a masonry. It has however to be reminded that in most cases of historical masonries, these mortars exhibit a very low strength, which would result in insignificant bond strengths of grouts in contact with such mortars. Otherwise, a relatively stronger mortar may be considered as a weak stone with high rugosity.

## 6.6 Selection of a Required $f_{gr,c}$-range, for Targeted $f_{wc}$-values

1. As stated in Sect. 6.3.1, the required compressive strength of the masonry $(f_{wc,requ.})$ was first calculated under the pertinent actions. The next step was to estimate (roughly though) a corresponding *required* compressive strength $f_{gr,c}$ of the grout to be injected, (although this is not a direct criterion of the structural effectiveness of the grout). This may be effectuated by means of masonry strength expressions offered in Sect. 6.3.2.1; thus, a corresponding $f_{gr,c}$-value is available, and an ample margin of scattering should be allowed.
   Consequently, a targeted range of required $f_{gr,c}$ appears.
   It is however repeatedly observed that this is only a roughly estimated $f_{gr,c}$-value. That is why, for the subsequent design procedure, a *range* of $f_{gr,c}$-values should be selected, close to the aforementioned estimated $f_{gr,c}$-value.
   Note: If, however, the necessary $f_{gr,c}$-value comes to be disproportionally high, the designer should in parallel take additional strengthening measures rather than grouting alone.
2. Based on this targeted range of $f_{gr,c}$-values, one is able to make use of the information included in the Tables of Sect. 6.4.4.2, in order to be assisted in selecting an appropriate *candidate composition*. Observing the appropriate Tables (from Tables 6.1, 6.2, 6.3, 6.4, 6.5, 6.6, 6.7, 6.8 and 6.9), the designer identifies those compositions which may (approximately though) satisfy the required $f_{gr,c}$-value within the ranges of grout strengths presented in these Tables. Thus, the nature and the number of binders, as well as the composition of the grout, may be roughly selected, temporarily though.
3. A further simplification of this procedure is possible by means of the following Table 6.10, which contains more condensed information. For four ranges of compressive strength (1.0–3.0 MPa, 3.0–6.0 MPa, 6.0–10.0 and 10.0–20.0 MPa), a series of corresponding grout compositions are given, including all types of grouts (one binder, two binders or three or more binders, based on hydraulic lime or cement with lime and pozzolan).

**Table 6.10** Candidate compositions of grouts corresponding to four ranges of compressive strength at 90 days

| Grout's compositions corresponding to ranges of compressive strength at 90 days ($f_{gr,c}^{90d}$) in MPa | | | |
|---|---|---|---|
| 1.0–3.0 (MPa) | 3.0–6.0 (MPa) | 6.0–10.0 (MPa) | 10.0–20.0 (MPa) |
| | | | L/MK/LSF/W:35/35/30/60 |
| NHL5/W: 100/82.5 | NHL5/W: 100/90–80 | | |
| NHL5/P/W: 90/10/80.2 | NHL5/P/W: 90/10/80 | | |
| NHL5/P/W: 80/20/85–80 | | | |
| NHL5/E/W: 70/30/90 | | | |
| NHL5/E/P/W: 63/30/7/87.5 | | | |
| | | | NHL5/L/MK/W:30/35/35/60 |
| NHL3.5$_Z$/W: 100/110–90 | NHL3.5$_Z$/W: 100/95–80 | | |
| NHL3.5$_z$/P/W: 80/20/120 | NHL3.5$_Z$/P/W: 80/20/95 | | |
| NHL2/W: 100/130 | | | |
| HL/W: 100/60 | | | |
| | CR/W: 100/70 | | |
| CA/W: 100/70 | CA/W: 100/55 | | |
| Unilit (0/0)/W: 100/80–70 | | | |
| CC/W: 100/60–65 | | | |
| | | | LI/W:100/?[a] |
| | | | C/L/W:80/20/85 |
| | | | C/P/W:60/40/85–75 |
| | C/L/W: 60/40/100 | C/L/W: 60/40/85 | |
| | | C/L/P/W:60/20/20/100 | C/L/P/W:60/20/20/75 |

(continued)

**Table 6.10** (continued)

| Grout's compositions corresponding to ranges of compressive strength at 90 days ($f_{gr,c}^{90d}$) in MPa | | | |
|---|---|---|---|
| 1.0–3.0 (MPa) | 3.0–6.0 (MPa) | 6.0–10.0 (MPa) | 10.0–20.0 (MPa) |
| | C/L/W: 50/50/95 | | |
| | C/L/P/W: 50/25/25/105 | | C/L/P/W: 50/25/25/80 |
| C/L/P/W: 50/20/30/125 | C/L/P/W: 50/20/30/120 | | |
| | | | C/FA/W:40/60/65 |
| C/L/P/W: 40/25/35/130 | C/L/P/W: 40/25/35/120–80 | C/L/P/W: 40/25/35/75 | |
| | | C/L/P/W: 40/30/30/110–90 | |
| | | C/L/P/W: 40/20–15/40–45/80 | |
| C/L/P/W: 30/30/40/140–130 | | | |
| | | | C/L/MK/W: 30/35/35/110 |
| | C/L/P/W: 30/25/45/80 | C/L/P/W: 30/25/45/80–72.5 | |
| | C/L/P/W: 30/20/50/80 | C/L/P/W: 30/20–10/50–60/90–85 | |
| | | | C/L/P/SF/W: 30/20–10/40–50/10/110 |
| | | | C/L/P/SF/W: 30/20–10/45–55/5/100 |
| C/L/Cl/W: 25/25/50/110 | | | |
| C/P/W: 20/80/120–115 | | | |
| C/L/P/W: 20/30/50/135 | | | |
| | C/L/P/W: 15/50/35/97 | | |
| | C/L/P/B/W: 13/43/26–22/18–22/98–93 | | |
| | C/L/P/W: 10/50/40/100 | | |

(continued)

**Table 6.10** (continued)

| Grout's compositions corresponding to ranges of compressive strength at 90 days ($f_{gr,c}^{90d}$) in MPa | | | |
|---|---|---|---|
| 1.0–3.0 (MPa) | 3.0–6.0 (MPa) | 6.0–10.0 (MPa) | 10.0–20.0 (MPa) |
| C/L/P/W: 10/30/60/50 | | | |
| | C/L/P/W: 10/20–15/70–75/85 | | |
| | C/L/P/B/W: 10/45/25/20/95–90 | | |
| | C/L/P/Cl/W:9/46/36/9/112 | | |
| | C/L/P/Cl/W:8/42/33/17/89 | | |

[a] *The water to solids ratio were not known to the authors of this book. The respective papers are given in the Appendix, Table 6.11).* *C* cement, *L* Lime, *NHL* natural Hydraulic Lime, *HL* Artificial Hydraulic Lime, *P* Pozzolan, *SF* silica fume, *FA* Fly ash, *MK* Metakaolin, *CR* Calx Romana (NHL), *CA* Calce Albazzana (NHL), *CC* Calce per Consolidamento (Premixed), *LI* Lime injection (Premixed), *LSF* Limestone Filler, *B* Brick Dust, *Cl* Clay and *W* water

It is worth to be noted that a "target strength range" of $f_{gr,c}$ may be achieved by more than one combination of binders, as it can be apparent in Table 6.10. Moreover, as one may see, the consecutive lines of the table follow a gradual increase of fineness of the compositions.

However, the aforementioned information is only a rough approximation, serving as a first guidance for the selection of the appropriate composition of the grout. Obviously, the designer by means of samples taken from the final composition, will check the values of the real strength of the grout and its variability as a function of the water-to-solids ratio and SP content (see Chap. 9). Besides, a successful grouting depends much more on the injectability of the grout and the application operations, than on any precision in strength values; That is why, even a large range of strength suffices.

4. These (possibly numerous) "candidate" compositions, should now be checked against the other performance requirements, i.e. durability, penetrability, fluidity and stability (see Chaps. 7 and 8).

## 6.7 Shrinkage

The shrinkage of grouts is expected to be relatively large. Indicatively, for a grout made of cement and Milos earth (a low reactivity natural pozolan), (Miltiadou et al. 1998), be it under rather arid conditions (50% of relative humidity), values between $30 \times 10^{-4}$ and $60 \times 10^{-4}$ were reported in 180 days. Similarly, binary grouts composed by cement with densified silica fume or hydrated lime, exhibited values between $10 \times 10^{-4}$ and $30 \times 10^{-4}$ in 28 days under 70% of relative humidity, or $40 \times 10^{-4}$ and $70 \times 10^{-4}$ under 50% of relative humidity (Paillère et al. 1993).

Compared with the typical shrinkage values valid for cement mortars $20 \times 10^{-4}$, grouts are recognized as more shrinkable—especially those which are richer in fines (as in the case of ternary grouts). Thus, a question about the consequences of this phenomenon within the masonry may be raised. However, practical evidence shows that this is not a serious problem: In fact, within a grouted masonry, the relative humidity of the air is rather high for the following reasons:

- relatively dry external air cannot easily penetrate in rather thick masonries,
- the air content enclosed in masonry is small, and evaporation takes place within the pores
- the injected liquids contain large quantities of water.

Besides, water absorption by masonry materials reduces the water content of the grout.

Moreover, within small and inclined voids in masonry, the final absolute value of displacement of shrinked grout surfaces is evidently very small. For all these reasons, even the aforementioned grouts resulted in completely satisfactory strengths of injected masonry elements (Miltiadou et al. 1993; Vintzileou and Tassios 1995;

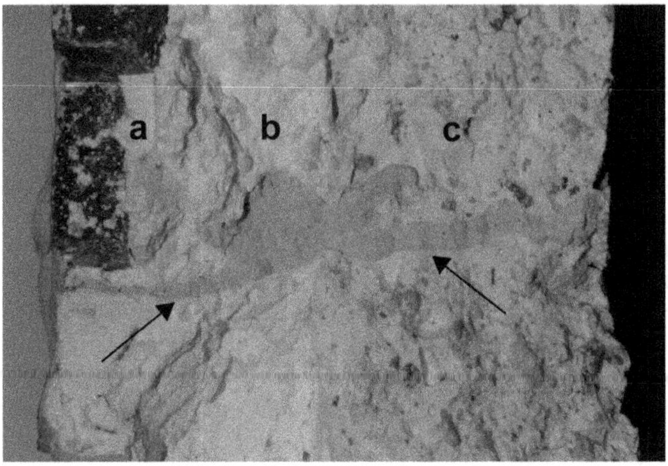


**Fig. 6.5** Grout in a crack of a mural mosaic's substratum mortar, observed on a piece selected after a diagonal compression test carried out on an injected masonry. This piece was cured in room conditions for 14 years: **a** Mosaic's tesserae.**b** First layer of substratum mortar with fine aggregates. **c** Second layer of substratum mortar with coarser aggregates (Total width of two layers ~5 cm)

Vintzileou and Miltiadou-Fezans 2008; Kalagri et al. 2010; Vintzileou et al. 2015; Mouzakis et al. 2017).

Figure 6.5 shows the grout filling a crack, produced in a mosaic substratum mortar, during the diagonal compression testing (Vintzileou and Miltiadou-Fezans 2008) of a wallette injected with a NHL5 grout. It is clear that the grout has filled the crack and no signs of shrinkage are present. The photo has been taken 14 years after the test was carried out.

However, in grouting regarding reattachment of frescoes and mosaics substrata or in situ repair of delaminated plasters and renderings, grout shrinkage may be of greater importance (Biçer-Şimşir et al. 2010; Tavares et al. 2010).

## Appendix: Data Regarding Grouts Compressive and Tensile Strength in Function of the Water to Solids Ratio

In the following Tables 6.11, 6.12 and 6.13 several numerical results of grout strengths are listed, as reported in the literature.

In Table 6.11 "One Binder" compositions are given. Although, the actual state of the art is in favour of binary and ternary compositions instead of plain cement grouts, plain cement compositions are also included in this table, as reference mixes and as well because they may be useful for other particular cases of grouting. Premixed solid phases are also included, although they may contain various combinations of binders, fillers and admixtures.

**Table 6.11** Compressive and tensile strength of One Binder grouts, as a function of the water to solids ratio (Premixed grouts based on limes are also included)

| Binder | Authors | Composition | W/S | At appr. 28 days | | At appr. 3 months | |
|---|---|---|---|---|---|---|---|
| | | | | $f_{gr,c}$ | $f_{gr,t}$ | $f_{gr,c}$ | $f_{gr,t}$ |
| Cement | Miltiadou 1990[(1)] | C100 | 0.50 | 34.00–25.00 | 3.00–4.00 (s) | | |
| | Van Germent et al. (1995) | C100 + 2% bentonite + 1.5% SP | 0.68 | 31.80 | 4.00 | | |
| | Miltiadou et al. (1998)[(2)] | C100 | 0.70 | 24.70 | 4.40 | | 4.20 |
| | Mirza et al. (2002) | C100 | 0.50 | 36.00 | | 41.00 | |
| | | C100 | 0.55 | 33.00 | | 34.00 | |
| | | C100 | 0.65 | 26.00 | | 29.00 | |
| | Saric-Coric et al. (2003) | C100 + HRWR and VMA admixtures (cubic specimens of grout with sand) | 0.40 | 31.00–41.00 | | | |
| | Mirza et al. 2013[(3)] | C100 | 0.80 | 11.00–29.00 | | | |
| | | | 1.20 | 7.00–12.00 | | | |
| | Uranjek et al. (2014) | C100 + 0.5% expansion additive | | | | 47.00 | 4.40, 1.60 (s) |
| | | Cement + 3% bentonite + 0.3% expansion additive | | | | 52.30 | 4.40, 1.40 (s) |
| Hydraulic lime and premixed solid phases based on limes [(4)] | Valluzzi (2000) | FEN X-B [(4)] | 0.50 | 3.23 | 0.35, 0.32 (s) | | |
| | | FEN X-A | 0.55 | 3.35 | 0.06 (s) | | |
| | | FEN X-A + Fluidifier (F) | 0.50 | 5.10 | - | | |
| | | FEN X-A + Retarder (R) | 0.55 | 3.21 | 0.20 (s) | | |
| | | FEN X-A + F + R | 0.55 | 3.65 | 0.28 (s) | | |
| | Corradi et al. (2002)[(5)] | Lime based hydraulic grout, kaolin calcined at low temperature, carbonates | ? | 8.00 | | | |
| | Miltiadou-Fezans et al. (2004a) | NHL3.5Z:100 | 0.85 | 1.86 | 1.16 | 3.54 | 1.52 |
| | | NHL3.5Z:100 | 0.90 | 1.74 | 0.80 | 4.10 | 1.92 |
| | | NHL3.5Z:100 | 0.95 | 2.00 | 0.63 | 4.10 | 1.92 |

(continued)

**Table 6.11** (continued)

| Binder | Authors | Composition | W/S | At appr. 28 days | | At appr. 3 months | |
|---|---|---|---|---|---|---|---|
| | | | | $f_{gr,c}$ | $f_{gr,t}$ | $f_{gr,c}$ | $f_{gr,t}$ |
| | | NHL3.5Z:100 | 1.00 | 1.12 | 0.53 | 2.36 | 1.20 |
| | | NHL3.5Z:100 | 1.10 | 0.98 | 0.54 | 1.21 | 0.97 |
| | Miltiadou-Fezans et al. (2006a) | NHL5:100 + SP:1 | 0.80 | 1.64 | 1.16 | | |
| | | NHL2:100 + SP:1 | 1.30 | 0.72 | 0.58 | 1.44 | 0.95 |
| | | Calx Romana | 0.70 | 2.59 | 1.14 | 3.92 | 1.32 |
| | | Albaria Calce Albazzana | 0.55 | 3.40 | 1.97 | 4.20 | 2.20 |
| | | Albaria Calce Albazzana | 0.70 | | | 2.40 | 1.00 |
| | | Unilit Fluid 0(4) | 0.80 | | | 1.94 | 1.50 |
| | | Unilit Fluid 0 (4) | 0.90 | 1.00 | 0.46 | | |
| | | Albaria Inezione(4) | 0.48 | | | 10.00 | 1.00 |
| | Oliveira et al. (2006)(5) | Commercial lime-based grout | ? | 17.60 | 0.29 | | |
| | Corradi et al. (2008)(5) | Albaria Iniezione 100 | ? | 7.00 | 3.00 | | |
| | Kalagri et al. (2010) | NHL5:100 | 0.80 | 2.06 | 1.10 | 4.88 | 1.75 |
| | | NHL5:100 + SP:1 | 0.80 | 2.82 | 1.90 | 4.50 | 2.52 |
| | | NHL3.5Z:100 + SP:0.7 | 0.80 | 3.10 | 1.65 | 4.67 | 2.19 |
| | | Calx Romana | 0.70 | 2.25 | 1.51 | 3.04 | 1.39 |
| | | Albaria Calce Albazzana | 0.70 | 1.69 | 1.02 | 2.60 | 0.88 |
| | | Unilit B Fluid 0 (4) | 0.70 | 1.53 | 1.27 | 2.56 | 1.53 |
| | Mazzon (2010) | FEN X-B (4) | | 12.80 | 3.80 | | |
| | Silva et al. (2012) | FEN-X/B (4) | | 12.50 | 2.80 | | |
| | Baltazar et al. (2014) (Strength values are approximately extracted from the graphs) | NHL5:100 + SP:0.6 | 0.45 | 5.40 | | | |
| | | NHL5:100 + SP:0.6 | 0.50 | 6.60 | | | |
| | | NHL5:100 + SP:0.6 | 0.55 | 4.80 | | | |
| | | NHL5:100 + SP:0.8 | 0.45 | 5.60 | | | |

(continued)

**Table 6.11** (continued)

| Binder | Authors | Composition | W/S | At appr. 28 days | | At appr. 3 months | |
|---|---|---|---|---|---|---|---|
| | | | | $f_{gr,c}$ | $f_{gr,t}$ | $f_{gr,c}$ | $f_{gr,t}$ |
| | | NHL5:100 + SP:0.8 | 0.50 | 6.80 | | | |
| | | NHL5:100 + SP:0.8 | 0.55 | 5.00 | | | |
| | | NHL5:100 + SP:1.2 | 0.45 | 5.50 | | | |
| | | NHL5:100 + SP:1.2 | 0.50 | 6.20 | | | |
| | | NHL5:100 + SP:1.2 | 0.55 | 5.00 | | | |
| | Papagianni and Pachta (2015) | HL | 0.61 | 1.76 | 1.15 | 2.25 | |
| | Yüzer et al.(2015) | Commercial hydraulic lime based grout | 0.60 | 2.00 | 0.80 | | |
| | Baltazar et al. (2017) | NHL5:100 + SP:0.8 | 0.50 | 7.00 | 1.50 | | |
| | | NHL5:100 + SP:0.8 + 1.5% linseed oil | 0.50 | 4.00 | 0.40 | | |
| | Luso and Lourenço (2016) (Strength values are approximately extracted from the graphs) | Mapei-Antique I ([4]) | 0.35 | 21.00 | 4.00 | 23.00 | 5.50 |
| | | Albaria Iniezione([4]) | 0.35? | 22.50 | 2.80 | 30.00 | 3.40 |
| | | Calce per Consolidamento ([4]) | 0.60–0.65 | 1.50 | | 2.50 | |
| | | Lime-Injection ([4]) | | 12.00 | 3.50 | 13.00 | 4.00 |
| | Giaretton et al. (2017) | Hydraulic lime based grout | 0.37 | 12.80 | 3.80 | | |
| | Nikolopoulou et al. (2018) | NHL5:100 | 0.90 | 1.03 | 0.67 | 3.14 | 1.61 |
| | | NHL5:100 | 0.925 | 0.98 | 0.64 | 3.02 | 1.52 |
| | Luso and Lourenço (2019) | Mape-Antique I ([4]) | 0.35 | 21.40 | 4.10 | | |

(1) Two different cements (CPA 55PM, CLK45-PM) were tested, using cylindrical specimens (d = 20 mm and h = 40 mm), both for compressive strength and for tensile strength, which was measured using the splitting test

(2) White Danish cement C55

(3) Two cements, 3 microfine cements, 4 blended Portland cements. The use of microfine cements, however, may offer further increased strengths

(4) Premixed grouts based on limes

(5) For some premixed grouts, the water to solids ratio was not known to the authors of this book. In some cases, the W/S ratio "suggested by the manufacturer" was used (see respective papers).

C = cement, NHL = natural Hydraulic Lime, HL = Artificial Hydraulic Lime, W = water

**Table 6.12** Compressive and tensile strength of Two Binders grouts as a function of the water to solids ratio

| Two Binders | Authors | Composition | W/S | At appr. 28 days | | At appr. 3 months | |
|---|---|---|---|---|---|---|---|
| | | | | $f_{gr,c}$ | $f_{gr,t}$ | $f_{gr,c}$ | $f_{gr,t}$ |
| Cement + Pozzolan | Miltiadou (1990)[(1)] | C90/DSF10 | 0.75 | 20.70–25.90 | 2.40–2.50 (s) | | |
| | | C75/DSF/25 | 0.75 | 22.30–31.00 | 2.50–3.00 (s) | | |
| | | C60/DSF40 | 1.00 | 16.60–21.90 | 2.10–2.50 (s) | | |
| | | C90/P10 | 0.75 | 21.50–20.70 | 2.10–2.70 (s) | | |
| | | C75/P25 | 0.75 | 18.60–14.60 | 1.80–1.90 (s) | | |
| | | C60/P40 | 1.00 | 10.70–9.30 | 1.20–1.50 (s) | | |
| | Tomaževič and Apih (1993) (The age of specimens is not reported; moreover, occasionally, sand was added) water/solid varies from 0.90–1.00 | C90/P10 | 0.90–1.00 | 30.30 | | | |
| | | C90/P10 | 0.70 | 32.50 | 1.90 | | |
| | | C90/P10 + 10 hydrophobic additive (a) | 0.70 | 19.70 | 1.60 | | |
| | | C70/P10 + 30 sand + 10additive (a) | ? | 6.80 | 0.60 | | |
| | | C70/P10 + 30 sand + 10 additive (a) | ? | 12.80 | 1.70 | | |
| | Miltiadou et al. (1998)[(2)] | C90/P10 | 0.70 | 19.80 | 3.90 | 24.20 | 4.60 |
| | | C75/P25 | 0.80 | 13.00, 12.03, 17.40, 17.40 | 3.50, 3.50, 4.50, 1.90 | 17.70, 18.20, 27.30, 27.30 | 3.90, 4.90, 4.80, 4.20 |
| | | C60/P40 | 0.85 | 8.80 | 2.80 | 13.10 | 3.50 |
| | | C20/P80 | 1.20, 1.15 | 1.40, 1.90 | 0.70, 0.80 | 1.84, 2.80 | 1.20 |
| | Mirza et al. (2002)[(3)] | C40/FA 60[(3)] | 0.50 | 16.00, 20.00, 15.00 | | 29.00, 26.00, 25.00 | |
| | | C40/FA60 | 0.55 | 14.00, 18.00, 12.00 | | 28.00, 20.00, 17.00 | |
| | | C40/FA60 | 0.65 | 12.00, 15.00, 9.00 | | 18.00, 16.00, 13.00 | |
| | Miltiadou-Fezans et al. (2003) | C60/P40 + SP0.7 | 0.65 | 25.54 | 6.96 | 26.02 | 5.57 |
| | | C60/P40 + SP0.7 | 0.75 | 17.30 | 4.66 | 18.33 | |
| | | C60/P40 + SP0.69 | 0.65 | 27.57 | 4.79 | 31.90 | 7.81 |

(continued)

**Table 6.12** (continued)

| Two Binders | Authors | Composition | W/S | At appr. 28 days | | At appr. 3 months | |
|---|---|---|---|---|---|---|---|
| | | | | $f_{gr,c}$ | $f_{gr,t}$ | $f_{gr,c}$ | $f_{gr,t}$ |
| Cement + Lime | Miltiadou (1990)[(1)] | C90/L10 | 0.75 | 16.10–16.70 | 2.10–2.10 (s) | | |
| | | C75/L25 | 0.75 | 13.60–10.00 | 1.80–1.70 (s) | | |
| | | C60/L40 | 1.00 | 3.20–3.40 | 0.70–0.90 (s) | | |
| | Miltiadou-Fezans et al. (2001a) | C60/L40 + SP0.7 | 0.35 | | | 6,80 | 0,88 (s) |
| | | C60/L40 + SP0.7 | 1.00 | | | 4,67 | 0,59 (s) |
| | | C50/L50 + SP0.75 | 0.95 | | | 3.50 | 0,43 (s) |
| | Toumbakari (2002) | C80/L20 | 0.85 | 14.60 | 5.60 | 19.5 | 4.50 |
| | Vintzileou and Adami (2009) | C80/L 20 | 0.85 | 14.60 | 3.40 | 17.9 | 4.50 |
| | Uranjek et al. (2014) | C75/L25 (+0.3 expansion additive + 3% bentonite) | ?[(4)] | | | 21.7 | 2.80, (1.40 s) |
| | Papagianni and Pachta (2015) | C23/L77 | 0.83 | 0.47 | 0.29 | 1.06 | 0.44 |
| | | C17/L83 | 0.93 | 0.29 | 0.25 | 0.75 | 0.29 |
| Hydraulic limes[(3] + Pozzolan | Miltiadou et al. (2004a) | NHL3.5Z:60/P:40 | 1.10 | 1.04 | 0.46 | | |
| | | NHL3.5Z:75/P:25 | 0.85 | 1.74 | 1.26 | | |
| | | NHL3.5Z 75/P:25 | 1.00 | 1.05 | 0.73 | | |
| | | NHL3.5Z:75/P:25 | 1.10 | 1.15 | 0.50 | | |

(continued)

**Table 6.12** (continued)

| Two Binders | Authors | Composition | W/S | At appr. 28 days | | At appr. 3 months | |
|---|---|---|---|---|---|---|---|
| | | | | $f_{gr,c}$ | $f_{gr,t}$ | $f_{gr,c}$ | $f_{gr,t}$ |
| | Miltiadou et al. (2006a), (2008), (2021) | NHL5:90//P:10 + SP:1 (lab) | 0.80 | 2.50 | 1.31 | | |
| | | NHL5:90//P:10 + SP:1 (in situ- pilot application different lab) | 0.80 | 1.50 | 1.01 | | |
| | | NHL5:90//P:10 + SP:1 (in situ during the works) | 0.80 | 2.02 | 1.01 | 5.26 | 2.80 |
| | | NHL5:90//P:10 + SP:1 (in situ during the works) | | 1.82 | 0.95 | 5.30 | 2.60 |
| | | NHL5:80/P:20 + SP:1,25 (lab) | 0.80 | 2.31 | 1.36 | 4.00 | 1.80 |
| | | NHL5: 80/P:20 + SP:1.5 (lab) | 0.80 | 2.11 | 1.22 | 4.05 | 1.67 |
| | | NHL5:80/P:20 + SP:1.5 (in situ- different lab) | 0.80 | 1.10 | 0.90 | | |
| | | NHL5:85/P15 + SP:1.25 (lab) | 0.80 | 2.14 | 1.41 | | |
| | | NHL5:85/P:15 + SP:1.25 (in situ- different lab) | 0.80 | 1.20 | 0.90 | | |
| | Moundoulas et al. (2009) | NHL3.5Z 80/P:20 | 0.95 | 2.30 | 2.35 | 4.32 | 0.37 |
| | | NHL3.5Z:80/P:20 | 1.20 | 1.20 | 0.88 | 1.59 | 0.27 |

(continued)

**Table 6.12** (continued)

| Two Binders | Authors | Composition | W/S | At appr. 28 days | | At appr. 3 months | |
|---|---|---|---|---|---|---|---|
| | | | | $f_{gr,c}$ | $f_{gr,t}$ | $f_{gr,c}$ | $f_{gr,t}$ |
| | | NHL3.5Z:80/P:20 + SP1:0.8 | 0.95 | 1.74 | 1.37 | 4.07 | 1.82 |
| | | NHL3.5Z:80/P:20 + SP2:1.7 | 0.95 | 1.61 | 0.79 | 3.74 | 1.67 |
| | | NHL3.5Z:80/P:20 + SP3:2.7 | 0.95 | 1.82 | 0.94 | 3.81 | 0.96 |
| | Artioli et al.(2011) | HL/P/sandy filler | ?(4) | 5.47 | 1.83 | | |
| | | HL/P/sandy filler, cured on site | ?(4) | 4.78 | 1.51 | | |
| | Badogiannis et al. (2012) | NHL5:100 + SP2:0.75 | 0.825 | 1.60 | 0.50 | 2.00 | 0.70 |
| | | NHL5:90/P1:10 + SF1:0.5 | 0.825 | 1.70 | 0.50 | 2.50 | 0.80 |
| | | NHL5:90/P1:10 + SP2:0.75 | 0.825 | 1.70 | 0.60 | 2.90 | 0.60 |
| | | NHL5:90/P2:10 + SP1:0.5 | 0.825 | 1.80 | 0.50 | 2.80 | 0.80 |
| | | NHL5:90/P2:10 + SP2:0.75 | 0.825 | 1.80 | 0.70 | 2.70 | 0.70 |
| | Baltazar et al. (2014) (Strength values extracted approximately from the graphs) | NHL5:98/SF:2 + SP:0.8 | 0.45 | 8.30 | | | |
| | | NHL5:98/SF:2 + SP:0.8 | 0.50 | 8.30 | | | |
| | | NHL5:98/SF:2 + SP:0.8 | 0.55 | 7.20 | | | |

(continued)

**Table 6.12** (continued)

| Two Binders | Authors | Composition | W/S | At appr. 28 days | | At appr. 3 months | |
|---|---|---|---|---|---|---|---|
| | | | | $f_{gr,c}$ | $f_{gr,t}$ | $f_{gr,c}$ | $f_{gr,t}$ |
| | | NHL5:94/SF:6 + SP:0.8 | 0.45 | 9.60 | | | |
| | | NHL5:94/SF:6 + SP:0.8 | 0.50 | 8.50 | | | |
| | | NHL5:94/SF:6 + SP:0.8 | 0.55 | 7.50 | | | |
| | | NHL5:90/SF:10 + SP:0.8 | 0.45 | 9.80 | | | |
| | | NHL5:90/SF:10 + SP:0.8 | 0.50 | 9.50 | | | |
| | | NHL5:90/SF:10 + SP:0.8 | 0.55 | 7.50 | | | |
| Hydraulic lime + Cement | Baltazar et al. (2017) | NHL5:60/C:40 + SP:0.6 | 0.45 | 17.70 | 5.00 | | |
| | | NHL5:60/C:40 + SP:0.6 + 1.5% linseed oil | 0.45 | 11.00 | 3.50 | | |
| Hydraulic lime + Earth | Nikolopoulou et al. (2018) | NHL5:80/E:20 | 0.85 | 0.72 | 0.63 | 2.13 | |
| | | NHL5:70/E:30 | 0.90 | 0.71 | 0.49 | 1.31 | |

(1) Two different cements (CPA 55PM, CLK45-PM) were tested, using cylindrical specimens (d = 20 mm and h = 40 mm), both for compressive strength and for tensile strength, which was measured using the splitting test. Densified silica fume, Santorini earth and Lime were used as ultrafine materials

(2) Four different pozzolan types were used, with white Danish cement C55

(3) Three types of Fly ash were used

(4) The water to solids ratio were not known to the authors of this book.

C = cement, L = Lime, NHL = natural Hydraulic Lime, HL = Artificial Hydraulic Lime, P = Pozzolan, SF = silica fume, DSF = densified silica fume, FA = Fly ash, E = earth, SP = superplasticizer and W = water

In Table 6.12 "Two Binders" compositions are reported, while compositions with "Three or More Binders" are presented in Table 6.13. For simplicity reasons, in all three Tables, strength, fineness and chemical composition of binders are not described here. For the assessment of hydraulicity of NHL the indications "5, 3.5, 3.5z or 2" are noted, according to EN 459.

As explained in Sect. 6.4.1 the role of the age of the grout is much more important than in the case of concrete. Here, the 28 days and 3 months strengths are listed. In all Tables, if not otherwise indicated, the tensile strength is determined using the flexural test. The indication(s) corresponds to splitting test results. Finally, in these tables, all kinds of grouts encountered in literature are listed, i.e., not only those able to ensure acceptable fluidity and stability for grouts on masonry.

**Table 6.13** Compressive and tensile strength of Three or More Binders grouts as a function of the water to solids ratio

| Three or more Binders | Authors | Composition | W/S | At appr. 28 days | | At appr. 3 months | |
|---|---|---|---|---|---|---|---|
| | | | | $f_{gr,c}$ | $f_{gr,t}$ | $f_{gr,c}$ | $f_{gr,t}$ |
| Cement-Lime/various types of pozzolans and fillers | Penelis et al. (1989) | C40/L50/P10 | 0.75 | 2.45 | 0.57 | | |
| | | C20/L50/P30 | 0.70 | 0.96 | 0.28 | | |
| | Miltiadou (1990)[1] | C70/L16.5/DSF13.5 + SP1.0 | 0.75 | 18.00- 17.20 | 1.70–1.80 (s) | | |
| | | C50/L27.5/DSF22.5 + SP1.66 | 1.00 | 12.50–10.00 (13) | 1.40–1.40 (s) | | |
| | Karaveziroglou et al.(1998) | C10/L30/P60 | 0.50 | 1.30 | 0.42 | 1.47 | 0.67 |
| | | C10/L30/P60 | 0.52 | 1.44 | 0.53 | 1.53 | 0.64 |
| | Miltiadou-Fezans (2000) | C50/L20/P30 | 1.20 | | | 4.45 | 2.05 (s) |
| | | C50/L20/P30 | 1.25 | | | 2.59 | 1.48 |
| | | C50/L18.8 + 1.2 $CaCO_3$/P30 | 1.25 | | | 2.60 | 1.36 |
| | | C40/L25/P35 | 1.20 | | | 3.44 | 1.66 (s) |
| | | C40/L25/P35 | 1.30 | 1.41 | 0.55 | 2.01 | 1.12 |
| | | C40/L23.5 + 1.5 $CaCO_3$/P35 | 1.30 | 1.17 | 0.57 | 2.15 | 1.12 |
| | | C30/L30/P40 | 1.30 | | | 2.79 | 0.98 (s) |
| | | C30/L20/P50 + SP1 | 1.00 | 1.74 | 0.94 | | |
| | | C30/L30/P40 | 1.40 | 0.74 | 0.30 | 1.35 | 0.70 |
| | | C30/L28.2 + 1.8 $CaCO_3$/P40 | 1.40 | 0.73 | 0.35 | 1.56 | 0.68 |

(continued)

**Table 6.13** (continued)

| Three or more Binders | Authors | Composition | W/S | At appr. 28 days | | At appr. 3 months | |
|---|---|---|---|---|---|---|---|
| | | | | $f_{gr,c}$ | $f_{gr,t}$ | $f_{gr,c}$ | $f_{gr,t}$ |
| | Miltiadou-Fezans et all. (2001a) | C60/L20/P20 + SP 0.6 | 0.75 | | | 11.10 | |
| | | C60/L20/P20 | 1.00 | | | 6.30 | |
| | | C50/L25/P25 + SP0.7 | 0.80 | | | 10.80 | |
| | | C50/L25/P25 | 1.05 | | | 4.90 | |
| | | C40/L30/P30 + SP0.7 | 0.90 | | | 7.10 | |
| | | C40/L30/P30 | 1.10 | | | 6.30 | |
| | Miltiadou -Fezans et al. (2001b) | C40/L25/P35 + SP0.7 | 0.80 | 3.96 | 1.86 | 5.50 | 2.10 |
| | | C40/L25/P35 + SP0.7 | 0.75 | 5.30 | 1.86 | 7.40 | 2.30 |
| | Toumbakari (2002) | C30/L17.5/P47.5 + 5 (SF) | 1.00 | 4.30 | 1.20 | 12.90 | 3.00 |
| | | C30/L14/P51 + 5 (SF) | 1.00 | 4.40 | 1.20 | 11.50 | 3.00 |
| | | C30/L11.7/P53.3 + 5 (SF) | 1.00 | 4.10 | 1.20 | 12.00 | 2.80 |
| | | C30/L17.5/P42.5 + 10 (SF) | 1.10 | 6.50 | 1.70 | 15.90 | 3.40 |
| | | C30/L14/P46 + 10 (SF) | 1.10 | 5.40 | 1.50 | 16.00 | 2.60 |
| | | C30/L11.7/P48.3 + 10 (SF) | 1.10 | 5.60 | 1.20 | 17.20 | 2.70 |
| | | C10/L22.5/P67.5 | 0.85 | 2.20 | 0.70 | 5.00 | 1.00 |

(continued)

**Table 6.13** (continued)

| Three or more Binders | Authors | Composition | W/S | At appr. 28 days | | At appr. 3 months | |
|---|---|---|---|---|---|---|---|
| | | | | $f_{gr,c}$ | $f_{gr,t}$ | $f_{gr,c}$ | $f_{gr,t}$ |
| | | C10/L18/P72 | 0.85 | 2.40 | 0.90 | 4.90 | 1.70 |
| | | C10/L15/P75 | 0.85 | 2.10 | 0.80 | 5.40 | 1.10 |
| | | C30/L17.5/P52.5 | 0.85 | 3.20 | 1.50 | 7.80 | 2.50 |
| | | C30/L14/P56 | 0.85 | 3.30 | 1.30 | 6.50 | 2.20 |
| | | C30/L11.7/P58.3 | 0.85 | 3.50 | 1.20 | 7.30 | 2.30 |
| | Miltiadou-Fezans et al. (2004b) | C20/L30/P50 | 1.35 | 0.60 | 0.31 | 2.76 | 1.98 |
| | | C30/L20/P50 | 1.00 | 1.75 | 0.95 | | |
| | Miltiadou-Fezans et al. (2004c), (2006b), (2007a), (2007b) | C40/L20/P40 | 0.80 | 4.16 | 2.32 | 8.00 | 3.50 |
| | | C40/L15/P45 | 0.80 | 4.25 | 2.26 | 7.00 | 3.20 |
| | | C30/L25/P45 (CEM II 42.5 N) | 0.80 | 2.99 | 1.25 | 5.00 | 3.20 |
| | | C30/L20/P50 (CEM II42.5 N) | 0.8 | 2.81 | 1.36 | 5.00 | 3.50 |
| | | C30/L25/P45/ + SP1.0 | 0.725 | 3.78 | 1.72 | 8.55, 9.23, 6.32, 7.62 | 2.84, 1.46, 1.33, 1.22 |
| | | C30/L25/P45 + SP0.85 | 0.75 | 3.75 | 2.67 | 7.46 | 3.80 |
| | Miltiadou-Fezans et al. (2006a) | C30/L25/P45/ + SP1.0 (Danish white) | 0.80 | 4.07 | 2.11 | 8.16 | 2.29 |
| | Adami and Vintzileou (2008) | C30/L35/MK35 | 1.10 | 9.90 | 2.00 | 13.60 | 1.00 |

(continued)

**Table 6.13** (continued)

| Three or more Binders | Authors | Composition | W/S | At appr. 28 days | | At appr. 3 months | |
|---|---|---|---|---|---|---|---|
| | | | | $f_{gr,c}$ | $f_{gr,t}$ | $f_{gr,c}$ | $f_{gr,t}$ |
| | | C30/L20/P50 | 0.90 | 3.30 | 1.70 | 7.90 | 1.90 |
| | Artioli et al. (2011) | C/HL/silty carbonate microfiller cured at standard conditions | ?(2) | 30.90, 37.70 | 1.79, 4.44 | | |
| | | C/HL/silty carbonate microfiller, cured on site | ?(2) | 32.82, 31.63 | 2.01, 2.20 | | |
| | Papagianni and Pachta (2015) | C10/L50/P40 | 1.00 | 1.73 | 0.75 | 4.79 | 0.30 |
| | | C15/L50/P35 | 0.97 | 2.51 | 1.04 | 5.39 | 1.24 |
| | | C12/L40/P28/20 limestone filler | 0.83 | 2.94 | 0.97 | | |
| | | C9/L46/P27/18brick dust | 0.90 | 1.72 | 0.84 | 4.75 | 1.18 |
| | | C13/L43/P26/18brick dust | 0.98 | 1.91 | 0.87 | 5.07 | 1.30 |
| | | C9/L45/P23/23brick dust | 0.96 | 1.46 | 0.54 | 3.06 | 0.60 |
| | | C13/L43/P22/22brick dust | 0.93 | 1.84 | 0.77 | 5.05 | 2.13 |
| | | C9/L46/P36/9 clay | 1.12 | 0.88 | 0.55 | 2.89 | 0.68 |
| | | C8/L42/P33/17 clay | 0.89 | 1.52 | 0.94 | 3.57 | 0.71 |
| | | C25/L25/50 clay | 1.11 | 0.66 | 0.23 | 1.22 | 0.28 |

(continued)

**Table 6.13** (continued)

| Three or more Binders | Authors | Composition | W/S | At appr. 28 days | | At appr. 3 months | |
|---|---|---|---|---|---|---|---|
| | | | | $f_{gr,c}$ | $f_{gr,t}$ | $f_{gr,c}$ | $f_{gr,t}$ |
| | Luso and Lourenço (2017a) | C30/L50/ /MK20 + SP4.0 | | 24.90 | 6.00 | | |
| | | C30/L17.5/ MK52.5 + SP3.3 | | 24.30 | 6.80 | | |
| | | C30/L35/ MK35 + SP5.5 | | 19.60 | 4.70 | | |
| | | C30/L35/ MK35 + SP3.33 | 0.60 | 21.50 | 3.50 | | |
| NHL5 or L / metakaolin/ fillers | Luso and Lourenço (2017a) (90 days strength values extracted approxi-mately from the graphs) | L35/NHL5:30/MK35 | 0.60 | 18.10 | 5.80 | ~ 21.00 | ~ 5.00 |
| | | L35/ MK35/Limestone filler30 | 0.60 | 17.50 | 4.90 | ~ 15.00 | ~ 3.00 |
| | Nikolopoulou et al. 2018 | NHL5 63/E30/P7 | 0.875 | 0.78 | 0.62 | 2.48 | |

(1) Two different cements (CPA 55PM, CLK45-PM) were tested, using cylindrical specimens (d = 20 mm and h = 40 mm), both for compressive strength and for tensile strength, which was measured using the splitting test. Densified silica fume, Santorini earth and Lime were used as ultrafine materials

(2) The water to solids ratio were not known to the authors of this book.

C = cement, L = Lime, NHL = natural Hydraulic Lime, HL = Artificial Hydraulic Lime, P = Pozzolan, SF = silica fume, DSF = densified silica fume, FA = Fly ash, E = earth, MK = metakaolin, SP = superplasticizer and W = water

# References


Adami CE, Vintzileou E (2008) Interventions to historic masonries: investigation of the bond mechanism between stones or bricks and grouts. RILEM Mater Struct 41(2):255–267

Adami CE, Vintzileou E (2010) The potential of medium and low mechanical properties hydraulic grouts for the retrofit of historic masonries. In: Proceedings of 8th international masonry conference, Dresden, 2010, Technische Universität Dresden and International Masonry Society

Adami CE, Vintzileou E., Moutselis M (2008) Shear behavior of interfaces between masonry units and hydraulic-lime grouts. In: Proceedings of RILEM HMC08, Historic Mortar Conference 2008, Portugal

Artioli G, Casarin F, Dalla Benetta M, da Porto F, Secco M, Valluzzi MR (2011) Restoration of historic masonry structures damaged by the 2009 Abruzzo earthquake through injection grouts. In: Proc. of 9th Australasian Masonry Conference, Queenstown, New Zealand 15–18 February 2011

Badogiannis E, Miltiadou-Fezans A, Kalagri A, Ipsilanti E, Dedeloudis C, Vintzileou E (2012) Mix design and performance evaluation of grouts with superfine natural pozzolan. In: Jasieńko J (ed) Structural Analysis of Historical Constructions, DWE Wrocław Poland, pp 773–780

Baltazar LG, Henriques FMA, Jorne F, Cidade MT (2014) Combined effect of superplasticizer, silica fume and temperature in the performance of natural hydraulic lime grouts. Construct Build Mater 50(2014):584–597

Baltazar LG, Henriques F, Cidade MT (2017) Grouts with improved durability for masonry consolidation: an experimental study with non-standard specimens. Key Eng Mater 747:480–487

Biçer Şimşir B, Rainer L (2013) Evaluation of lime-based hydraulic injection grouts for the conservation of architectural surfaces a manual of laboratory and field test methods. Getty Conservat Instit Los Angeles 2013:105

Biçer-Şimşir B, Griffin I, Palazzo-Bertholon B, Rainer L (2010) Lime-based injection grouts for the conservation of architectural surfaces. Studies in Conservation, June 2010. DOI: 10.1179/sic. 2010.55 Supplement-1.3

Binda L, Modena C, Baronio G (1993) Strengthening of masonries by injection technique. In: Proceedings of 6° NaMC, vol I, Philadelphia, pp 1–14

Binda L, Modena C, Baronio G, Abbaneo S (1997) Repair and investigation techniques for stone masonry walls. Constr Build Mater 11(3):133–142

Binda L, Baronio G, Tiraboschi C, Tedeschi C (2003) Experimental research for the choice of adequate materials for the reconstruction of the Cathedral of Noto. Constr Build Mater 17:629–639

Binda L, Modena C, Baronio G, Gelmi A (1994) Experimental qualification of injection admixtures used for repair and strengthening of stone masonry walls, In: Proceedings 10th International brick/block masonry conference on Calgary, Vol. 2, pp 539–548

Bras A, Henriques MS (2012) Natural hydraulic lime based grouts—the selection of grout injection parameters for masonry consolidation. Constr Build Mater 26(2012):135–144

Corradi M, Borri A, Vignoli A (2002) Strengthening techniques tested on masonry structures struck by the Umbria-Marche earthquakes of 1997–1998. Constr Build Mater 16:229–239

Corradi M, Tedeschi C, Binda L, Borri A (2008) Experimental evaluation of shear and compression strength of masonry wall before and after reinforcement: Deep repointing. Constr Build Mater 22:463–472

Egermann R (1993) Investigation on the load bearing behavior of multiple leaf masonry. In Proceedings of IABSE symposium structural preservation of the architectural heritage. Rome, Italy, pp 305–312

EN 1771 (2004) Products and systems for the protection and repair of concrete structures-Test methods-Determination of injectability and splitting test.

EN 197–1 (2000) Part 1: Composition, specifications and conformity criteria for common cements

EN 459–1 (2015) Building lime. Definitions, specifications and conformity criteria

Ferragni D, Forti M, Malliet J, Mora P, Teutonico JM, Torraca G (1984) Injection Grouting of mural paintings and mosaics. In: Proceedings of symposium international adhesives and consolidants, Paris, 2–8 Sept., pp. 110–116.

Van Gemert D, Ladang C, Carpentier L, Geltmeyer B (1995) Consolidation of the Tower of St. Mary's Basilica at Tongeren. Int. Journal for Restoration of Buildings and Monuments, Vol. 1, no. 5, 1995, pp. 371–392

Giaretton M, Valluzzi MR, Mazzon N, Modena C (2017) Out-of-plane shake-table tests of strengthened multi-leaf stone masonry walls. Bull Earthq Eng 15:4299–4317

Griffin I (2004) Pozzolanas as additives for grouts. Stud Conserv 49(2004):23–34

Holmstrom I, (1982) Mortars, cements and grouts for conservation and repair. Some urgent needs for research. In: Proceedings of international symposium mortars, cements and grouts used in the conservation of historic buildings, 3–6 November 1981, Rome, ed. ICCROM, Rome, pp 19–24

Jorne F, Henriques FMA, Baltazar LG (2015) Injection capacity of hydraulic lime grouts in different porous media. Mater Struct 48:2211–2233. https://doi.org/10.1617/s11527-014-0304-9

Kalagri A, Miltiadou-Fezans A, Vintzileou E (2010) Design and evaluation of hydraulic lime grouts for the strengthening of stone masonry historic structures. Mater Struct 43:1135–1146

Karaveziroglou M, Papayianni I, Penelis G (1998) Mortars and grouts in restoration of Roman and Byzantine monuments. In: Biscontin G, Moropoulou A, Erdik M, Delgado Rodrigues J (eds) Compatible materials for the protection of European cultural heritage, PACT 55 1998, Technical Chamber of Greece, pp 219–245

Luso E (2019) Lourenço P (2019) Mechanical behaviour of two leaf masonry wall-strengthening using different grouts. J Mater Civ Eng 31(7):04019096

Luso E, Lourenço PB (2016) Experimental characterization of commercial lime-based grouts for stom]ne masonry consolidation. Construct Build Mater 102:216–225

Luso E, Lourenço PB (2017a) Experimental laboratory design of lime-based grouts for masonry consolidation. Int J Architect Heritage https://doi.org/10.1080/15583058.2017.1354095

Luso E, Lourenço PB (2017b) Bond strength characterization of commercially available grouts for masonry. Constr Build Mater 144(2017):317–326

Mazzon N (2010) Influence of Grout injection on the dynamic behaviour of stone masonry buildings. PhD Thesis, University of Padova

Miltiadou A, Durville JL, Martineau F, Massieu E, Serrano JJ (1993) Etude mécanique de mélanges cailloux-mortier-influence de l'injection de coulis. Bulletin de liaison, Laboratoire des Ponts et Chaussées-183- janv.- févr. 1993. Réf. 3677:75–84

Miltiadou A, Papakonstantinou E, Zambas K, Panou A, Frantzikinaki K (1998) Structural restoration of the columns of Parthenon Opisthodomos with hydraulic grouts of high injectability. Research report, Archives of the Conservation Service of Acropolis Monuments, p 78 (in Greek)

Miltiadou AE (1990) Étude des coulis hydrauliques pour la réparation et le renforcement des structures et des monuments historiques en maçonnerie. Thèse de Doctorat de l'Ecole Nationale des Ponts et Chaussées. Pub. LCPC in Collection Etudes et recherches des Laboratoires des Ponts et Chaussées, série Ouvrages d'art, OA8 ISSN 1161–028X, LCPC, Décembre 1991, Paris, France, p 278

Miltiadou-Fezans A (1998) Criteria for the design of hydraulic grouts injectable into fine cracks and evaluation of their efficiency. In: Biscontin G, Moropoulou A, Erdik M, Delgado Rodrigues J (eds) Compatible materials for the protection of European cultural heritage, PACT 55 1998, Technical Chamber of Greece, pp 149–163

Miltiadou-Fezans A (2000) Design of the hydraulic grouts compositions for the structural restoration of Milos catacombs, Greece. Research report, Laboratory of Restoration Materials and Techniques, Directorate for the Restoration of Byzantine and Post-Byzantine Monuments, Hellenic Ministry of Culture and Sports (in Greek)

Miltiadou-Fezans A, Anagnostopoulou S, Gatzionis K (2001a) Investigations for the design of grouts for the structural restoration of Eptapyrgion Fortress in Thessaloniki, Greece. Researsh report, Laboratory of Restoration Materials and Techniques, Directorate for the Restoration of

Byzantine and Post-Byzantine Monuments, Hellenic Ministry of Culture and Sports, December 2001 (in Greek)

Miltiadou-Fezans A, Anagnostopoulou S, Gatzionis K, Kordoulas A (2001b) Laboratory and in situ tests for the design of mortars and grouts for the structural restoration of the NE range of Cells of Hosios Loucas Monastery in Boeotia, Greece. Researsh report, Laboratory of Restoration Materials and Techniques, Directorate for the Restoration of Byzantine and Post-Byzantine Monuments, Hellenic Ministry of Culture and Sports, November 2001 (in Greek)

Miltiadou-Fezans A, Kalagri A, Kordoulas A (2003) Technical report of grouting interventions in the building SPAP II in Olympia and in the Municipal market of Pyrgos in Ileia, Peloponese, Greece. Researsh report, Laboratory of Restoration Materials and Techniques, Directorate for the Restoration of Byzantine and Post-Byzantine Monuments, Hellenic Ministry of Culture and Sports, October 2003 (in Greek)

Miltiadou-Fezans A., Kalagri A., Anagnostopoulou S. (2004a) Research and investigations for the design of grouts, mortars and pastes for the restoration of the Ancient Theatre of Megalopolis, Greece. Research report, Directorate for Technical Research on Restotation of the Hellenic Ministry of Culture and Sports, October 2004 (in Greek)

Miltiadou-Fezans A, Kalagri A, Anagnostopoulou S (2004b) Research and investigations for the design of grouts for the restoration of the Ancient Theatre of Epidaure, Greece. Research report, Directorate for Technical Research on Restotation of the Hellenic Ministry of Culture and Sports, October 2004 (in Greek)

Miltiadou-Fezans A, Kalagri A, Anagnostopoulou S (2004c) Research and investigations for the design of grouts for the restoration of ranges of cells and the East Gate of Dahni Monastery, Attica, Greece. Research report, Directorate for Technical Research on Restotation of the Hellenic Ministry of Culture and Sports, 2004 (in Greek)

Miltiadou-Fezans A, Papakonstantinou E, Zambas K, Panou A (2005) Design and application of hydraulic grouts of high injectability for structural restoration of the column drums of the Parthenon Opisthodomos. In: Brebbia CA, Torpiano A (eds) Structural studies, repairs and maintenance of heritage architecture IX. WIT Press, Southampton, pp 461–472

Miltiadou-Fezans A, Kalagri A, Triantafyllou M (2006a) Research report for the design of hydraulic grouts for the structural restoration of the Katholikon of Dafni Monastery, Attica, Greece. Comparative study of hydraulic lime-based grouts. Research report, Directorate for Technical Research on Restoration, Hellenic Ministry of Culture and Sports, p 68 (in Greek)

Miltiadou-Fezans A, Anagnostopoulou S, Kalagri A (2006b) Restoration mortars and grouts for consolidation of the Cells of internal yard of Dafni Monastery. 1st Conference on Restoration, Society for Research and Promotion of Scientific Restoration of Monuments (ETEPAM), Thessaloniki (in Greek)

Miltiadou-Fezans A, Kalagri A, Delinikolas N (2007a) Design of hydraulic grout and application methodology for stone masonry structures bearing mosaics and mural paintings: the case of the Katholikon of Dafni Monastery. In: Arun G (ed) Proc. of Int. Symposium Studies on Historical Heritage, Antalya, Turkey, 17–21 September, 2007, pp 649–656

Miltiadou-Fezans A, Kalagri A, Savvidou M (2007b) Research and investigations for the design of grouts for the structural restoration of the Church of the Assumption of Virgin Mary in Tegea Arkadias, Greece. Research report, Directorate for Technical Research on Restotation of the Hellenic Ministry of Culture and Sports (in Greek)

Miltiadou-Fezans A, Kalagri A, Kakkinou S, Ziagou A, Delinikolas N, Zarogianni E, Chorafa E (2008) Methodology for in situ application of hydraulic grouts on historic masonry structures. The case of the Katholikon of Dafni Monastery». In: D' Ayala D and Fodde E (eds), Structural analysis of historic construction, preserving safety and significance, Proceedings of 6th international conference on SAHC, 2–4 July, Bath, CRC Press/Balkema, Taylor and Francis Group, London UK, Vol. II, pp 1025–1033

Miltiadou-Fezans A, Delagrammatikas M, Kalagri A, Vassiliou P (2021) Evaluation of performance of matured hydraulic grouts: strength development, microstructural characteristics and durability issues. In: Roca P, Pelà L, Molins C (Eds) Proceedings of 12th International Conference on

Structural Analysis of Historical Constructions SAHC21, on line event 29th Sept to 1st Oct 2021, International Centre for Numerical Methods in Engineering (CIMNE), Barcelona, Spain, pp 2480–2491

Mirza J, Mirza MS, Roy V, Saleh K (2002) Basic rheological and mechanical properties of High-volume fly ash grouts. Constr Build Mater 16:353–363

Mirza J, Saleh K, Langevin MA, mirza S, Bhutta MAR, Tahir M Md, (2013) Properties of microfine grouts at 4° C, 10° C and 20° C. Constr Build Mater 47(2013):1145–1153

Moundoulas P, Morout P., Miltiadou A, Aggelakopoulou E, Bakolas S, Kouloumbi N, Moropoulou A (2009) The impact of fluidifiers on hydraulic grouts. In: Proceedings of 11th paints symposium research and technology of paints, varnishes and inks on the eve of 2010, Athens, 7–8/5/2009, pp 278–304

Mouzakis C, Adami CE, Karapitta L, Vintzileou E (2017) Seismic Behaviour of timber-laced stone masonry before and after interventions: shaking table testes on two-storey masonry model. Bull Earthquake Eng. https://doi.org/10.1007/s10518-017-0220-9

Nikolopoulou V, Adami CE, Karagiannaki D, Vintzileou E, Miltiadou-Fezans A (2018) Grouts for strengthening two- and three-leaf stone masonry, made with earthen mortars. Int J Architect Heritage Conservat Anal Restorat 13(5):663–678

Oliveira DV, Lourenço PB, Garbin E, Valluzzi M.R, Modena C (2006) Experimental investigation on the structural behaviour and strengthening of three-leaf stone masonry walls. In: Proceedings of international conference on structural analysis of historical constructions, New Delhi, Nov 2006, pp 817–826

Paillère AM, Serrano JJ, Buil M (1986) Possibilités offertes par l'emploi d'ultrafines siliceuses dans les coulis d'injection à base de liants hydrauliques. Bulletin De Liaison Des Laboratoires Des Ponts Et Chaussées, Paris, no. 141:23–25

Paillère A-M, Serrano JJ, Miltiadou A (1993) Formulation des coulis hydrauliques pour l'injection de fines fissures et cavités dans les structures dégradées en béton et maçonnerie. Bull. de Liaison Labo. Ponts et Chaussées, 186-juil.-août pp 61–78, Ref. 3676

Papayianni I (2015) Experimental study on the performance of lime-based grouts used in consolidating historic masonries. Mater Struct 48(7):2111–2121

Penelis G, Karaveziroglou M, Papayianni I (1989) Grouts for repairing and strengthening old masonry structures. In: Brebbia CA (ed) Structural repair and maintenance of historical buildings. Computational Mechanics Publications, Southampton, UK, pp 179–188

Saric-Coric M, Khayat KH, Tagnit-Hamou A (2003) Performance characteristics of cement grouts made with various combinations of high-range water reducer and cellulose-based viscosity modifier. Cem Concr Res 33:1999–2008

Schuller MP, Atkinson RH, Borgsmiller JT (1994) Injection grouting for repair and retrofit of unreinforced masonry. In: Proceedings of 10th international block masonry conference. Calgary, Canada, July 5–7, 1994

Silva B (2012) Diagnosis and strengthening of Historical masonry structures: Numerical and experimental analyses. Ph.D. Thesis, University of Brescia, April 2012, p 407

Silva B, Pigouni AE, Valluzzi MR, da Porto F, Modena C (2014) Calibration of analytical formulations predicting compressive strength in consolidated three-leaf masonry walls. Construction and Building materials, vol 64, August 2024, pp 28–38. https://doi.org/10.1016/j.conbuildmat.2014.04.044

Tassios TP (2004) Rehabilitation of three-leaf masonry. In: Evoluzione nella sperimentazione per le costruzioni, Seminario Internazionale. Centro Internationale di Aggiornamento Sperimentale—Scientifico (CIAS), 2004

Tassios TP, Chronopoulos M (1986) Aseismic dimensioning of interventions on low-strength masonry buildings. In: Proceedings of Middle East and Mediterranean Regional Conference on Earthen and Low-strength masonry in seismic areas, Middle East University, August 31st to September 6th, Ankara

Tavares ML, Veiga MR Fragata A (2010) Grouting mortars for consolidation of historical renders showing loss of adhesion. In: Proceedings of 2nd Historic Mortar Conference HM2010 and RILEM TC 203-RHM Final Workshop September 2010, Prague, Chech Republic, pp 743–752

Tomazevic M, Apih V (1993) The strengthening of stone-masonry walls by injecting the masonry friendly grouts. In European Earthquake Engineering, Vol 7, No 2 Patron Bologna, pp 10–20

Toumbakari EE (2002) Lime-Pozzolan-Cement grouts and their structural effects on composite masonry walls. Ph. D. Thesis, Department of Civil Engineering, Katholieke Universiteit Leuven, Sept. 2002, p 364

Toumbakari EE, Van Gemert D, Tassios TP, Tenoutasse N (1999) Effect of mixing procedure on injectability of cementitious grouts. Cem Concr Res 29(1999):867–872

Toumbakari EE, Vintzileou E, Tassios TP, Van Gemert D (2007) Shear behaviour of masonry units and lime-pozzolan-cement grout interfaces. In: Proceedings of 10th NAMC, The Masonry Society, St Louis, Missouri, pp767–776

Uranjek M, Žarnić R, Bokan-Basiljkov V, Basiljkov V (2014) Seismic resistance of stone masonry building and effect of grouting. Gradevinar 66(8):715–726. https://doi.org/10.14256/JCE.1031.2014

Valluzzi MR (2000) Comportamento meccanico di murature storiche consolidate con materiali e tecniche a base di calce. PHD Th., Univ. of Trieste, p 276

Valluzzi MR, da Porto F, Modena C (2004) Behaviour and modelling of strengthened three-leaf stone masonry walls. Materials and Structures, Vol 37, April 2004, pp 184–192

Van Rickstal F (2000) Grout injection of masonry, scientific approach and modeling. Departement of Civil Engineering, Katholieke Universiteit Leuven, May 2000, p 197

Vintzileou E (2011) Three-leaf masonry in compression, before and after grouting: a review of literature. Int J Architect Heritage 5(4–5):513–538

Vintzileou E, Miltiadou-Fezans A (2008) Mecahnical properties of three-leaf stone masonry grouted with ternary or hydraulic lime-based grouts. Engineering structures 30 92008) 2265–2276

Vintzileou E, Adami CE (2009) The bond mechanism in stone- or brick-to-grout interfaces Strain 45:400–409

Vintzileou E, Tassios TP (1995) Three leaf stone masonry strengthened by injecting cement grouts. J Struct Eng ASCE 121(5):848–885

Vintzileou E, Mouzakis C, Adami CE, Karapitta L (2015) Seismic Behavior of three-leaf stone masonry buildings before and after interventions: Shaking table tests on two-storey masonry model. Bull Earthquake Eng. https://doi.org/10.1007/s10518-015-9746-x

Yüzer N, Oktay D, Ulukaya S, Gökyğit-Arpaci EY (2015) Mechanical behaviour of two-wythe brick masonry walls injected with hydraulic lime grout. In Proceedings of International Conference REHAB2015, 22–24 July 2015, Porto, Portugal

# Chapter 7
# Durability

**Abstract** This Chapter is related to durability issues. Physical effects are considered first, referring to the consequences of water introduced in masonry by the grouting (freezing and dissolution of soluble phases). Subsequently, chemical effects are considered, such as sulphate reactions, alkali-silica reactions, possible chlorides' attack and leaching. The chapter ends with a brief presentation of literature results of durability tests, and with a guide for the selection of binders vs durability.

## 7.1 Introduction

Compared with concrete and steel structures, masonry walls are in general more durable; and the same is expected for grouted masonry walls.

However, since an irreversible intervention like grouting has been decided, it is reasonable to take the necessary measures so that the modified structure remains durable after this intervention.

In what follows, a brief description of durability issues is undertaken, together with suggested design and/or construction methods, in order to ensure durability after grouting.

Durability issues can be distinguished into those attributed mostly to the effects of physical phenomena and those attributed purely to chemical reactions.

## 7.2 Physical Effects

In the case of grouted masonry walls, physical effects like wear, abrasion and erosion and extreme high temperatures (fire) do not need to be examined, since this kind of vulnerability of masonry is not affected by grouting or because grouting may have positive consequences on masonry durability.

Similarly, any shrinkage of the thickness of the "veins" of grouting material inside masonry, cannot produce structural problems to it.

 
A. Miltiadou-Fezans and T. P. Tassios, *Mix-Design and Application of Hydraulic Grouts for Masonry Strengthening*, Springer Tracts in Civil Engineering,
https://doi.org/10.1007/978-3-030-85965-7_7

The physical effects however, which are related to the action of water within the pores of grouted masonry, remain to be examined, as follows.

### 7.2.1 Water Introduced During Grouting

As it is known, internal free water content in masonry may be harmful, and as it is known, large quantities of water are introduced into the masonry wall during grouting. Water to binder ratio of typical hydraulic binder mortar is about 40–45%, while a grout usually contains 80% or even more water. Hydration reactions require only a small fraction (about 25% of it)—but the excess water is needed to ensure the injectability of the grout (adequate fluidity without however jeopardizing the stability of the suspension, see Chaps. 3 and 4). This water excess is introduced deep into the wall; its evaporation has to take place through the pores of the existing materials. However, before its evaporation, two are our main concerns about this water.

**Freezing**: During the execution of grouting, freezing temperatures may be destructive for masonry; in fact, the injected large quantities of water within the body of masonry, could be subjected to freezing expansion, with adverse cracking results. Consequently, under expected freezing conditions, grouting operations should be suspended. Recommendations for concreting-works under low temperature conditions (EN 206: 2013) are useful in this respect. Similarly, in case of already executed grouting under doubtful low temperature conditions, defence measures similar to those used in concrete structures may be taken. In all cases, conservatism is suggested, taking into account the slower setting time and rate of strength evolution in most grout compositions used for historical masonry consolidation, compared to OPC concrete ones.

**Dissolution of soluble phases**: Water-soluble phases existing in the wall masonry may be dissolved by the introduced water; their ions may migrate through the pores and crystallize, with adverse effects.

It is also noted that grouts with disproportionately long setting-times allow absorption mechanism to take up larger quantities of the grout-water in the masonry, whereas possible rapid evaporation further reduces its water content. The reduction of the available water may hinder the hydration of the grout. Moreover, if the initial masonry materials are chemically vulnerable, the water absorption may increase the overall area, where adverse substances can be dissolved and allow the leaching of some chemical components contained in masonry. Additionally, if grouting takes place under potential freezing conditions, the risk of freezing damage would be increased, both in terms of time and of affected area.

In this connection, it is noted that after the setting and hardening of the grout, overall permeability of masonry walls is reduced, thus reducing the risks of external waters to penetrate. However, the low water permeability of masonry, ensured by grouting and by appropriate surface pointing, does not substantially reduce air permeability of the wall, necessary to facilitate escape of vapour through masonry.

**Earth-based mortars' expansion**: In masonries containing earthen mortar and/or adobe bricks, clay minerals (such as montmorillonite and others) may expand when absorbing water, producing damages. It has to be noted however that, in case of stone masonries, such damage may be tolerated, especially if the mortar joints are not very thick. Some first experimental results (Nikolopoulou et al. 2018) of grouting in two and three-leaf stone masonries built with earthen mortars, have shown that grouting may be a promising technique for strengthening such structures, provided that the grout is carefully designed, and the execution of works are carried out following a meticulous methodology (see Chap. 10). In the case of adobe walls, grouting may be used for repairing such walls, but further research is necessary, in order to determine the appropriate materials and compositions of grouts, as well as their properties and their application methodology (see pertinent literature works Jäger and Fuchs 2003; Papayianni 2006; Chaundry 2007; Vargas et al. 2008; Silva et al. 2009, 2012; Blondet et al. 2012; Papayianni and Pachta 2015a; Illampas et al. 2017).

### *7.2.2 Fluctuation of Moisture*

As it is known, masonry walls may be submitted to adverse wetting-and-drying cycles, leading to a possible accumulation of water-soluble salts within the pores of masonry materials, with the following consequences: If crystallisation of these salts occurs at the outer surface of the walls ("efflorescense"), mostly aesthetic problems are produced; while, internal crystallisation and growth of salts "subflorescense" may result in cracking and flaking.

Consequently, it is of paramount importance that grouting should not contain substances (such as e.g. sulphates or soluble alkalis) potentially contributing to the production of such salts (see Sects. 7.3.1 and 7.3.2). This is even more important when such substances already exist within the existing masonry materials.

## 7.3 Chemical Effects

The following possible harmful effects of chemical nature should be considered in the case of grouted masonry.

### *7.3.1 Sulphate Reactions*

Sulphate attack typically occurs when the cementitious materials are exposed to water that contains a high concentration of dissolved sulphates. In fact, durability is usually a concern when the sulphate content of the water is >200 mg/l (BS EN 206: 2013). The two most common types of sulphate attack are: (i) physical attack, where the sulphates in water penetrate the surface of hardened material, crystallize, and expand, disrupting it, and (ii) chemical attack, where the sulphate salts react with the paste, causing it to dissolve, soften, and erode.

During sulphate attack, both the calcium hydroxide and the aluminate phases in the binder matrix react with sulphate to form insoluble substances with much greater volumes.

There are two main mechanisms of sulphate attack on cementitious materials, namely the Late Ettringite[1] Formation (LEF) and the Thaumasite[2] Sulphate Attack (TSA). Development of Ettringite needs high alumina content, while development of thaumasite is connected to the presence of carbonates.

Consequently, penetration of sulphates from external sources (e.g. in the case of vicinity of some curative mineral waters) may cause adverse chemical reactions, mainly with Portland cement components, resulting in the production of expansive substances and subsequent microcracking and spalling. Under such conditions, the use of appropriate hydraulic binders in the grout composition, as well as the increase of tightness of the face of the wall are recommended.

In order to minimize the risk of sulphate attack, a dense grout paste with a relatively low permeability and a binder less susceptible to sulphate reactions should be beneficial. This can mainly be achieved by either using sulphate resistance binders (e.g. characterized by a reduced $C_3A$, gypsum and alkalis content) and/or pozzolans, so that there is less portlandite [calcium hydroxide (CH)]) for the destructive reaction.

Seawater is a frequent cause for the accumulation of sulphates. Seawater exposure provides potential for various types of exchange reactions and types of deterioration to occur simultaneously. The main concern with seawater is the presence of magnesium sulphate. Leaching actions remove lime and calcium sulphate, while, under high temperature, reaction with magnesium sulphate leads to the formation of calcium sulphoaluminate; this, may cause expansion, rendering the material more open for further attack and leaching (Lea 1970). Note that, in coastal areas, highly permeable masonries may accumulate in their body magnesium sulphate components through the years.

Another source of sulphates is Portland cement, which typically contains 3–5% gypsum (calcium sulphate dihydrate); it may be present in historic masonries due to previous interventions or additions.

A more serious situation may appear in the case of initially in-masonry used building mortars based on sulphates, as is the case with old gypsum mortars used in some regions. In such a case, the introduction of aluminates may lead to LEF, while the introduction of carbonates (namely $CaCO_3$) may lead to TSA. Characteristic is the case of the Bell-tower of Neauphle-le-Chateau in France (reported by De Lépinay and Caillault 1999), where the application of a hydraulic lime-based grout

[1] Ettringite is a hexacalcium aluminate trisulfate hydrate formed by the reaction of calcium aluminate with calcium sulfate (gypsum).

[2] Thaumasite is composed of calcium silicate, carbonate, sulfate and water, and can be formed directly through the reaction of calcium silicate hydrate (C–S–H) with calcite in the presence of moisture and unbound sulfate ions, and indirectly by the reaction of ettringite with C–S–H and carbonates/bicarbonates.

to the masonry walls containing gypsum mortar, led to severe TSA and complete destruction of the historic masonry, especially in places that were exposed to high humidity and low temperatures.

### 7.3.2 Alkali-Silica Reaction (ASR)

Depending on the local building traditions and available materials, stones or mortar-sand of a given masonry might contain active silica minerals, potentially subject to ASR, leading to the formation of a sodium or potassium silicate gel. The formation of this hydroscopic gel induces stresses to the aggregates and surrounding binders, resulting in dissociation of the aggregates and to cracking. As it is known, such alkalis are contained in most cements (except the so called low-alkali cements), as well as in some pozzolans. ASR is usually associated with the presence of amorphous or poorly crystalized silica and usually present in igneous rock of relatively recent formation; however slow rate ASR may occur in the presence of well crystalized silica rich aggregates. Consequently, in such cases, the use of low alkali cements or even better the use of other low alkali hydraulic binders without Portland cement is recommended. Other common source of accumulation of alkali are sea water and fertilizers, in costal and agricultural environments.

The complex mechanism of deterioration due to the evolution of ASR, involves the presence of portlandite (calcium hydroxide). Thus, special attention has to be given in case of a clay mortar masonry: First, we should examine if this clay contains active silica and alkali, because grouting with hydraulic materials, even of low alkali content (which inevitably contain calcium hydroxide) may result in the evolution of ASR.

Addition of a finely ground pozzolanic material is a common mitigation measure against ASR induced damage; the active silica in the surface of pozzolan particles will react both with portlandite and with the alkali, and disperse alkali silicate gel in a micro scale, which does not allow the evolution of stresses.

### 7.3.3 Chlorides

Chloride attack is a concern because chloride reacts with the calcium hydroxide in the paste to form soluble products, which can be leached away, resulting in a possible loss of material's strength. Chlorides can also form salt crystallization products in the pores when the material is subjected to wetting and drying cycles (such as in groundwater fluctuations). These salt deposits can eventually cause cracking and disintegration.

The addition of finely ground pozzolan, which reduces available calcium hydroxide, is a remedy measure when chloride ion presence is expected (e.g. structures in coastal areas).

Temperature and chloride concentration, control the chemical reactions involved in chloride attack. When temperatures is higher than 40 °C and chloride concentrations are above 10000 mg/L (seawater has a chloride ion concentration of about 19400 mg/L), ettringite can decompose to form Friedel's salt and gypsum (below 20 °C trichloride forms instead) (Luna et al. 2006). In case that such high concentrations of chloride may be expected (i.e. in structures in the direct vicinity of the sea, where salt-spray and/or tidal circles lead to accumulation of salts, wetted by geothermal spring water or near certain anthropogenic activities), care should be taken to use a denser, less permeable and low alumina grout in conjunction with proper pointing.

Moreover, chlorides can severely attack metallic (and in particular iron) elements that may exist in the masonry as reinforcement, joints, braces, anchors etc.

As a general rule, grouting reduces the penetration of chlorides into the masonry. However, precautions should be taken that chlorides are not introduced by the grout itself (e.g., through the use of mixing water that is not in accordance with concrete regulations EN1008: 2002, inappropriate superplasticizers or other admixtures, and not properly washed sand in case of sand containing grouts).

### 7.3.4 Leaching

Leaching involves the attack of water on the calcium hydroxide and calcium silicate hydrate (CSH) present in hydrated binders. Solid hydrates of cement paste are more stable at pH above 12–13, but at a lower pH the hydration phases no longer remain stable and thus dissolve. The pore solution of a typical portland cement paste is highly alkaline, so that the leaching process starts by removing alkalis ($Na^+$ and $K^+$), followed by dissolution of portlandite (CH), and subsequently by the leaching of calcium from silicates (e.g., CSH) (Holt 2008). Aluminate phases are also affected (Luna et al. 2006). Leaching is highly dependent on the permeability of the hardened material and may be a slow process: 5–10 mm leached depths have been found in concretes submerged for 100 years in standing natural water (Holt 2008).

Much more intense is the effect of leaching when there is water flow that washes out ions, prevents the solution within the pores or fine cracks to reach saturation and to retain high alkaline pH.

Grouting should inhibit or greatly reduce such "free" move of water within the masonry, but appropriate pointing and capping are most important for protecting both the grout and other inner masonry materials from leaching.

## 7.4 Brief Presentation of Main Literature Results on Grout's Durability Testing

Although there is a plethora of studies on the durability of masonry materials (stones, ceramic tiles, mortars and binders) (Ritchie 1955; Lewin 1981; Tabasso et al. 1984; Rodriguez-Navarro et al. 2000; Cardell et al. 2003; Groot 2016; Lubelli et al. 2018), the literature focusing on testing the durability of grouts and grouted masonry elements is rather limited.

Ferragni et al. as early as 1984, specified that the amount of extractable sodium and potassium ions contained in the raw materials should be as small as possible (e.g., not more than 120 milliequivalents per kg of mixture), but also the soluble calcium should be kept reasonably low (e.g., not more than 60 milliequivalents per kg of mixture).

Miltiadou (1990) studied the resistance in aggressive environment of various grouts. To this end, the expansion of grout specimens conserved during 28 days at 70% RH and then immersed in sea water or water containing high percentages of sulphates (according to NF P18-837), was measured at specified time intervals. The binary grouts consisted of cement (three different types were tested: CPA, CPA-PM, CLK) combined with hydrated lime or a natural pozzolan, namely Santorini earth, or a densified silica fume; the ternary grouts consisted by combinations of cement with densified silica fume and hydrated lime. It has to be noted that the binary grouts prepared with densified silica fume or Santorini earth did not suffered excessive expansion at 180 days, in all types of aggressive water used. On the contrary, the use of hydrated lime in combination with cement led to the destruction of specimens, especially those immerged in water with high sulphate content (Paillère et al. 1993). It has to be noted however, that in the case of ternary grouts, where the use of hydrated lime is combined with densified silica fume, the behaviour of the grouts was improved (Paillère et al. 1993).

Karaveziroglou et al. (1998) studied the influence of temperature and moisture extremes on mortars and grouts made with traditional materials such as lime, pozzolan, and cement, used as binding agents, as well as sand and crushed brick used as aggregates. Prismatic specimens were subjected to wetting-drying, boiling-drying and freezing-thawing cycles. Length changes and strength loss were measured to compare various mortar and grout compositions. They concluded that the high cement content, although advantageous for strength development, does not contribute to better volume stability of the materials.

Toumbakari et al. (1999) examined, among other, the behavior of ternary cement-lime-pozzolan grout mixtures exposed to sulfate. The grouts were prepared with ultrasound mixing and subsequently, they were introduced in a Hobart mixer, together with sand and gypsum in constant ratios. The results of the expansion of prismatic specimens indicate that the mortars expanded continuously and stabilise only after 120–180 days, accompanied by a decrease of the dynamic modulus of elasticity.

Miltiadou-Fezans et al. (2005) studied the durability (with capillary absorption tests: RILEM II-6) of a series of binary compositions of 75% white Aalborg cement and 25% of natural pozzolan (4 different natural pozzolans were examined: Milos earth, Santorini earth, pozzolan of Milos and pumice) in the framework of the research undertaken for the design of the optimum grout composition to be injected in the cracked drums of the columns of the Parthenon Opisthodomos of the Acropolis of Athens. In all cases, enhanced durability was observed, both under the action of sulphate rich water and sea water. Grouts containing Santorini earth presented the best performance.

Regarding the behaviour of hydraulic lime-based grouts, a comparative study was carried out by Kalagri et al. (2010); it was proved that the resistance of grouts in suplhates depends a lot on the chemical and mineralogical composition of the hydraulic lime used. The use of St Astier NHL5 gave the best results, together with a ternary grout composed by 30% of white Aalborg Cement, 45% of natural pozzolan and 25% of hydrated lime.

Biçer-Şimşir (2016) has pointed out the importance of studying the resistance of grouts to soluble salts at very early ages, and highlighted the need for establishing an adequate test to serve this purpose. The author has presented a preliminary study to this end by injecting grouts into clean and salt laden crushed brick, and observing the behaviour of such composite specimens in comparison with the behaviour of grout-only specimens, when subjected to cycles of immersion in sulphate salts baths and drying. The grout-only specimens exhibited faster and higher levels of deterioration than the specimens prepared by injecting grouts into a crushed brick system, which is more representative of wall conditions in the field.

The performance of matured hydraulic lime grouts, associated or not with natural pozzolan, was examined by Miltiadou-Fezans et al. (2021), using specimens as old as 14 years. It was proved that the replacement of a small percentage of NHL5 by a natural pozzolan of the order of 10%, has a favorable effect both on strength and durability, while larger addition of pozzolan (20%) may deteriorate the NHL5 performance. The optimum percentage to be replaced depends on the raw material characteristics, and needs to be examined in a more systematic way.

## 7.5 Guidance for the Grout Design Versus Durability

Grouting is an irreversible intervention that affects the entire volume of the structure, where is applied. Thus, great attention should be given to the proper design of the grout in order to meet the durability requirements and ensure retreatability (Van Balen 2000). Keeping in mind that it is the grouted masonry and not only the grout itself that should be durable and able to maintain its integrity over time, the adequate grout composition strongly depends on the existing masonry materials and their pathology, as well as on the expected performance of the grouted materials.

On the basis of the topics discussed and of literature results, some guidance for the preliminary selection of the type of the binder of the grout for a successful grouting intervention may be formulated.

**Detailed knowledge of the existing masonry materials and environmental characteristics** is of paramount importance, as already presented in Sect. 1.2 Basic data necessary for the design of grout. The most common tests to be carried out are summarized in what follows (Van Balen et al. 2005; Middendorf et al. 2005a, b; van Hees et al. 2012):

- *Chemical and mineralogical composition of the different masonry materials should be studied.* A combination of techniques such as X-ray Fluorecence (XRF), X-ray Diffraction (XRD), Optical Microscopy (OM) of thin sections, Scanning Electron Microscopy coupled with Energy (or better Wavelength) Dispersion Spectroscopy (SEM/EDS or WDS), Fourier Transform Infrared Spectroscopy (FTIR) and Raman Microscopy, as well as thermal methods (DSC, DTA, TG) and wet chemistry techniques, should provide adequate information for the grouting design needs.
- *The water-soluble phases should be identified and quantified.* Wet chemistry techniques may be used, but best results can be achieved by Ion Chromatography (IC).
- *Porosity and pore size distribution is also very important* in order to determine the masonry materials accessibility to weathering by crystallization of water soluble salts. Water absorption test and Mercury Intrusion Porosimetry (MIP) should address this aspect.

**Geographic location and environmental conditions of structure** should be studied. Proximity to the sea and salt-spray, industrial or urban pollutants, as well as intense agricultural activities are parameters which should be taken into consideration in the grout design process. Information regarding the risks of freezing and thawing, sulphate attack and water evaporation rates is valuable.

**General rules for the preliminary selection of grout materials**. The general rule of chemical compatibility adopted for restoration materials should be taken into account. Consequently, materials are selected to have chemical affinity with the original ones. As a rule of thumb, Portland cement should be avoided in historic masonry structures, or cement content should be kept low.

Thus, hydraulic lime-based grouts (neat hydraulic lime or in combination with a small quantity of pozzolan) or ternary grouts with a low cement content, combined with pozzolanic materials and hydrated lime in adequate percentages, are usually the two possible alternatives.

In all cases, avoid using materials containing large quantities of water-soluble phases. Thus, the use of Portland cement with high gypsum and alkali content has to be avoided; instead, the use of low sulphate and alkali content cement is recommended. To this end, white cement and low alumina cement should be preferred, as they usually contain smaller amounts of gypsum.

Regarding natural hydraulic limes (NHL), attention should be given to have low alkali and aluminates content; most important for grouting work is a low content of free lime (calcium hydroxide), which is met in eminently hydraulic limes (NHL5). The replacement of a small percentage of NHL with finely ground pozzolan is intented to mitigate free lime, but testing should be performed in order to establish the appropriate percentage to be used in the composition.

Pozzolans should also to be selected having a low alkali content and appropriately fine particle size distribution (indicatevely max d < 45 μm), directly affecting the pozzolanic reaction, and the penetrability of the grout, as presented in Chap. 2. It has to be noted that natural pozzolans available in the market, may sometimes have a high content in coarser grains. Thus, to ensure penetrability, the pozzolan to be used in a grout has to be finer than the $d_{85}$ of the hydraulic lime or cement (as reported in Chap. 2). In case of very fine cements or NHL, observing the grading criteria for penetrability, pozzolans have also to be carefully selected, to ensure that durability, as well as strength is served, without jeopardizing penetrability (see Sect. 2.6).

Generally, it is recommended to select a prominently hydraulic over air setting character of the binder of the grout.

It has to be noted, however, that when frescoes, mosaics or other decorative elements are present, the use of hydraulic lime-based grouts or even lime-pozzolan based grouts, may be preferable over ternary grouts containing cement (see Ferragni et al. 1984; Kalagri et al. 2010; Miltiadou-Fezans et al. 2008; Bicer-Simsir et al. 2010; Papayianni et al. 2010; Papayianni and Pachta 2015b; Biçer-Şimşir 2016).

In the case of masonry containing gypsum mortars, research is still necessary for the design of adequate grouts, since even the use of hydraulic lime-based compositions may be detrimental under specific site conditions. A survey of the appropriate literature is recommended, as well as an extended research program before any intervention.

As far as masonry built with clay mortars or clay blocks is concerned, specific research has to be carried out taking into account recent research on the subject (Jäger and Fuchs 2003; Papayianni 2006; Chaundry 2007; Vargas et al. 2008; Silva et al. 2009, 2012; Blondet et al. 2012; Papayianni et al. 2010; Papayianni and Pachta 2015a; Illampas et al. 2017; Nikolopoulou et al. 2018).

**Basic criteria for the raw materials**. The criteria proposed in the literature (Regourd 1982) and in standards (e.g., NF 15-319 and NF 15-317) for cement to be used in structures resistant in sulphate or marine environment respectively (see Table 7.1), may guide the selection of this material in case of binary and ternary grouts.

The acceptable content of soluble alkalis (which may be present in cement or hydraulic lime or pozzolanic materials) estimated in "equiv. $Na_2O$" has to be lower than 0.6%, as demonstrated by the early work of Stanton (1940).

For pozzolans the total content of alkalis in "equiv. $Na_2O$" has to be lower than 2% (ASTM C 618; Lee 1986).

In Table 7.2 the potentially damaging components and their contents on Portland cement (OPC) and natural hydraulic limes (NHL) are presented, together with a synopsis of their potentially damaging effects. The comparison of these contents may better justify the selection of NHL in some cases, when frescoes, mosaics or other decorative elements are present. However, for the strengthening of marble architectural members or masonry structures, experimental results have shown that adequate binary or ternary grouts present also satisfactory behaviour against the

**Table 7.1** Criteria for the chemical characteristics of cement in high sulphate and marine environment (Regourd 1982; NF P 15-317; NF P 15-319)

| Chemical characteristics | Requirements for Sulphate resistance (NF P 15-319) | Requirements for resistance in Marine environment (NF P 15-317) |
|---|---|---|
| $\%C_3A$<br>$\%C_4AF$ | $C_3A < 5.0\%$<br>$C_4AF + 2C_3A < 20\%$ | (i) If $C_3S < 50\%$, $C_3A < 10\%$<br>(ii) If $C_3S > 50\%$, $C_3A + 0.27 \cdot C_3S < 23.5\%$ |
| $SO_3$ | < 2.3% | < 2.5% |
| MgO | < 4.0% | < 3.0% |
| $Al_2O_3$ | | < 8.0% |
| LOI | < 3.0% | |
| Insoluble | < 0.75% | |

$C_3A = (Al_2O_3, 3CaO)$, $C_4AF = (Al_2O_3, Fe_2O_3, 4CaO)$, $C_3S = (SiO_2, 3CaO)$

**Table 7.2** Synopsis of potentially damaging components in binders and their potential damaging effects; comparison of their usual content on OPC and NHL (taking into account the chemical characteristics of ordinary commercial binders)

| Potentially damaging components | | % Content on | | Potential damaging effects |
|---|---|---|---|---|
| | | OPC | NHL | |
| Tricalcium Aluminate | $C_3A$ | 3–10+ | < 1 | It reacts with sulphates and water, producing sulphate attack and causing extensive cracking and possible failure. It also reacts with sea salts and affects bricks/stones |
| Tetracalcium Aluminoferite | $C_4AF$ | 8–10+ | Nil | It reacts with gypsum, causing expansion |
| Sulphates | $SO_3$ | 1.3–3.5 | < 1.5 | It contributes to sulphate attack |
| Alkalis | $Na_2O/K_2O$ | 0.2–1.3 | < 0.5 | It reacts with the silicates in cement and sand, producing gradual disintegration (ASR—Alkali-Silica Reaction) |

action of salts, when raw materials with low sulfate and alkali content are used (such as Danish white cement in combination with appropriate pozzolans and hydrated lime) (Miltiadou-Fezans et al. 2005; Kalagri et al. 2010).

**Mixing water**. The standards for the use of mixing water (e.g., EN 1008:2002) and aggregates in concrete should be met, and the grouting intervention should not take place in extreme environmental conditions.

**Note**: Some more special durability problems might appear with the use of some superplasticizers. However, this subject is not adequately covered in literature.

# References


ASTM C618-19 Standard Specification for Coal Fly Ash and Raw or Calcined Natural Pozzolan for Use in Concrete

Biçer-Şimşir B (2016) Preliminary study on the sulfate resistance of lime-based hydraulic injections grouts for the conservation of architectural surfaces. In: Papagianni I, Stefanidou M, Pachta V (eds) Proceedings of 4th historic mortars conference, Santorini, 10–12 Oct 2016, pp 548–555

Biçer-Şimşir B, Rainer L (2013) Evaluation of lime-based hydraulic injection grouts for the conservation of architectural surfaces. The Getty Conservation Institute, Los Angeles, p 105

Biçer-Şimşir B, Griffin I, Palazzo Bertholon B, Rainer L (2010) Lime- based injection grouts for the conservation of architectural surfaces. Studies in conservation, June 2010. https://doi.org/10.1179/sic.2010.55. (Supplement-1.3)

Blondet M, Vargas J, Groenenberg R (2012) Evaluation of the efficacy of mud injection to repair seismic cracks on adobe structures via full-scale shaking table tests. In: Proceedings of the 15th world conference on earthquake engineering, Lisbon, Portugal, 24–28 Sept 2012

Cardell C, Delalieux F, Roumpopoulos K, Moropoulou A, Auger F, Van Grieken R (2003) Salt-induced decay in calcareous stone monuments and buildings in a marine environment in SW France. Constr Build Mater 17:165–179

Chaundry C (2007) Evaluation of grouting as a strengthening technique for earthen structures in seismic areas: case study of Chiripa. M.Sc. Thesis, University of Pensylvania, EUA

De Lépinay H, Caillault P-Y (1999) Saint-Nicolas Neauphle-le-Château (Yvelines)-le mariage impossible des maçonneries montées au plâtre et des coulis de ciment. In: Proceedings of Journée technique internationale: La consolidation dans les maçonneries anciennes: les coulis d'injection, French National Committee of ICOMOS, 14th December

EN 206:2013 Concrete—specification, performance, production and conformity

EN 1008:2002 Mixing water for concrete. Specification for sampling, testing and assessing the suitability of water, including water recovered from processes in the concrete industry, as mixing water for concrete

Ferragni D, Forti M, Malliet J, Mora P, Teutonico JM, Torraca G (1984) Injection grouting of mural paintings and mosaics. In: Proceedings of symposium international of adhesives and consolidants, Paris, 2–8 Sept 1984, pp 110–116

Groot G (2016) Repair mortars for historic masonry; effects of the binder choice on durability, HERON 61(1)

Holt E (2008) Durability of low-pH injection grout: a literature survey. Technical report, VTT Technical Research Centre of Finland

Illampas R, Silva RA, Charmpis DC, Lourenço PB, Ioannou I (2017) Validation of the repair effectiveness of clay-based grout injections by lateral load testing of an adobe model building. Constr Build Mater 153(30):174–184

Jäger W, Fuchs C (2008) Reconstruction of the Sistani House at Bam Citadel after the collapse due to the earthquake 2003. In: D'Ayala D, Fodde E (eds) Structural analysis of historic construction. Taylor & Francis Group, London, pp 1181–1187

Kalagri A, Miltiadou-Fezans A, Vintzileou E (2010) Design and evaluation of hydraulic lime grouts for the strengthening of stone masonry historic structures. Mater Struct 43:1135–1146

Karaveziroglou M, Papayianni I, Penelis G (1998) Mortars and grouts in restoration of Roman and Byzantine monuments. In: Biscontin G, Moropoulou A, Erdik M, Delgado Rodrigues J (eds) Compatible materials for the protection of European cultural heritage, PACT 55 1998, Technical Chamber of Greece, pp 219–245

Lea FM (1970) The chemistry of cement and concrete. Edward Arnold Ltd., London

Lee C (1986) Available alkalis in fly ash and their effects on alkali-aggregate reaction. Retrospective Theses and Dissertations, 8264. https://lib.dr.iastate

Lewin SZ (1981) The mechanism of masonry decay through crystallization. In: Proceedings of the conference conservation of historic stone buildings and monuments, Washington, D.C., 2–4 Feb 1981

Lubelli B, Cnudde V, Diaz-Goncalves T, Franzoni E, Van Hees RPJ, Ioannou I, Menendez B, Nunes C, Siedel H, Stefanidou M (2018) Towards a more effective and reliable salt crystallization test for porous building materials: state of the art. Mater Struct 51(55)

Luna M, Arcos D, Duro L (2006) Effects of grouting shotcreting and concrete leachates on backfill, geochemistry, ISSN 1402 3091, SKB Rapport R 06 107

Middendorf B, Hughes JJ, Callebaut K, Baronio G, Papayianni I (2005a) Investigative methods for the characterisation of historic mortars—part 1: mineralogical characterisation. Mater Struct 38(8):761–769

Middendorf B, Hughes JJ, Callebaut K, Baronio G, Papayianni I (2005b) Investigative methods for the characterisation of historic mortars—part 2: chemical characterisation. Mater Struct 38(8):771–780

Miltiadou AE (1990) Étude des coulis hydrauliques pour la réparation et le renforcement des structures et des monuments historiques en maçonnerie. Thèse de Doctorat de l'Ecole Nationale des Ponts et Chaussées. Publ. by LCPC in Collection Etudes et recherches des Laboratoires des Ponts et Chaussées, série Ouvrages d'art, OA8 ISSN 1161-028X, LCPC, Décembre 1991, Paris, France, p 278

Miltiadou-Fezans A, Papakonstantinou E, Zambas K, Panou A, Frantzikinaki K (2005) Design and application of hydraulic grouts of high injectability for structural restoration of the column drums of the Parthenon Opisthodomos. In: Brebbia CA, Torpiano A (eds) Structural studies, repairs and maintenance of heritage architecture IX. WIT Press, Southampton, pp 461–472

Miltiadou-Fezans A, Kalagri A, Kakkinou S, Ziagou A, Delinikolas N, Zarogianni E, Chorafa E (2008) Methodology for in situ application of hydraulic grouts on historic masonry structures. The case of the Katholikon of Dafni Monastery. In: D'Ayala D, Fodde E (eds) Proceedings of the 6th international conference on structural analysis of historic construction, 2–4 July, Bath, CRC Press/Balkema, Taylor and Francis Group, vol I, pp 1025–1033

Miltiadou-Fezans A, Delagrammatikas M, Kalagri A, Vassiliou P (2021) Evaluation of performance of matured hydraulic grouts: strength development, microstructural characteristics and durability issues. In: Roca P, Pelà L, Molins C (Eds) Proceed. 12th International Conference on Structural Analysis of Historical Constructions SAHC21, on line event 29th Sept to 1st October 2021, International Centre for Numerical Methods in Engineering (CIMNE), Barcelona, Spain, pp 2480–2491

NF P 18-837 Produits spéciaux destinés aux constructions en béton hydraulique: produits de calage et/ou scellement à base de liants hydrauliques: essai de tenue à l'eau de mer et/ou à haute teneur en sulfates

NF P 15-317 Liants hydrauliques - Ciments pour travaux à la mer, Septembre 2006

NF P 15-319 Liants hydrauliques - Ciments pour travaux en eaux à haute teneur en sulfates, Septembre 2006

Nikolopoulou V, Adami C-E, Karagiannaki D, Vintzileou E, Miltiadou-Fezans A (2018) Grouts for strengthening two- and three-leaf stone masonry, made with earthen mortars. Int J Archit Herit: Conserv Anal Restor 13(5):663–678 doi 10.1080/15583058.2018.1463414

Paillère AM, Serrano JJ, Miltiadou A (1993) Formulation des coulis hydrauliques pour l'injection de fines fissures et cavités dans les structures dégradées en béton et maçonnerie. Bull. de Liaison Labo. Ponts et Chaussées, 186-juil.-août.pp 61–78, Ref. 3676

Papayianni I (2006) Restoration of earth blocks buildings in vernacular arvhitecture, materials and techniques. In: Proceedings of international conference on heritage protection, 18 Jan 2006, Paris, France

Papayianni I, Pachta V (2015a) Constructional characteristics and restoration aspects of historic earth block structures. Eng Struct Mater 3(2):113–121. https://doi.org/10.17515/resm2016.72st0726

Papayianni I, Pachta V (2015b) Experimental study on the performance of lime-based grouts used in consolidating historic masonries. Mater Struct 48(7):2111–2121

Papayianni, Stefanidou M, Pachta V (2010) Grouts for injection of historical masonries: influence of the binding system and other additions on the properties of the matrix. In: Proceedings of 2nd historic mortars conference & Rilem TC 203-RHM repair mortars for historic masonry, Final workshop, 22–24 Sept 2010, Prague, Rilem, pp 1123–1134

Regourd M (1982) La résistance du béton aux altérations physiques et chimiques. In: Baron J, Sauterey R (eds) Le béton hydraulique: connaisance et paratique, Presses de l'École Nationale des Ponts et Chaussées, pp 513–530

RILEM TC 25-PEM (1980) Test method no II-6: Water absorbion coefficient (capillarity), RILEM, Bagneux, France

Ritchie T (1955) Study of efflorescence produced on ceramic wicks by masonry mortars. J Am Ceram Soc 38:362–366

Rodriguez-Navarro C, Doehne E, Sebastian E (2000) How does sodium sulfate crystallize? Implications for the decay and testing of building materials. Cem Concr Res 30(2000):1527–1534

Silva RL, Schueremans L, Oliveira DV (2009) Grouting as a repair/strengthening solution for earth constructions. In: 1st WTA International PhD symposium at Leuven, Belgium

Silva RL, Schueremans L, Oliveira DV, Dekoning K, Gyssels T (2012) On the development of unmodified mud grouts for repairing earth constructions: Rheology, strength and adhesion. Mater Struct 45(10):1497. https://doi.org/10.1617/s11527-012-9853-y

Stanton TE (1940) Expansion of concrete through reaction between cement and aggregate. Proc Am Soc Civ Eng 66(10):1781–1811

Tabasso M, Diana S, Sammuri P (1984) Evaluation of mortars for use in conservation from the standpoint of release of soluble salts. In: ICOM Committee for Conservation, 7th Triennial Meeting, Copenhagen

Toumbakari EE, Van Germent D, Tassios TP (1999) Multi-blend cementitious injection grouts for repair and strengthening of masonry structures. In: Proceedings of the international conference specialist techniques and materials for concrete construction, held at the University of Dundee, Scotland, UK on 8–10 Sept 1999

Van Balen K (2000) Compatibility and retreatability as criteria completing the Venice Charter's criteria of reversibility in monument conservation. J Sophia Asian Stud 18(18):211–221

Van Balen K, Papayianni I, van Hees R, Binda L, Waldum A (2005) Introduction to requirements for and functions and properties of repair mortars. Mater Struct 38(282):781–785

Van Hees R, Van Balen K, Bicer-Simsir B, Binda L, Elsen J, von Konow T, Lindqvist JE, Maurenbrecher P, Papayanni I, Subercaseaux M, Tedeschi C, Toumbakari E-E, Thompson M, Valek J, Veiga R (2012) RILEM TC 203-RHM: repair mortars for historic masonry. From problem to intervention: a decision process. Mater Struct 45(9):1295–1302. https://doi.org/10.1617/s11527-012-9917-z

Vargas J, Blondet M, Cancino C, Ginocchio F, Iwaki C, Morales K (2008) Experimental results on the use of mud-based grouts to repair seismic cracks on adobe walls. In: D'Ayala D, Fodde E (eds) Proceedings of 6th international conference on structural analysis of historic constructions 2–4 July, Bath, UK, Taylor & Francis Group, London, vol 2, pp1095–1100

# Chapter 8
# Optimisation of Grout Performances

**Abstract** This Chapter refers to some inevitable *contradictory* requirements of a grout, especially related to ultrafines' content (penetrability against fluidity), superplasticizer's content (fluidity against stability), type and content of binders to achieve sufficient bonding with masonry materials, without however imparting to the masonry disproportionally high stiffness (strength against ductility) and without jeopardizing durability (strength against durability) or level of grouting pressure to avoid local rupture of very low masonry strength. Optimization of grout performances is needed in most cases. To this end, a simple first procedure is proposed following a step-by-step selection of mix-design parameters (Sect. 8.2). First, the type of binder is selected, based on strength and durability demands. An appropriate percentage of fines is then selected, based mainly on penetrability criteria, in combination with strength and durability aspects as well. Using experimental results already presented in Chapters 2–6, diagrams combining penetrability, fluidity, stability and strength characteristics versus W/S ratio are traced. Thus, a "usable" range of W/S-values appears; its compatibility with the required grout-strength range will be checked. Normally, several satisfactory solutions are thus found. Otherwise, the necessary correcting measures are discussed in the chapter, for each specific case, mainly by means of increased percentages of fines and addition of superplasticizers. This way, the rationality of the proposed methodology is understood. Thus, the designer is in better position to handle the final design of the grout, by means of a better knowledge of the interplay of the intervening numerous parameters.

## 8.1 Introduction

Let us recall the main requirements for a successful grout.

1. Penetrability of the grains into the fissures and voids of the masonry.
2. Fluidity of the suspension to allow its flow throughout the body and under the minimum possible pressure.
3. Stability of the grout in-time and in-flow, avoiding segregation and extensive bleeding, both resulting in possible blocking.

 
A. Miltiadou-Fezans and T. P. Tassios, *Mix-Design and Application of Hydraulic Grouts for Masonry Strengthening*, Springer Tracts in Civil Engineering,
https://doi.org/10.1007/978-3-030-85965-7_8

4. Appropriate strength (tensile strength, mainly) of the grout, in order to adhere with the constituents of the masonry, without however imparting to the masonry disproportionately high stiffness, especially when decorative elements (frescoes or mosaics) are present. Thus, among candidate compositions having similar tensile strength (see Chap. 6), the composition with the lower compressive strength may be selected to avoid undesirable stiffening of the injected masonry.
5. Durability, avoiding internal chemical incompatibility and undesirable influences of external environmental harmful causes.

The mix-designer avails of the following materials and means: a white or grey cement or hydraulic lime, one or two ultrafine materials, a commercial superplasticizer and a colloidal high turbulence mixer. And we will assume that she/he has determined a width of the apertures (nominal width $W_{nom}$) to be filled by the grout (see Chap. 5).

Furthermore, the designer, taking into account the characteristics of the existing constituent materials of the masonry to be injected and the environmental conditions (as well as the expected performances of the masonry after its repair or strengthening), has estimated the target values for the necessary compressive and tensile strength of the grout (see Chap. 6), and has decided about the durability requirements. Thus, an initial selection of the type of the basic binder, and the type of grout may be made, on the basis of available literature and experience (see Chap. 7).

The designer should now appropriately select the composition of solids and the water-to-solids ratio of the mix, in order to satisfy all the aforementioned requirements. But some of them may be *contradictory* to each other. Here are some undesirable cases of this nature:

- Too much of ultrafines (to ensure penetrability or strength requirements), may induce unacceptably low fluidity and a need to increase water-content; thus, the resulting strength-reduction might be also unacceptable.
- At this stage, a generous addition of a superplisticizer may disproportionately increase the risk of extensive bleeding.
- Besides, a very low strength masonry (unable to resist the needed grouting pressures), may require the use of extremely fluid grouts—of possibly very low strength or high bleeding.

It becomes apparent that an *optimisation* should be sought each time between the various desirable performances of a grout (Miltiadou-Fezans and Tassios 2016). To this end, a better understanding is needed of the way each design-parameter *interacts* with the others, in order to shape the final performances of the grout. It is believed that thanks to such an understanding, the aforementioned contradictory effects will be avoided.

## 8.2 The Interaction Between Design-Parameters

Before starting the design procedure itself, it is suggested to follow a *thought-experiment*, as the one presented in this section, in order to better understand the way, the grout-constituents (solid phase, water, superplasticiser) interact between themselves, in order to shape the grout properties (fluidity, stability and strength).

This thought-experiment may be carried out as follows:

1. A class of voids' *aperture* ($W_{nom}$) is selected, usually based on an engineering decision (see Chap. 5).
2. Furthermore, a target value of the *compressive strength* $f_{wc,s}$ of the wall after repair and strengthening is given. In doing so, however, possible adverse effects of high compressive strengths on decorative elements of the wall should also be considered: Lower strength plain masonry walls fail with more distributed cracks (i.e., more easily repairable), as opposed to higher strength, thus more brittle, walls. It is therefore suggested to select the lowest structurally required compressive strength of wall, as a target for the design of the intervention.
3. On the other hand, taking into account the possible existence of soluble salts in the interior of the structure, as well as the conditions of the environment, the durability requirements are defined in order to avoid any kind of efflorencenses, as well as any undesirable chemical reaction.
4. Based on experience and /or the information included in Chaps. 6 and 7, the type of the basic binder (cement or hydraulic lime) is selected; the type of grout (one binder, ternary or binary) is also selected, respecting the target $f_{wc,s}$-value after grouting and possible durability issues.
5. An appropriate *nature* of fines should be selected, based on the following criteria:

   - To be fine enough as to serve penetrability (see Chap. 2).
   - To improve stability (see Chap. 4).
   - To reduce the mass-percentage of active binders (mainly of cement), in order (i) to avoid possible brittleness, due to unnecessarily high compressive strength, (ii) to reduce the mass of *sulphur* compounds of cement (see Chaps. 6 and 7).
   - To increase the specific surface of the solids, which now may serve as water suppliers to the binders, in order to achieve satisfactory hydration.
   - To enhance possible favourable pozzolanic reactions with other materials.

6. An appropriate volume *percentage of fines* ($f_{vp}$) should now be selected to fulfil penetrability requirements, starting with a temporary minimum value:

$$f_{vp} \cong \frac{1 - W_{nom}}{n \cdot d_{85}}$$

   where n ~ 5, and $d_{85}$ denotes the grain diameter of the cement or hydraulic lime used, corresponding to 85% passing (see Sect. 2.5).

   - This value will be now expressed in terms of masses ($f_{mp}$).

$$f_{mp} = \frac{\rho_{mf} \cdot f_{vp}}{(1 - f_{vp})\rho_b + f_{vp}\rho_{mf}}$$

where, $\rho_{mf}$ = average density of fines, and $\rho_b$ = density of the basic binder, usually cement (C) or natural hydraulic lime (NHL), its mass being equal to $(1 - f_{mp})$.

- The average "specific surface" $S_A$ of the solid phase is now calculated.
- For this composition of solids (cement "C" or "NHL" and fines "$f_{mp}$"), a series of Fluidity Factor Tests is supposed to be carried out, and the corresponding "$F_l$ vs. $(W/S)$" curve is traced (like the one shown in Fig. 8.1). In Fig. 8.1a the basic properties of injectability are considered, in order to define a first usable interval of $(W/S)$ ratio values:

A *fluidity factor line (i)* is drawn (Fig. 8.1a) as a function of $(W/S)$, for a mix containing a minimal value $f_{mp}$ of fine solids, serving penetrability.
The horizontal part of this line *(ii)* illustrates *the maximum fluidity* of the aforementioned mix.
The *minimum acceptable fluidity factor* for the target $W_{nom}$, as required by "Eq. 3.4", is illustrated by the horizontal line *(iii)*.
The vertical line *(iv)* represents the *maximum acceptable* $(W/S)$ *ratio* ($w_2$) in order to avoid instability, be it because of excessive bleeding ("Eqs. 4.23, 4.30 and 4.31") or harmful segregation ("Eqs. 4.41 and 4.42").

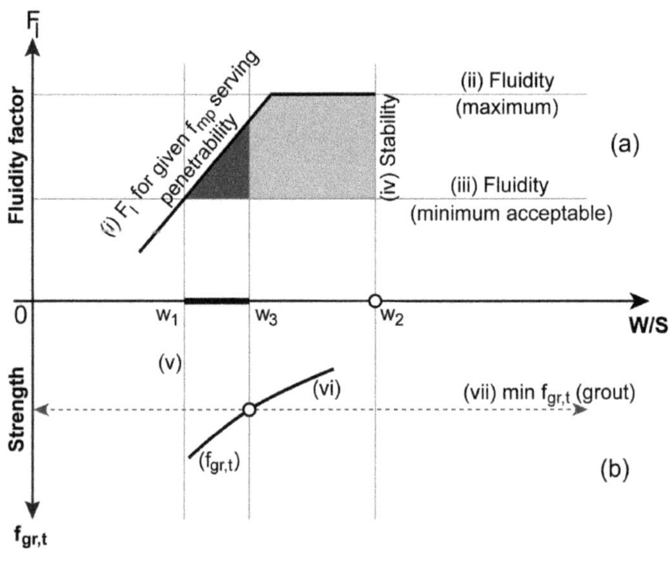


**Fig. 8.1** Fluidity and W/S ratios as affected by the four basic properties of a grout mix, resulting in an interval of W/S ratios $w_1$ to $w_3$ usable for the mix design. Adapted from Miltiadou-Fezans and Tassios (2016)

The *crossing point of lines (i) and (iii)* corresponds to a *minimum* $(W/S) = w_1$ value. Water-to-solids ratios higher than $w_1$ are needed to ensure minimum fluidity of the given penetrable mix.

In Fig. 8.1b, a *tensile strength curve (vi)* of this grout is drawn as a function of (W/S)-values: It is reminded that a minimum tensile strength (min $f_{gr,t}$) should be respected in order to produce a satisfactory bonding between grout and masonry constituents (so that a selected compressive strength $f_{wc,s}$ of grouted masonry be achieved).

This *minimum tensile strength (min $f_{gr,t}$)* is illustrated by the horizontal *line (vii)*. Thus, a new $(W/S)$ limit value $w_3$ appears: the designer is free to select any intermediate value $w_1 < (W/S) < w_3$ to satisfy injectability and strength requirements.

In conclusion, under the conditions of this thought-experiment, water-to-solids ratios between $w_1$ and $w_3$-values, result in grouts satisfying all requirements.

Yet, Fig. 8.1 seems to be a somehow complicated representation. It is believed, however, that it very much helps the understanding of the *interplay* of the numerous variables of the phenomenon, ensuring an easier final design of the grout.

Nevertheless, in real cases, such a clear successful solution may not be possible; the following sections deal with such cases.

## 8.3 Increase of Fines to Improve Compatibility Between Stability and Fluidity

Because of possible instability, it cannot be excluded that the water content $w_2$ might be lower than $w_1$ (of Fig. 8.1). If this is the case, a further increase of fines $f_m$ ($>f_{mp}$) will be needed, so that $w_2' > w_1'$ (Fig. 8.2).

A detailed study of such an increase of fines is shown in what follows, only as a demonstration of the intervening procedures.

To illustrate this procedure, we will take the case of grouts presented on Fig. 3.7 (see Chap. 3), where, a given cement C23 ($S_A = 4100$ cm$^2$/g, $d_{85} = 20$ μm) was used, satisfying the grading rules for a $W_{nom} = 108$ μm (corresponding to the 0.63–1.25 mm sand column), without necessarily any substitution by ultrafines ($f_{mp} = 0$). Nevertheless, for stability, durability and strength reasons, such a substitution may be needed, and to this end a Santorini earth was chosen (SE1, max d < 20 μm). Fluidity measurements were carried out (FFT method), in order to estimate the necessary W/S ensuring injectability in a milieu with $W_{nom} = 108$ μm.

Using Fig. 3.7, the Fig. 8.3 has been drawn; On a "$f_m$ vs. $(W/S)$" diagram (Fig. 8.4), we will try to visualise various phenomena:

1. In Fig. 8.3 the line (i) connects approximately the breaking points of the $F_l$ lines of Fig. 3.7. This line (i) of Fig. 8.3 is "translated" to line (1) in Fig. 8.4.
2. For $W_{nom} = 108$ μm, from "Eq. 3.4" we calculate min $F_l = 1.2 \times 10^3$ mm/s, line (ii) in Fig. 8.3. For this min $F_l$-value we read on Fig. 8.4 the corresponding

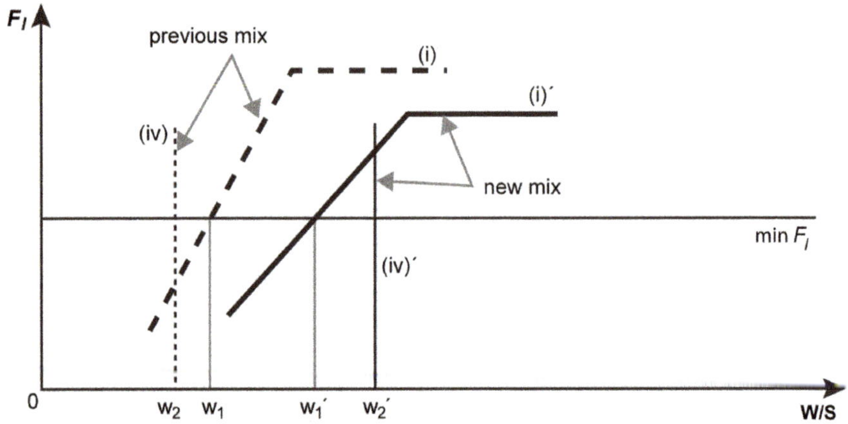


**Fig. 8.2** Consequences of mix modification, if further fines are added: stability is enhanced, and a new interval of W/S ratios ($w_1', w_2'$) appears to be usable [(iv)and (iv)′ lines represent critical bleeding]. Adapted from Miltiadou-Fezans and Tassios (2011)

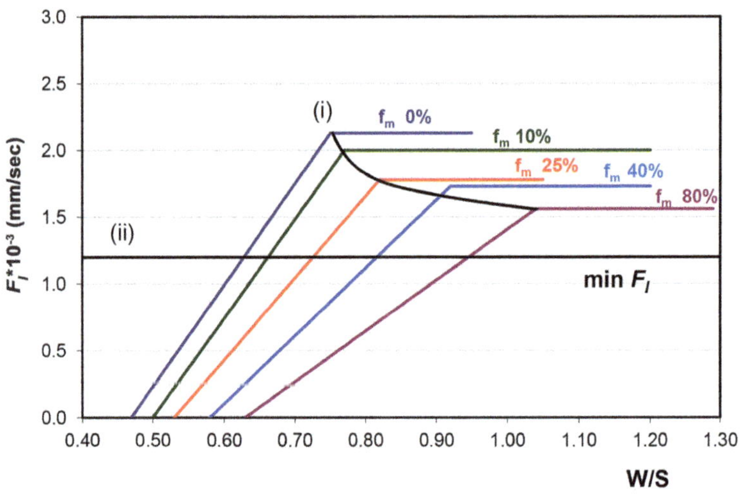


**Fig. 8.3** On Fig. 3.7, two additional lines (i) and (ii) are traced, as explained in the text

$(W/S)$ values for various $f_m$-percentages. Thus, a "respective" line (2) is traced on Fig. 8.4.

3. Subsequently, for average $S_A$-values read out in Table 4.1 for various (SE1) percentages, we calculate the critical bleeding related $(W/S)_{bl,crit}$-values "Eq. 4.23", and we report them in Fig. 8.4 as line (3).
4. In conclusion, the coordinates of points located between lines (1), (2) and (3) of Fig. 8.4 correspond to "permissible solutions".

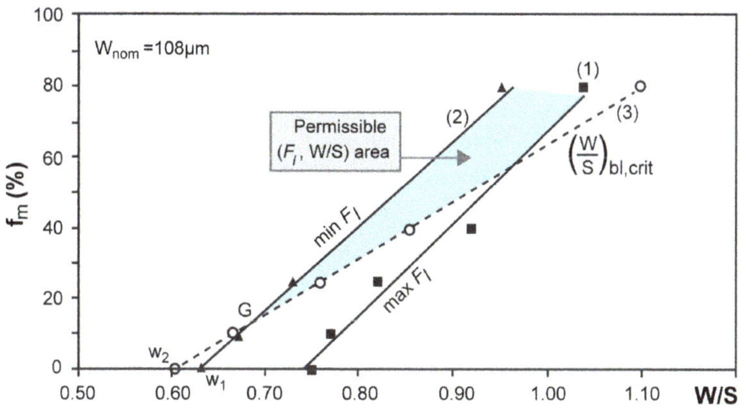


**Fig. 8.4** A graph illustrating the conditions governing the selection of an appropriate ultrafines' percentage $f_m$ of Santorini earth, to substitute a very fine commercial cement. Adapted from Miltiadou-Fezans and Tassios (2011)

5. However, the lower part of this area (near point G) may suffer of unacceptable *uncertainties*: Because of scattering of measurements, the exact location of lines (2) and (3) is not known; thus, in this particular example, it seems that solutions with $(W/S) > 0.75$ (i.e., $f_m > 0.30$) should be retained.
6. At this stage, strength considerations may offer additional criteria about the quantity of SE1 needed:
   (a) Low SE1 percentage (i.e., high cement content) results in higher strength if needed.
   (b) If a high strength is not needed, a lower cement content will be used, i.e., a higher SE1 percentage.

By means of this example, it was shown that compatibility between stability and fluidity requirements may be succeeded by means of an appropriate increase of the percentage of fines. So, in this particular example, it is noted that, although the very fine cement initially selected was able to ensure penetrability (with $f_{mp} = 0\%$), it was now proved that stability and fluidity requirements lead to the use of at least 30% of ultrafines, unless additional fines are possibly needed to reduce strength.

## 8.4 Increase of Fines to Improve Stability Itself

In the preceding paragraphs, the usefulness of the presence of ultrafine materials was again made apparent: Thanks to them, the specific surface $S_A$ of the solid phase is increased (see Table 8.1) and, as a consequence, harmful bleeding may be avoided, and a higher fluidity factor may be ensured, without however threatening the stability of the mix. On the other hand, accidental $(\boldsymbol{W}/\boldsymbol{S})$ increase in jobsite will not result

**Table 8.1** Specific surface $S_A$ of solid phase increased, thanks to the addition of ultrafine materials [Lime (L), two types of Santorini earth, (SE and SE1)] and corresponding (*W/S*) ratios resulting in bleeding initiation (b = 0%) and in critical bleeding (b = 5%) values

| Solid phase of the grout | $S_A$ [a] ($cm^2/g$) | W/S | |
|---|---|---|---|
| | | b = 0% | b = 5% |
| 100% C10 | 3489 | 0.43 | 0.52 |
| 90% C10 + 10% SE | 4490 | 0.65 | 0.79 |
| 75% C10 + 25% SE | 5992 | 0.74 | 0.89 |
| 60% C10 + 40% SE | 7493 | 0.87 | 1.06 |
| 90% C10 + 10% L | 4940 | 0.73 | 0.84 |
| 75% C10 + 25% L | 7117 | 0.87 | 1.01 |
| 60% C10 + 40% L | 9293 | 0.97 | 1.19 |
| 100% C23 | 4100 | 0.55 | 0.64 |
| 90% C23 + 10% SE1 | 4555 | 0.64 | 0.74 |
| 75% C23 + 25% SE1 | 5225 | 0.72 | 0.84 |
| 60% C23 + 40% SE1 | 5898 | 0.80 | 0.92 |
| 20% C23 + 80% SE1 | 7697 | 0.89 | 1.2 |

[a] Calculated average fineness of the solid phase of the grout

in unacceptable bleeding. Figure 8.5 clearly illustrates these two advantages of the use of ultrafines:

- An increase of specific surface from 3500 to 7000 $cm^2/g$, results in increasing of "safe" against bleeding ($W/S$) values from approximately 0.5 to 1.00.

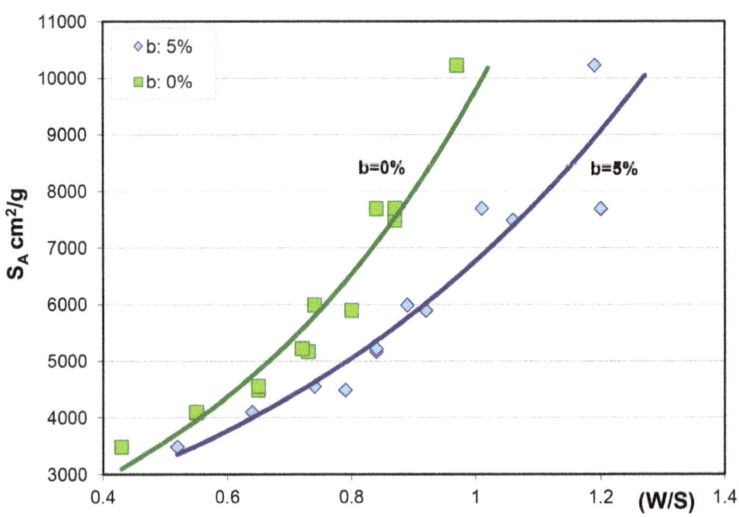


**Fig. 8.5** Water to solid ratios: (i) initiating bleeding (b = 0%) and (ii) leading to critical bleeding, as a function of the increase of the specific surface of solids

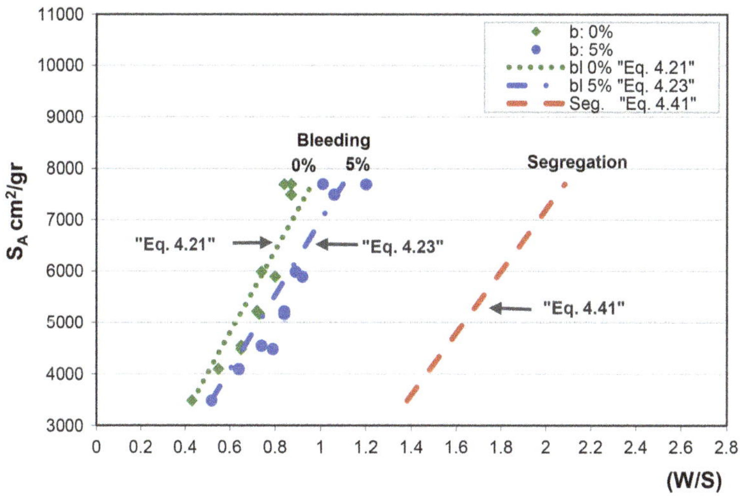


**Fig. 8.6** The water to solid ratios leading to the appearance of unacceptable segregation are quite higher than those of critical bleeding (s. Table 4.4). Adapted from Miltiadou-Fezans and Tassios (2011)

- Similarly, if $S_A = 3500\ cm^2/g$, a $(W/S)$ increase by only 0.10, would increase bleeding from 0 to 5%, whereas for $S_A = 7000\ cm^2/g$, the same consequence would be possible with a W/S increase by 0.20. Increased safety margins are apparent.

On Fig. 8.6 now, another critical line is added, illustrating the appearance of unacceptable segregation (as defined in Sect. 4.5). It is seen that normally, when a superplasticizer is not used, segregation is not critical.

## 8.5 Modifications to Obtain a Minimum Tensile Strength

Another parameter has also to be considered; a minimum tensile strength (min $f_{gr,t}$) of the grout has to be observed for satisfactory bonding between grout and masonry constituents. To this end, on Fig. 8.1b a maximum water content $w_3$ was found. Similarly, for the new mix decided in Sect. 8.3 (see Fig. 8.2), the new value $w_3'$ (Fig. 8.7), happily enough, is larger than $w_1'$; thus, a practical solution is feasible.

If, however, $w_3' < w_1'$, appropriate modifications should be made to the previous mix in order to decrease the $w_1'$-value or to further increase $w_3'$-value:

- Lower $w_1'$-values could be taken by decreasing the percentage of fines, although penetrability and stability requirements do not offer much space for such a solution.
  Similarly, lower $w_1'$-values may also be achieved by adding a superplasticizer.

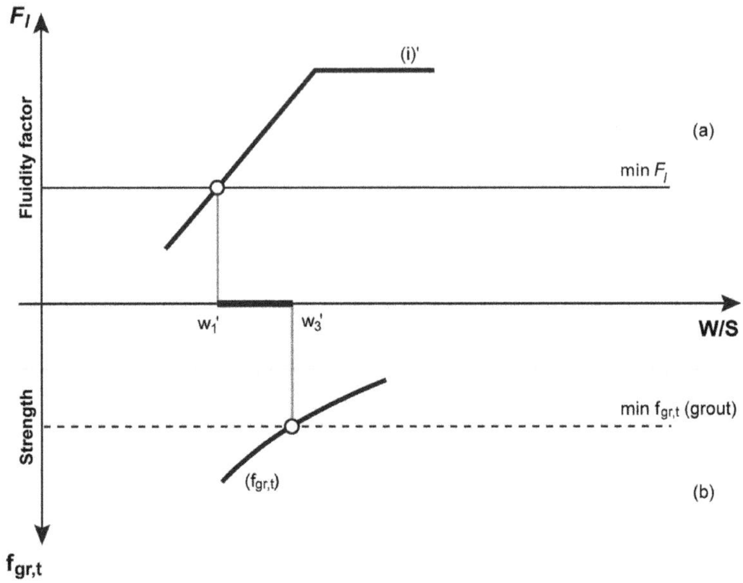


**Fig. 8.7** Limit W/S-values ($w_1'$, $w_3'$) observing the inequality $w_1' < w_3'$

- Higher $w_3'$-values would necessitate higher active binder percentage or the use of a more resistant binder, in order to succeed the necessary tensile strength value $f_{gr,t}$ of the grout.

However, as far as the increase of active binder is concerned, the following points should be reminded:

- In case of binary solids i.e., of cement (or hydraulic lime) and pozzolanic material, we should try to exploit the maximum possible percentage of pozzolan corresponding to the full binding of totally available lime.
- In case of ternary solids compositions, additional trial mixes are needed, modifying the dosage of each constituent.

## 8.6 Addition of Superplasticizer

As far as the addition of SP is concerned, two cases are presented in what follows.

1. The first case covers a binary composition of natural hydraulic lime (NHL3.5z) and ultrafine pozzolanic material (volcanic gaia), without and with superplasticizer (1% Rheobuild 5000) (Moundoulas et al. 2009).
   The natural hydraulic lime used ($d_{85} = 20.2$ μm, $d_{99} = 48.2$ μm see Sect. 2.3), satisfied the penetrability grading rules for a $W_{nom} = 175$ μm, without any need to substitute part of the basic binder with ultrafines ($f_{mp} = 0$). The combination

of the natural hydraulic lime with the pozzolanic material, without and with SP, was examined to study mainly durability and strength characteristics.

In Fig. 8.8, full lines correspond to the results without superplasticizer, while dotted lines represent the results with superplasticizer. As it is seen on Fig. 8.8, after the addition of the superplasticizer, all lines "min $F_l$", "max $F_l$" and "critical bleeding" moved to the left in such a way that the "permissible area" I was shifted to the place II, leaving ample space for a profitable selection of parameters. Thus, an almost 10% reduction of mixing water was feasible. Moreover, bleeding was not critical; and the addition of the superplasticizer proved to be profitable.

2. In the case depicted in Fig. 8.9, a ternary grout was used, i.e., a combination of cement 30% and fines constituted of a pozzolanic material-Lava Antica 45% and lime 25% (Miltiadou-Fezans et al. 2007). The white cement used ($d_{85}$ = 41 μm, $d_{99}$ = 83 μm), did not satisfied the grading rules for a $W_{nom}$ = 175 μm. Thus, at least 15% of cement had to be substituted with ultrafine materials to satisfy penetrability requirements (min $f_{mp}$ = 15%, see Fig. 8.9).
3. In this case, the addition of a superplasticizer (1% CHEM SPLM) was rather detrimental (see Fig. 8.9): Bleeding became very critical, and the "permissible area" I was drastically restricted in II, too close to min $F_l$, with doubtful consequences on the required performances.

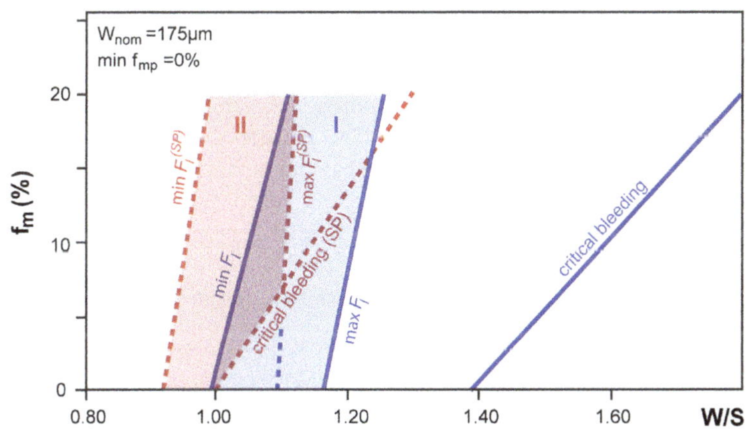


**Fig. 8.8** Illustration of the conditions governing the selection of ultrafines' percentage $f_m$ of pozzolanic material for a binary grout (natural hydraulic lime and pozzolan-volcanic earth), without and with superplasticizer. Permissible areas are II and I, with and without superplasticizer correspondingly. In this case, bleeding was not critical at all; thus, the addition of the superplasticizer proved to be profitable in an effort to reduce $w_1'$ (see Fig. 8.7)

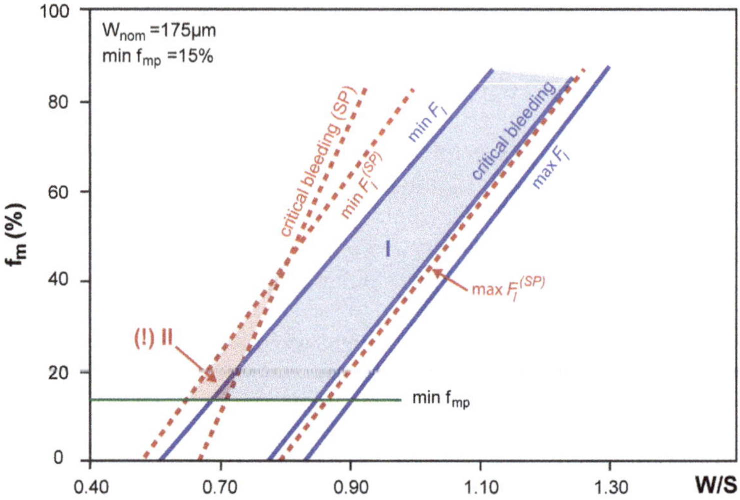


**Fig. 8.9** Illustration of the conditions governing the selection of ultrafines' percentage $f_m$ (in the case of ultrafines consisting of pozzolanic material and lime), for a ternary grout without or with superplasticizer. In this case, the addition of the superplasticizer had clearly adverse consequences

## 8.7 Conclusion

As a conclusion, it seems that the systematic quantitative knowledge of basic properties of grouts (such as penetrability, fluidity, and stability), together with some practical numerical predictions regarding these properties, allow for a better understanding of the otherwise *too* complicated behaviour of these grouts, especially when some strength and durability aspects have also to be considered. The mix-designer may find profitable to visualise these properties on appropriate graphs (like those presented in Figs. 8.1, 8.2, 8.3, 8.4, 8.5 and 8.6), and facilitate her/his poly-parametric decision making. It is hoped that this way, grout mix-design becomes more rational and less fortuitous.

## References


Miltiadou-Fezans A, Tassios TP (2011) Practical guidelines for the design of hydraulic grouts and the quality control during their application in historic masonry structures. In: Proceedings of one day seminar on restoration techniques, materials and application problems, Society for Research and Promotion of Scientific Restoration of Monuments (ETEPAM), Thessaloniki, Greece, 10 Nov 2010, pp 51–66 (in Greek)

Miltiadou-Fezans A, Tassios TP (2016) Holistic methodology for the mix design of hydraulic grouts in strengthening historic masonry structures. In: Papagianni I, Stefanidou M, Pachta V (eds), Proceedings of 4th Historic Mortars Conference (HMC2016), 10–12 Oct 2016, Santorini, Greece, pp 580–587

Miltiadou-Fezans A, Kalagri A, Savvidou M (2007) Research and investigations for the design of grouts for the structural restoration of the Church of the Assumption of Virgin Mary in Tegea Arkadias. Research report, Directorate for Technical Research on Restoration of the Hellenic Ministry of Culture and Sports (in Greek)

Moundoulas P, Morout P, Miltiadou A, Aggelakopoulou E, Bakolas S, Kouloumbi N, Moropoulou A (2009) The impact of fluidifiers on hydraulic grouts. In: Proceedings of 11th paints symposium research and technology of paints, Varnishes and inks on the eve of 2010, Athens, 7–8/5/2009, pp 278–304

# Chapter 9
# Practical Guidelines for the Mix Design of Grouts in Masonry Strengthening

**Abstract** In this Chapter, the scientific approach followed in this book finds its justification: The rational and detailed examination of all properties of a grout, offers now the possibility to follow a practical step-by-step procedure of mix-design, permitting to handle numerous parameters in a logical sequence. Thus, this chapter contains more practical guidelines for the mix-design of grouts used in masonry strengthening. The use of *Tables* and *empirical formulae* included in previous Chapters, greatly facilitate the selection of (i) the type of the binders and the final grading of the solids, (ii) the minimum acceptable fluidity factor, depending on the finest discontinuities width class ($W_{nom}$), (iii) the zero-bleeding and the critical-bleeding (W/S) expressions (with or without superplasticizers), as a function of the calculated average specific surface $S_A$ of the solid phase, and (iv) the limit value W/S against segregation. Subsequently, a practical procedure for the selection of the final (W/S) ratio of the mix is described, respecting all the aforementioned limits. Corresponding remedy-measures are presented in case such a complete respect is not feasible. Moreover, for masonries of minor historical importance and with representative $W_{nom} \geq 0.25$ mm, a Table is offered, containing approximate compositions of grouts for three different required grout strength ranges. Experimental verification of the required grout-performances will be in any case necessary.

**Preamble**

In Chap. 8 an understanding of the interplay of the numerous parameters of grout-design was sought. Now, a more practical step-by-step procedure for the mix-design will be presented, making use of previously formulated design criteria and empirical formulae, predicting some grout properties.

An optimisation of occasionally contradictory requirements will also be achieved, following a trial-and-error method when needed.

The following design-steps are suggested.

New numbering of previous equations is followed in this chapter.

 
A. Miltiadou-Fezans and T. P. Tassios, *Mix-Design and Application of Hydraulic Grouts for Masonry Strengthening*, Springer Tracts in Civil Engineering,
https://doi.org/10.1007/978-3-030-85965-7_9

## 9.1 Selection of Binders

A preliminary selection of the *type* of binders will first be based on *durability* considerations only (Chap. 7). Next, *strength* considerations will be taken into account. To this end, the value of the required compressive strength $f_{gr,c}$, of the grout has to be known. This will be estimated by means of the data of Sect. 6.3.2; note that the compressive strength of masonry before ($f_{wc,0}$) and after grouting ($f_{wc,s}$) were given by the Structural Engineer. Based on the required range of $f_{wc,s}$-values after grouting, a range of necessary grout-strength values $f_{gr,c}$ may be estimated (by means of "Eq. 6.1" or "Eq. 6.3"), accounting for the scattering of data. It is assumed however, that the targeted masonry strength should not be disproportionately higher than its initial value. In some cases of rubble three-leaf masonry, an indicative upper limit was suggested to be as high as twice the initial strength of the wall.

Now, Table 6.10 offers the possibility of selecting a preliminary type of binders (together with their provisional percentages), respecting the estimated range of $f_{gr,c}$-values. Note that Table 6.10 offers the possibility to select more than one candidate compositions; and this will greately facilitate the final decisions of the designer. However, only the compositions observing the durability requirements will be retained for further consideration.

Besides, the respective W/S-value read in Table 6.10, is only a simple indication.

Note: Because of the fundamental significance of the tensile rather than the compressive strength of the grout, it is customary to take note of the estimated $f_{gr,t}$ as well, by means of the data of Sect. 6.4.4.3.

## 9.2 Selection of a $W_{nom}$-value

For a given masonry, a nominal value of the effective width of its discontinuities ($W_{nom}$) has to be roughly estimated. To this end, the suggestions included mainly in Sect. 5.4 are helpful; and it is reminded that thanks to the concept of such a nominal effective width of voids to be filled, the subsequent steps of grout-mix design are greatly facilitated.

## 9.3 Checking the "Fineness" of the Binders' Mixture

After the selection of the types of binders (and their provisional percentages) made in step 9.1, the designer is now invited to decide the use of the respective binders available in the market. Subsequently, the fineness of the mixture of these binders will be checked as follows:

Based on the nominal value of the width of the effective discontinuities, the criteria regarding the grading of the mixture of these binders are applied, in order to ensure *penetrability*, as per Chap. 2:

$$d_{85} < \frac{W_{nom}}{5} \tag{9.1}$$

and

$$d_{99} < \frac{W_{nom}}{2} \tag{9.2}$$

The grain diameters corresponding to 85% and 99% passing of the entire solid phase of the grout are numerically determinated, based on the knowledge of grading of each binder and its percentage.

The verification of the "Eq. 9.1" and "Eq. 9.2" should be carried out for all the candidate compositions indicated by the Table 6.10 (after durability considerations), making use of the gradings of the available binders. Obviously, that composition is selected which better respects the aforementioned inequalities. If this is the case, the selected binders' mixture satisfies the requirements of "Eqs. 9.1 and 9.2", and we go to step 9.5.

If not, the correcting measures described in step 9.4 are needed.

## 9.4 Additional Ultrafine Materials

1. Before any calculation, an elementary control of the available ultrafines, already participating in the selected composition, should be made: Is it in fact an ultrafine? Or (as it accidentally may happen in practice), this product may contain exceptionally coarse grains? The designer should exclude such a possibility.
   In any event, the ultrafines (UF) of the materials participating in the selected compositions, should satisfy the following condition (Sect. 2.5): Their maximum grain size should be smaller than the $d_{85}$ of the basic hydraulic binder to be partially substituted (as it will be explained hereafter):

$$d_{99}^{UF} < d_{85}^{B} \tag{9.3}$$

2. When for the selected binders, "Eq. 9.1" was not satisfied, suitable ultrafine materials are selected in order to substitute a percentage of the basic hydraulic binder (Cement or Hydraulic lime), calculated in volume, as follows:

$$\min f_{vp} \approx 1 - \frac{W_{nom}}{5 \cdot d_{85}} \tag{9.4}$$

Note: This value will be now expressed in masses, $f_{mp} = (\rho_{mf} \cdot f_{vp})/[(1 - f_{vp})\rho_b + f_{vp}\rho_{mf}]$ where, $\rho_{mf}$ = average density of fines, and $\rho_b$ = density of the basic binder.
In any event, "Eq. 9.2" should always be observed.
It should be noted, however, that the nature of the selected ultrafine materials (usually natural or artificial pozzolan and/or lime), as well as their proportions, may affect the strength and the durability requirements, on which see Chaps. 6 and 7.

## 9.5 Expected Fluidity Factor and Minimum Water-to-Solids Ratio

Now the minimum acceptable value of fluidity factor is approximately *calculated* (see Sect. 3.3):

$$\min F_l \approx \left[1.2 - 45 \cdot (W_{nom} - 0.1)^2\right] \cdot 10^3 \text{ [mm/s]}, \text{ for } 0.1 < W_{nom} < 0.2 \text{ [mm]} \quad (9.5)$$

The estimation of the corresponding $(W/S)$-ratio may be facilitated taking into account the zero-bleeding $(W/S)$-values, which can be estimated with the following expressions (Sect. 4.4):

Without superplasticizer

$$(W/S)_{0,bl} \cong 1.25 \cdot 10^{-4} \cdot S_A \ [\text{cm, g}] \quad (9.6)$$

where $S_A$ denotes the calculated average specific surface of the solid phase.

If a superplasticizer is used, this limit $(W/S)$-value will be reduced; e. g., in the case of melamine formaldehyde resin (Chryso GT),[1] the following expression could be useful.

$$(W/S)_{0,bl,SP} \cong \left(1.25 - 0.5 \cdot SP^{\%}\right) \cdot 10^{-4} \cdot S_A \ (\pm 0.10) \ [\text{cm, g}] \quad (9.7)$$

In practice, it is observed that these zero-bleeding values are systematically lower (roughly by 0.05–0.10) than those corresponding to $(W/S)$ resulting in min $F_l$-values. Thus, one way or another, a *lower bound* limit of $(W/S)$ is approximately found.

[1] For other superplasticizers, a separate preliminary experimental investigation is needed, as described in Sect. 4.4 for bleeding and Sect. 4.5 for segregation.

## 9.6 Maximum Water-to-Solids Ratio to Ensure Stability

In order to determine the appropriate quantity of mixing water ensuring *stability* ($w_2$ of Fig. 8.1), the maximum acceptable $(W/S)$-values of the grout can be estimated, as follows:

1. Against critical bleeding ($b < 5\%$):

   Without superplasticizer

$$(W/S)_{bl,crit} = 1.1 \cdot 1.25 \cdot 10^{-4} \cdot S_A + 0.04 \ [\mathrm{cm, g}] \tag{9.8}$$

where $S_A$ denotes the calculated average specific surface of the solid phase.

If a superplasticizer is used, this limit value will be reduced; e. g., in the case of melamine formaldehyde resin (Chryso GT),[2] the following expressions could be useful.

$$W/S)_{bl,SP,crit} \cong 1.1\left(1.25 - 0.5 \cdot SP^{\%}\right) \cdot 10^{-4} \ S_A + 0.04 \ [\mathrm{cm, g}] \tag{9.9}$$

With the use of the same superplasticizer, in case that DSF is used as ultrafine, the following expression may be used (valid only for a binary grout "75% Cement C10 + 25% Densified silica fume" and the above-mentioned superplasticizer:

$$(W/S)_{bl,SP,crit.} = 1.55 \cdot \left(1 - \frac{SP^{\%}}{4.30}\right) \quad \text{for SP} \geq 0.20\,\% \tag{9.10}$$

2. Against segregation

   Without superplasticizer

$$(W/S)_{segr} \cong 0.80 + \left(\frac{S_A}{6000}\right) \ [\mathrm{cm, g}] \tag{9.11}$$

Without superplasticizer, in case that DSF is used as ultrafine

$$(W/S)_{segr} \cong 1.60 + \left(\frac{S_A}{30000}\right) \ [\mathrm{cm, g}] \tag{9.12}$$

If a superplasticizer is used, this limit value will be reduced; e. g., in the case of melamine formaldehyde resin (Chryso GT),[3] the following expressions could be useful (if others than DSF ultrafines are used):

---

[2] See Footnote 1.

[3] See Footnote 1.

$$(W/S)_{segr} \cong 0.80 + \left(\frac{S_A}{6000}\right) - 1.7 \cdot (SP)^{(\%)} \ [\text{cm, g}] \tag{9.13}$$

With the use of the same superplasticizer, in case that DSF is used as ultrafine, the following expression may be used.

$$(W/S)_{segr} \cong 1.60 + \left(\frac{S_A}{30000}\right) - 1.7 \cdot (SP)^{(\%)} \ [\text{cm, g}] \tag{9.14}$$

Thus, the experimental investigation will be carried out with $(W/S)$-ratios approximately lower than the aforementioned values.

## 9.7 Experimental Examination of the Candidate Composition

Now we are ready to initiate our experimental investigation regarding the *fluidity* and the *stability* of the candidate composition of the mix, as it was selected on the basis of the approximations described in the previous six steps. To this end, an initial $(W/S) = w_0$ ratio is selected as follows:

- First, an approximated $w_1$-value (see Fig. 8.1) is found by means of "Eq. 9.6" or "Eq. 9.7", *plus* say 0.10. This is a fluidity-related limit value.
- Second, the lowest of the stability-related $(W/S)$-values of Chap. 4 is selected as $w_2$-value (Fig. 8.1) from "Eqs. 9.8–9.14".

If by bad chance, $w_2 < w_1$, go to step 9.8.
If $w_1 < w_2$, take the mean value of the aforementioned $(W/S)$ ratios

$$w_0 = \frac{w_2 + w_1}{2}$$

In case $w_2$ and $w_1$ differ more than 0.20, use just the minimum (*W/S*) plus 0.10.

Subsequently, using the estimated $w_0$ find *experimentally*[4] the following quantities:

- $F_{lo}$ (Fluidity Factor, using the experimental procedure of Sect. 3.2.3).
- $b_0$ (normalized bleeding value, using the experimental procedure of Sect. 4.4).
- check the thickness of a distinct denser sediment on the bottom of a container, as described in Sect. 4.5.4.

Check, correspondingly if

$$\mathrm{F}_{l_0} > \min \mathrm{F}_l \text{ (estimated using "Eq. 9.5")} \tag{9.15a}$$

[4] To this end, a high turbulence colloidal mixer has to be used, since the ultrasound mixing method is not yet generally applied under worksite conditions.

$$b_0 < b_{crit} = 0.05 \tag{9.15b}$$

$$sediment \leq 1\,\text{mm} \tag{9.15c}$$

1. If all conditions (9.15a, 9.15b and 9.15c) are observed, then a final solution may be sought as in Sect. 9.9.
2. If, however, one or more of the inequalities 9.15a, 9.15b and 9.15c are not observed, proceed as follows (when the use of a superplasticizer is not yet decided):

   – If $F_{l_0} < \min F_l$, a reasonable increase of $(W/S)$ ratio should be made (see Fig. 9.1), to satisfy "Eq. 9.15a", under the condition that "Eq. 9.15b" and "Eq. 9.15c" will be also observed, as checked by means of a new experiment.
   – If $b > b_{crit}$, (either because of the initial conditions or after the endeavours described in Fig. 9.1), substantial modifications of the initial mix may be needed: In most cases, additional ultrafines (or another more suitable ultrafine material) are needed, together with appropriate changes in $(W/S)$ ratios. In some cases, a selection of a more suitable mixing method may also be considered.

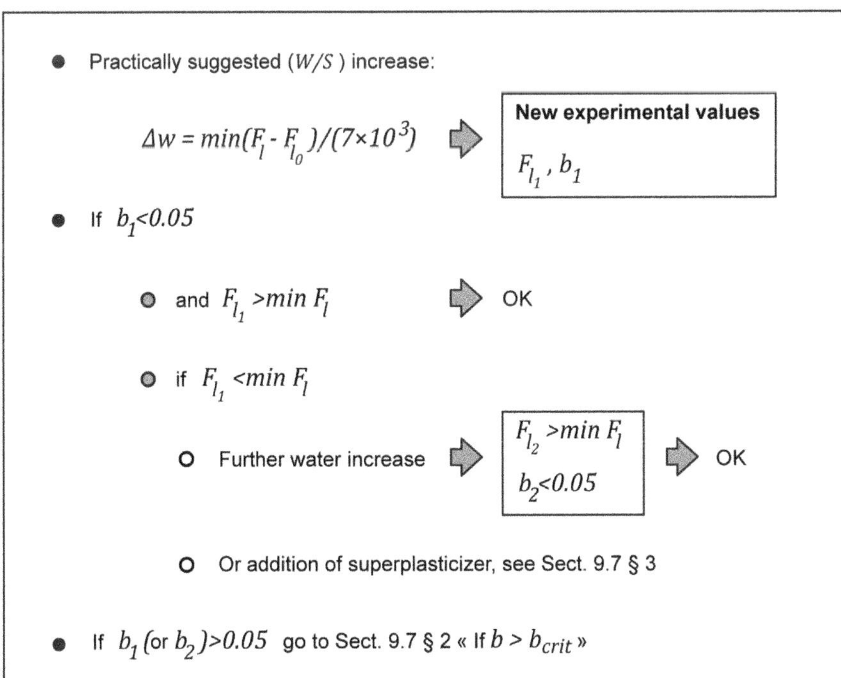


**Fig. 9.1** Schematically suggested procedure for reinstatement of satisfactory fluidity. Reworked from Miltiadou-Fezans and Tassios (2016)

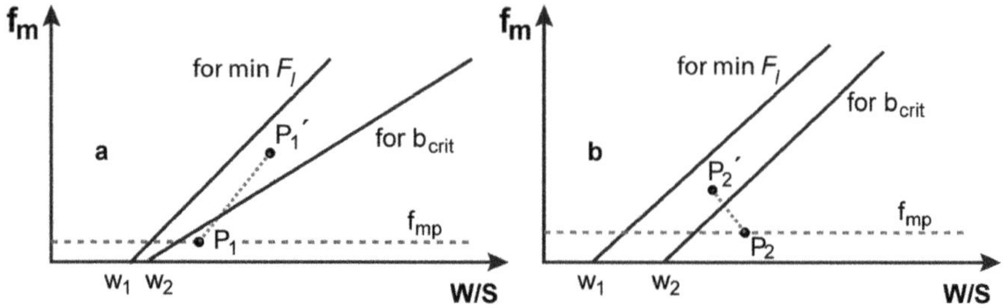


**Fig. 9.2** Qualitative illustrations, assisting decision making in case of initial trial mixes violating bleeding limitations, in two indicative cases **a** and **b**. Reworked from Miltiadou-Fezans and Tassios (2011)

To this end, the following remarks may assist decision making: In order to better introduce the pertinent concepts, suppose that relationships between $f_m$[5] and $(W/S)$ resulting in min $F_l$ and $b_{crit}$-lines are available (as it was the case in Figs. 8.3 and 8.8).

For instance, in a case like the one illustrated in Fig. 9.2a, the point $P_1$ representing the unsuccessful initial trial mix should move to $P_1'$, i.e., both "$f_m$" and "*W/S*" values should be *increased*. In a different case, like the one illustrated in Fig. 9.2b, increase of "$f_m$" should be accompanied by a relative decrease of the apparently disproportionately high-water content used. These or other possible modifications should be experimentally checked again.

In this respect, the following more practical suggestions may be offered to the mix designer:

(a) When an initial trial mix fails to observe bleeding limitation, an *increase* of content of ultrafine materials should be effectuated, for the same water content. If bleeding is sufficiently reduced, without however violating the required minimum fluidity factor value (min $F_l$), the solution is retained.

(b) If, however (because of the aforementioned modification of the mix), an insufficient $F_l$-value has occurred, two possible interventions remain: an appropriate increase of water content or a smaller increase of ultrafines. Alternatively, at this stage the use of a superplasticizer may be decided. The corresponding bleeding value should be *experimentally* checked again.

In all cases segregation should always be checked.

- In the rare cases where the thickness of the distinct denser sediment is larger than approximately 1 mm, the following remedial measures may be taken: Reduce $(W/S)$ value or increase of $f_m$, provided that $F_l$-values will continue to observe "Eq. 9.15a".

  However, it has to be noted that happily enough, in the absence of a superplasticizer, segregation does not appear before bleeding becomes critical (see Fig. 8.6).

[5] It is supposed that always $f_m \geq f_{mp}$ (Sect. 8.2).

Now that, thanks to these modifications, all inequalities (9.15) are observed, go to Step 9.9.

3. In many cases however, the satisfaction of conditions 9.15a, 9.15b and 9.15c cannot be achieved without the use of an appropriate superplasticizer. If so, the following cases could be considered.

   - In the case of low content of ultrafines, *critical bleeding* may occur before a satisfactory fluidity factor is achieved. In such cases the addition of an appropriate small percentage of a superplasticizer may be tried: The corresponding reduction of $(W/S)$-ratio, may allow the increase of fluidity factor above the level of min $F_l$ required by "Eq. 9.5". Figure 9.3 schematically illustrates this phenomenon.
   - An additional benefit of the use of an appropriate superplasticizer is the expected reduction of water content, resulting also in an increase of the tensile strength of the grout. This increase is frequently required, as explained in Sect. 8.2 and Fig. 8.1.
   - If a densified silica fume is used, the addition of a suitable superplasticizer cannot be avoided, anyway, in order to achieve satisfactory deflocculation.
   - If a superplasticizer is to be used, some preliminary experimental investigations are indispensable, because of the *extreme variety* of available commercial products and the sensitivity of their results on fluidity and stability. Similarly, such an investigation is required in order to match the type of the superplasticizer and the nature of the solids of the mix.

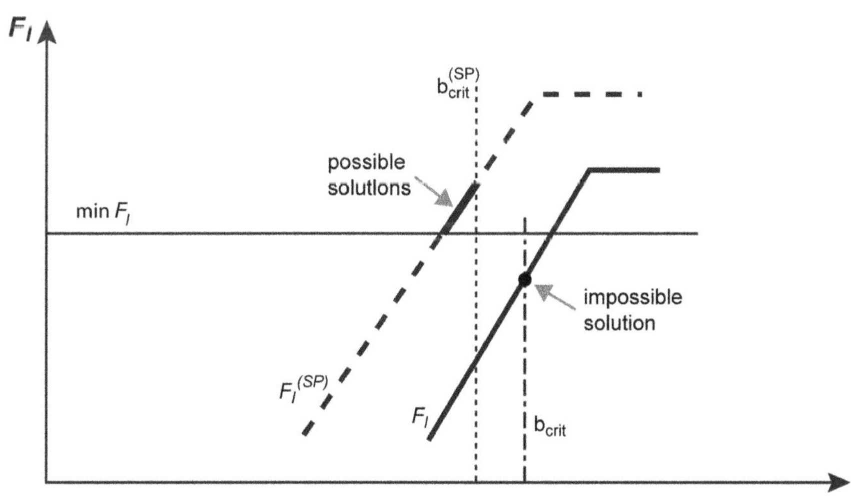


**Fig. 9.3** If bleeding is critical, without achieving a satisfactory $F_l$-value, the addition of a superplasticizer may offer satisfactory solutions with $F_l$ > min $F_l$ and acceptable bleeding. Reworked from Miltiadou-Fezans and Tassios (2016)

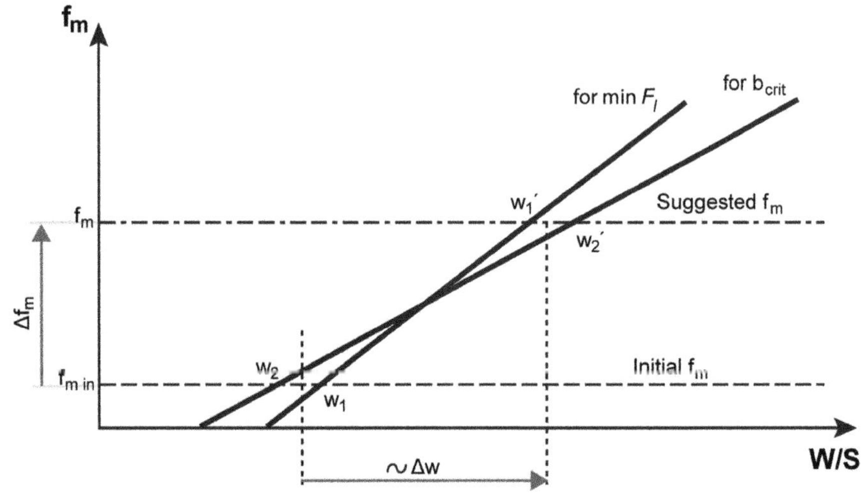


**Fig. 9.4** Qualitative illustration of the necessary additional quantities $\Delta f_m$ and $\Delta w$. Adapted from Miltiadou-Fezans and Tassios (2016)

- Finally, the possible consequences of the superplasticizers on the durability of the grout should also be considered (Chap. 7).

## 9.8 Early Critical Bleeding

If $w_1 > w_2$, proceed as follows:

1. Because of the inevitable uncertainties of the equations mentioned in Sect. 9.6, if the difference between $w_1$ and $w_2$ is not significant (e.g., around 0.05), try again a value $w_0 = \frac{(w_1 + w_2)}{2}$: In case of satisfactory results concerning the inequalities (9.15), proceed to Sect. 9.9.
2. Otherwise, the most decisive modification of the mix is to add ultrafines (see Fig. 8.2) of an appropriate nature. Thus, bleeding may be avoided, whereas, with a simultaneous increase of $(W/S)$-ratio, fluidity factor would not be jeopardized. The necessary additional quantities $\Delta f_m$ and $\Delta w$ are qualitatively visualised[6] in Fig. 9.4.

Now that $w_{1'} < w_{2'}$, proceed as in Sect. 9.7.

[6] As for example in Figs. 8.4 and 8.9.

## 9.9 Strength Evaluation

Since the composition of the grout satisfying all injectability[7] requirements is found, it remains to measure the mechanical strengths of this trial grout. In other words, the selected $(W/S)$-ratio should be sufficiently low to ensure the required strength of the grout.

To this end, a rough estimatiom of this strength may be found by means of a splitting test of the cylindrical specimens contained in the "sand column" after its successful filling with grout.

Precise strength values, however, are found by means of grout specimens (preferably cylindrical, see also Sect. 6.4.3), to measure both the compressive and tensile strengths, in 28, 90 and 360 days. Normally, only the 90-days strength is of direct importance, in order to check the suitability of the selected $(W/S)$-ratio. Earlier strengths, however, are useful, in order to rapidly estimate the expected 90-days strength by means of check-tests in shorter intervals. On the other hand, one year's tests are needed to check possible in-time deterioration of strength, as observed in some cases of very ultrafine binders.

In the not frequent case that the strength requirement is not satisfied, the mix-design has to be restarted anew.

## 9.10 Possible Simplification

Under specific conditions, however, the entire mix-design procedure may be substantially simplified.

Such is the case of masonries of minor historical importance and with a $W_{nom} \geq 0.25$ mm. In such cases, the preliminary design may be based on Table 9.1 (identical with Table 6.10). Selecting one of these compositions observing the required strength-ranges of the grout ($f_{gr,c}$), may be a good practice.

Nevertheless, a pilot application of grouting should be organized at the worksite: if the selected composition satisfies the criteria described in Sect. 10.5.2, the selection is retained. Otherwise, the complete design procedure of Sects. 9.1–9.9 should be followed.

## 9.11 Worksite Conditions

Before the initiation of the execution of the works, it is necessary in all cases to produce final trial mixtures by using the *worksite* equipment, as well as the materials which are going to be used in situ. During the work, continuous control of the mixtures is required, as indicated in Chap. 10.

[7] I.e., penetrability, fluidity and stability.

**Table 9.1** Roughly suggested grout compositions for $W_{nom} > 0.25$ mm and for four ranges of strength at 90 days (identical with Table 6.10)

| Grout's compositions corresponding to ranges of required compressive strength at 90 days ($f_{grc}^{90d}$) in MPa | | | |
|---|---|---|---|
| 1.0–3.0 | 3.0–6.0 | 6.0–10.0 | 10.0–20.0 |
| | | | L/MK/LSF/W:35/35/30/60 |
| NHL5/W: 100/82.5 | NHL5/W: 100/90-80 | | |
| NHL5/P/W: 90/10/80.2 | NHL5/P/W: 90/10/80 | | |
| NHL5/P/W: 80/20/85-80 | | | |
| NHL5/E/W: 70/30/90 | | | |
| NHL5/E/P/W: 63/30/7/87.5 | | | |
| | | | NHL5/L/MK/W:30/35/35/60 |
| $NHL3.5_Z$/W: 100/110-90 | $NHL3.5_Z$/W: 100/95-80 | | |
| $NHL3.5_z$/P/W: 80/20/120 | $NHL3.5_Z$/P/W: 80/20/95 | | |
| NHL2/W: 100/130 | | | |
| HL/W: 100/60 | | | |
| | CR/W: 100/70 | | |
| CA/W: 100/70 | CA/W: 100/55 | | |
| Unilit (0/0)/W: 100/80-70 | | | |
| CC/W: 100/65-60 | | | |
| | | | LI/W:100/?[a] |
| | | | C/L/W:80/20/85 |

(continued)

**Table 9.1** (continued)

| Grout's compositions corresponding to ranges of required compressive strength at 90 days ($f_{grc}^{90d}$) in MPa | | | |
|---|---|---|---|
| | | | C/P/W:60/40/85-75 |
| | C/L/W: 60/40/100 | C/L/W: 60/40/85 | |
| | | C/L/P/W:60/20/20/100 | C/L/P/W:60/20/20/75 |
| | C/L/W: 50/50/95 | | |
| | C/L/P/W: 50/25/25/105 | | C/L/P/W: 50/25/25/80 |
| C/L/P/W: 50/20/30/125 | C/L/P/W: 50/20/30/120 | | |
| | | | C/FA/W:40/60/65 |
| C/L/P/W: 40/25/35/130 | C/L/P/W: 40/25/35/120-80 | C/L/P/W: 40/25/35/75 | |
| | | C/L/P/W: 40/30/30/90-110 | |
| | | C/L/P/W: 40/20-15/40-45/80 | |
| C/L/P/W: 30/30/40/140-130 | | | |
| | | | C/L/MK/W: 30/35/35/110 |
| | C/L/P/W: 30/25/45/80 | C/L/P/W: 30/25/45/80-72.5 | |
| | C/L/P/W: 30/20/50/80 | C/L/P/W: 30/20-10/50-60/90-85 | |
| | | | C/L/P/SF/W: 30/20-10/40-50/10/110 |
| | | | C/L/P/SF/W: 30/20-10/45-55/5/100 |
| C/L/Cl/W: 25/25/50/110 | | | |
| C/P/W: 20/80/120-115 | | | |
| C/L/P/W: 20/30/50/135 | | | |

(continued)

**Table 9.1** (continued)

| Grout's compositions corresponding to ranges of required compressive strength at 90 days ($f_{grc}^{90d}$) in MPa | | | |
|---|---|---|---|
| | C/L/P/W: 15/50/35/97 | | |
| | C/L/P/B/W: 13/43/26-22/18-22/98-93 | | |
| | C/L/P/W: 10/50/40/100 | | |
| C/L/P/W: 10/30/60/50 | | | |
| | C/L/P/W: 10/20-15/70-75/85 | | |
| | C/L/P/B/W: 10/45/25/20/95-90 | | |
| | C/L/P/Cl/W:9/46/36/9/112 | | |
| | C/L/P/Cl/W:8/42/33/17/89 | | |

[a] *The water to solids ratio were not known to the authors of this book. The respective papers are given in the Appendix, Table 6.11). C* cement, *L* lime, *NHL* natural hydraulic lime, *HL* artificial hydraulic lime, *P* pozzolan, *SF* silica fume, *FA* fly ash, *MK* metakaolin, *CR* calx romana (NHL), *CA* calce albazzana (NHL), *CC* calce per consolidamento (premixed), *LI* lime injection (premixed), *LSF* limestone filler, *B* brick dust, *Cl* clay and *W* water

**Note** To a possibly not yet experienced reader, the described procedure for the design of a correct grout may appear too long and complicate. But it should not be forgotten that the final product aimed at by this design, depends on 6 performance requirements: penetrability, fluidity, bleeding, segregation, strength and durability—some of them being contradictory to each other, needing optimisation.

Mathematically thinking, it is expected that the solution of a system of 6 equations with 6 variables, would not be a short procedure; besides, after appropriate modifications, the system should be solved several times, in order for an overall optimum solution to be achieved. With this analogy, the proposed relatively long procedure for grout design, may be better justified. Otherwise, without some disciplined method, such a design may become chaotic.

Obviously, experienced designers will continue to employ their own design method; but it is believed that, even experienced designers, thanks to the preceding long analysis may better recognise the nature of some phenomena.

## References

Miltiadou-Fezans A, Tassios TP (2011) Practical guidelines for the design of hydraulic grouts and the quality control during their application in historic masonry structures. In: Proceedings of One day Seminar on Restoration Techniques, materials and application problems, Society for Research and Promotion of Scientific Restoration of Monuments (ETEPAM), Thessaloniki, Greece, 10 Nov 2010, pp 51–66 (in Greek)

Miltiadou-Fezans A, Tassios TP (2016) Holistic methodology for the mix design of hydraulic grouts in strengthening historic masonry structures. In: Papagianni I, Stefanidou M, Pachta V (eds) Proceedings of 4th historic mortars conference (HMC2016), 10th–12th Oct 2016, Santorini, Greece, pp 580–587

# Chapter 10
# Practical Recommendations for the Execution of Grouting

**Abstract** In this Chapter a set of recommendations is presented for the *execution* of grouting, regarding grout tubes installation, description of equipment needed, in situ preparation of the grout and in situ control of injectability characteristics, as well as in-time evolution of grout's strength. Moreover, methods of in-situ checking of injection procedure are described, together with the data that should be recorded during grouting operations, and their evaluation. The chapter concludes with the assessment of the efficiency of grouting; overall quality management is finally described, together with a detailed presentation of laboratory and in situ non-destructive control tests.

## 10.1 Introduction

The purpose of this Chapter is to present practical guidelines, focusing on the following: (i) the information regarding the preparation of the masonry and the installation of grout entrance (and possibly exit) tubes, (ii) the main requirements for the in situ grouting equipment, (iii) the basic information concerning the preparation of the grout in situ, (iv) the in situ controls of injectability characteristics of the grout, as well as of the in-time evolution of strength and porosity or other characteristics of the grout, (v) the injection procedure itself, and (vi) the data that have to be recorded during grouting execution, and their evaluation. Finally, (vii) post intervention assessment of grouting efficiency will be discussed, together with (viii) the overall quality management needed.

It has to be made clear that masonry grouting with its ambitious purposes, regarding strength and durability, is a delicate operation which was correctly assimilated with a medical surgery operation. Consequently, all stages of this operation have to be meticulously conceived, executed and controlled. Otherwise, the idea that masonry grouting is a simple pouring of an "empirically known" liquid mixture, cannot at all result in the expected radical improvement of masonry.

 
A. Miltiadou-Fezans and T. P. Tassios, *Mix-Design and Application of Hydraulic Grouts for Masonry Strengthening*, Springer Tracts in Civil Engineering,
https://doi.org/10.1007/978-3-030-85965-7_10

## 10.2 Preparation of Masonry and Installation of Injection Tubes

### *10.2.1 Masonry Survey*

Before the application of grouts, the faces of masonry elements should be adequately examined.[1] In case that the masonry is rendered and the rendering is not to be preserved, this rendering should be removed, to allow for a detailed survey of the masonry and its pathology to be carried out.

Such a detailed survey of the geometry and the pathology of the faces, as well as a knowledge regarding the interior spaces of the wall, is necessary for the following reasons:

- Finding out if the walls consist of monolithic or two or three-leaf masonry
- Detailed evaluation of damage
- Decision making for repair and strengthening interventions (re-pointing, deep-repointing, stitching, grouting, etc.)
- Selection of the position of the grout entrance (and possibly exit) tubes.

Whereas face-survey methods are well known, the interior of a masonry element may be examined by one or all of the following techniques (see also Chap. 5):

- Taking-off of blocks at appropriate places
- Taking of cores of adequate diameter
- Endoscopical observation
- Application of Non-Destructive methods (see also Sect. 10.8 of this Chapter).

In Fig. 10.1 the survey of the external East façade of the Katholikon of Daphni Monastery, Attica, Greece and its pathology is presented (Miltiadou-Fezans et al. 2004).

Regarding the selection of the position of the grout tubes, an in-space correlation of masonry geometry and cracks position between the two (or even four) faces of the masonry element, may offer very useful data for the design of injection tubes positioning.

Figure 10.2 shows the survey of cracks of the internal façade of the East wall of the Katholikon of Daphni Monastery (Miltiadou-Fezans et al. 2003). Although this drawing was made before the removal of the internal renderings, a simple comparison of external and internal crack pattern proved that the most of the cracks went through the masonry thickness. Such a correlation of external and internal cracks' position, will be taken into account when the locations of grouting-tubes will be decided.

[1] As proposed by Ashurst (1989), Ashurst and Dimes (1990), Baronio et al. (1992), Schuller et al. (1994), Binda et al. (1997), Valluzzi (2003), Miltiadou-Fezans and Tassios (2009), Miltiadou-Fezans et al. (2008a, 2014).

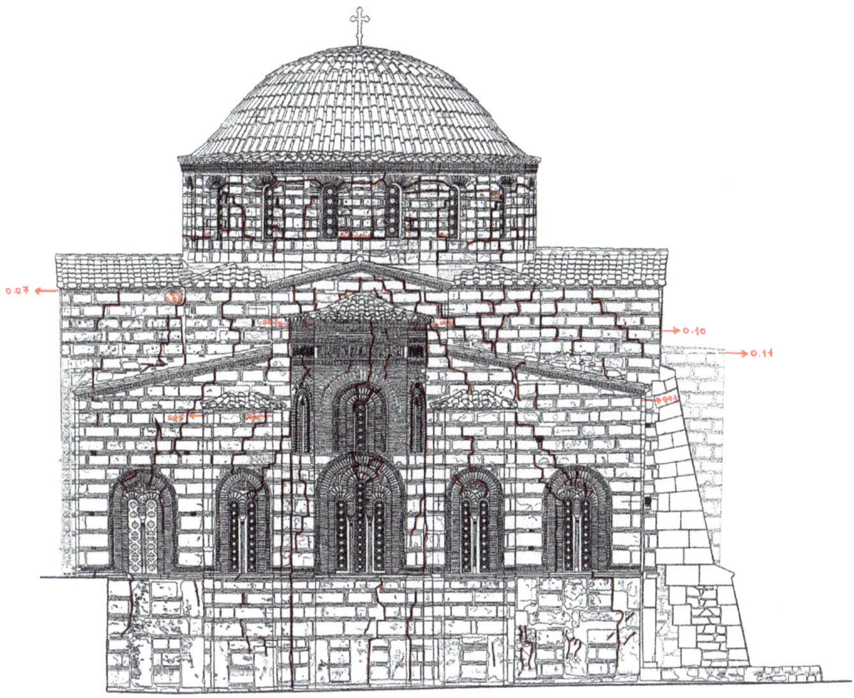


**Fig. 10.1** Survey of the geometry of blocks and joints, as well as of the crack pattern of masonry of the *external* façade of the East wall of the Katholikon of Daphni Monastery, Attica, Greece. Adapted from Miltiadou-Fezans et al. (2004)

### *10.2.2 Cleaning of Loose Material and Sealing of Cracks and Voids*

In order to apply grouting and fill internal discontinuities of various widths, an adequate level of pressure has to be attained inside the grout flow-paths in the interior of the masonry.

To build up this pressure, it is necessary to prevent uncontrollable leakage of the grout; thus, sealing of all masonry cracks and surface voids is absolutely necessary.[2]

To this end, cleaning carefully the masonry is the first step; hidden fine cracks and delaminations should also be uncovered. In all cracked and disintegrated areas of masonry, loose material (within and on both sides of each crack) should be removed, and the region of cracks should be thoroughly cleaned, to allow for their openings

[2] As already pointed out in the literature i.e., for example Ashurst and Dimes (1990), Schuller et al. (1994), Van Rickstal (2000).

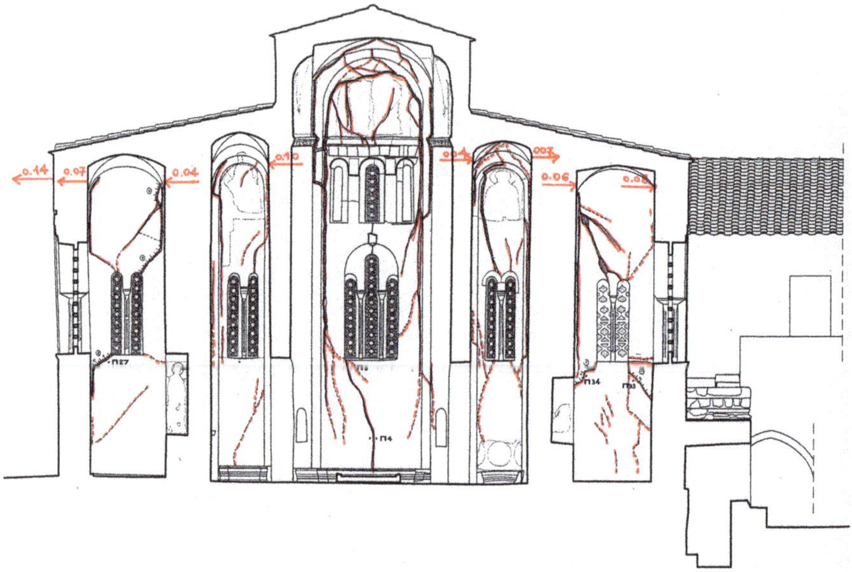


**Fig. 10.2** Survey of the crack pattern of the *internal* face of the East wall of the Katholikon of Daphni Monastery, Attica, Greece. Simple comparison with the external façade reveals that most of the cracks go through the thickness of the wall. Adapted from Miltiadou-Fezans et al. (2003)

to be visible.[3] Subsequently, sealing of cracks will be carried out after the drilling of injection holes along the cracks and after the insertion of injection tubes, as discussed below.

When the building mortar near the surface of masonry is generally in poor condition, deep re-pointing is necessary to allow for efficient injection. In case of local deficient building mortars, disintegrated joints should be raked out, followed by local joint re-pointing. The mix proportions of the mortar used for re-pointing should be prescribed by the structural design, taking into account the characteristics of the existing materials, ensuring air and vapour movement through masonry width, as well as accounting for the need to develop the necessary strength to resist the injection pressure.

Care must also be taken to seal the intrados of masonry openings (doors and windows) or possibly other internal or external cavities (niches, cupboards etc.), since otherwise leakage will occur in such areas.

In case of internal cavities like chimneys and gutters, which should not be filled with grout, it is recommended to fill them before grouting with very thin sand (having a $W_{nom}$ smaller than the one used for the design of the grout). In most cases, a maximum grain diameter < 0.5 mm, may be an effective solution. After grouting, the

[3] In many cases of heavily damaged masonries, the Structural design may prescribe stitching of cracks or even removal and replacing of deficient blocks, or even a removal and rebuilding of an entire part of masonry. Such interventions should be undertaken before grouting application.

**Fig. 10.3** The gutter at the external North façade of Hagia Eirini Church in Aiolou St. in Athens, Greece (see circled area) was filled with fine sand to avoid grout penetration, thus allowing for serving its initial function after restoration works

sand will be removed, allowing the internal cavities to serve their initial function. In Fig. 10.3 the gutter at the external North façade of Hagia Eirini Church in Aiolou St. in Athens is presented; this gutter was filled with very fine sand before grouting operation (Miltiadou-Fezans 2004).

### *10.2.3 Drilling the Holes—Grid of Injection Tubes*

Holes are drilled at adequate distances to form a grid, allowing both for the grout to be injected and for the air to escape. Moreover, these holes allow for a possible overflow to be "channelled" through these holes and eventually to be controlled.

Rotary bits are used; it has to be noted, however, that such bits have to be used carefully. As underlined also by Schuller et al. (1994) "rotary bits have the tendency to force dust and drill cuttings into the hole being drilled, thus sealing the crack from grout penetration". To avoid such difficulties, the aforementioned Authors recommend the use of a hollow-core drill bit, with vacuum chuck; this apparatus allows residue to be removed during drilling.

When opening these holes, percussion-drilling should never be used.

The following criteria apply in selecting the position of the drilled holes in which grout tubes will be fixed:

- The distance of consecutive mortar-joints, both horizontal and vertical, (since holes for grout-tubes are normally drilled through such joints)
- The thickness of the wall: Exceptionally thin walls need closer grout-tubes
- The injectability of the finally available grout, as compared to the actual permeability of the masonry
- The in-space correlation between possible cracks on inner and outer faces of the wall
- In the case of a multi-leaf masonry, the need for several tubes to penetrate the shell/infill interface, could also influence the distance between consecutive grout tubes, as well as the length of the internal part of each tube: some tubes should reach the middle of the infill material and other tubes should reach the interfaces between the external leaf and the infill material.

Nevertheless, modifications of initial decisions are expected after the pilot-application suggested in Sect. 10.5.2. Denser pattern may be used in some particular zones, as well as additional holes can be drilled in some locations if necessary.

As a rough approximation, however, the following practical rules may be useful.

It is suggested to drill holes preferably following the closest pattern, i.e., the equilateral triangular pattern,[4] forming rhombus of unequal diagonals, on both faces of masonry or cracked area (Fig. 10.4). Installation of tubes on only one face is also proposed in the literature, to avoid taking away of internal renderings. This may be the case when the renderings have an aesthetic value (see also Sect. 10.2.6). But, in the case of usual renderings, their removal is essential in order to inspect and survey the masonry, and carry out all necessary works for grouting application.

Depending on the walls geometry this staggered grid may be installed having the longer diagonal in the vertical direction, as presented in Figs. 10.4 and 10.5, or in the horizontal direction (Ashurst and Dimes 1990; Schuller et al. 1994).

In case the masonry is cracked and the width of cracks is small, holes should also be drilled at distances along the cracks, in order to ensure their direct filling with grout. In areas the cracks are wide enough, tubes may be inserted directly in the crack (Fig. 10.4a).

The location of holes on back face should *not* coincide with their location on the front face.[5] More specifically, whenever possible, the nodes of the back face grid should be arranged at mid-distances of nodes of the front face grid, both horizontally and vertically (Fig. 10.4a, b).

---

[4] Also proposed by Caleca e de Vecchi (1990) (as mentioned in Valluzzi 2003, p. 34), Van Rickstal (2000), Valluzzi et al. (2004).

[5] As proposed as well by Ashurst (1989).

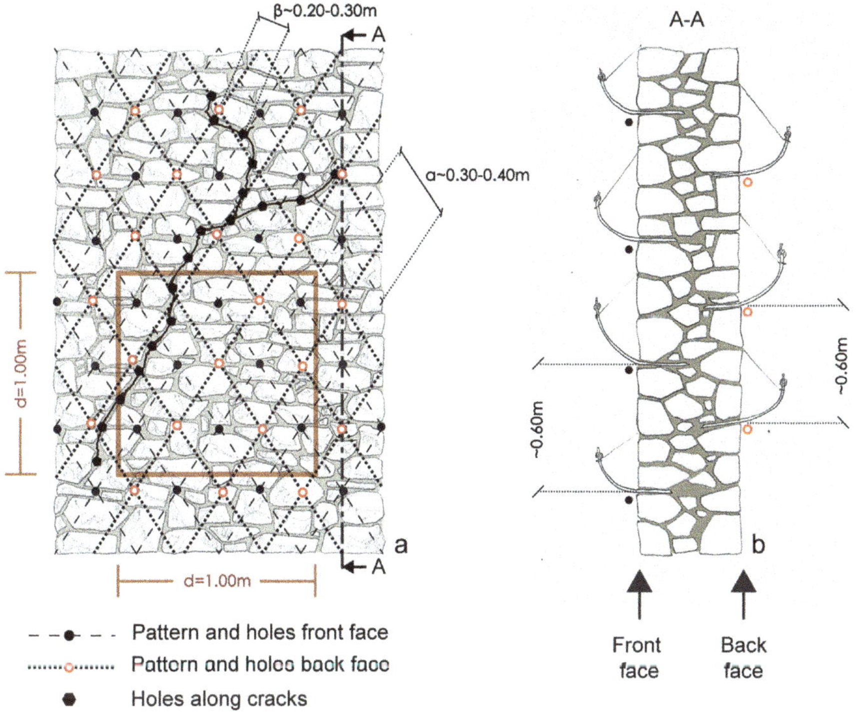


**Fig. 10.4** Schematic presentations: **a** The grid of grout holes on a front wall face, noted with filled circles for holes and dashed lines. On the same drawing the grid of the back face is shown with empty circles for nodes and dotted lines. **b** The grouting holes and tubes on a wall section (AA on **a**). In both **a** and **b**, the nodes of the back face grid are arranged at mid-distances of the nodes of the front face grid, both horizontally and vertically

Holes are normally drilled in mortar joints and inclined downwards, if possible.

The distance of consecutive holes depends on the construction type, the thickness of masonry, as well as on its pathology. In case of a very dense crack pattern the distance of the holes may be larger than in case of a less cracked structure.

That is why it is difficult to state rules of general validity.

Nevertheless, the installation of small number (2–4) of tubes per $m^2$ used in the past is inadequate, as experimentally studied by Baronio et al. (1992) in masonries with cracks of various widths and voids irregularly distributed. A possible explanation is given by Van Rickstal (2000), referring to the probability that the grout reaches a void of relatively larger volume, where lower grout pressure is established with adverse consequences for the filling of fine discontinuities.

Thus, a distance of injection tubes of 25 cm was proposed by Baronio et al. (1992), while Valluzzi (2003) has installed the injection tubes on wallettes of 40 cm width in a distance of 25–30 cm, only in one face of the wall. Some distant tubes were also installed in the other face, to control the progress of grouting. Similar procedure and grout tube distance was used by other researchers as well (Silva et al. 2014;

**Fig. 10.5** Grid of injection tubes in various masonries: **a** Rubble stone-brick masonry of the NE Range of Cells of Hosios Loukas Monastery, Boeotia, Greece. Adapted from Miltiadou-Fezans (2003). **b** Byzantine masonry of the Katholikon of Daphni Monastery, Attica, Greece, built according to the enclosed brick system (stones with bricks around them): white arrows indicate the special horizontal wires installed to facilitate the upwards fixing of tubes after injection. Adapted from Miltiadou-Fezans et al. (2008a). **c** Rubble stone masonry of a building in Thiva, Greece: yellow (circle) and black (rectangle) tape was used at the end of the tubes to indicate their in-masonry depth. Courtesy of Structural Design Office Epilysi

Giaretton et al. 2017; Luso and Lourenço 2019). Smaller distances (0.20–0.25 m) are proposed by Binda et al. (2003) for the injection of the walls of Noto Cathedral. In other Laboratory experimental works on wallettes (to a scale of almost 2:3, in which tubes were installed on both faces of the wall), the distance of holes was 0.2–0.3 m (Vintzileou and Tassios 1995; Toumbakari et al. 2005) and 0.15–0.20 (Vintzileou and Miltiadou-Fezans 2008).

In case of grouting application in real scale structures, using *grouts of adequate injectability* and *tubes on both faces* of the wall, a distance of holes approximately equal to 0.30–0.40 m is thought to be satisfactory (Schuller et al. 1994; Miltiadou-Fezans et al. 2014). Nevertheless, this distance should not be lower than the wall thickness.

Thus, for 0.3–0.4 m length of each triangle side in the aforementioned equilateral triangles pattern, the distance of the tubes in one direction (for example horizontally) would be also 0.3–0.4 m, whereas in the other direction (for example, vertically) it would be 0.50–0.70 m.

In such a configuration, a number of approximately 8–10 tubes are installed per square meter on each face (Fig. 10.4a), plus those directly inserted in large cracks. In case the distance of holes is smaller (example 0.25–0.30 m), 11–12 holes per $m^2$ may be necessary (Valluzzi et al. 2004).

However, as already stated, the size of blocks and their geometry, as well as the masonry pathology or the existence of surface decorations, may impose larger or smaller distances of holes.

### *10.2.4 Cleaning of Drilled Holes*

It is noted that, after drilling, a thorough cleaning of each hole should be carried out in order to remove dust and debris, by means of air current (air under pressure).

In all cases, during this operation, by observing the exit of the air out of adjacent holes, one may possibly record the several intercommunications between them. This is a very helpful indication for the existence of interconnected discontinuities, allowing to predict the expected grout internal movement. In areas where such an intercommunication is not detected, some corrective measures should be taken to possibly improve it. Additional holes should be drilled in one or all faces of the wall at various depths.

In Fig. 10.6a the internal façade of a masonry pier of the drum of the cupola of the Katholikon of Daphni Monastery is shown; since, the pier bears in the most of its surface a mural mosaic, the installation of masonry grouting tubes on its internal face, was possible only at its upper and lower part, as presented in Fig. 10.6a. In the area with the mural mosaic, fine tubes were installed by the competent Conservators (see Sect. 10.2.6). The cleaning of the holes with air in pressure revealed many intercommunication paths between the most of the holes of the upper part of the pier. In Fig. 10.6b the drawing of the survey of the positions of the holes of the injection tubes on the upper part of the pier is presented: full circles represent holes with a

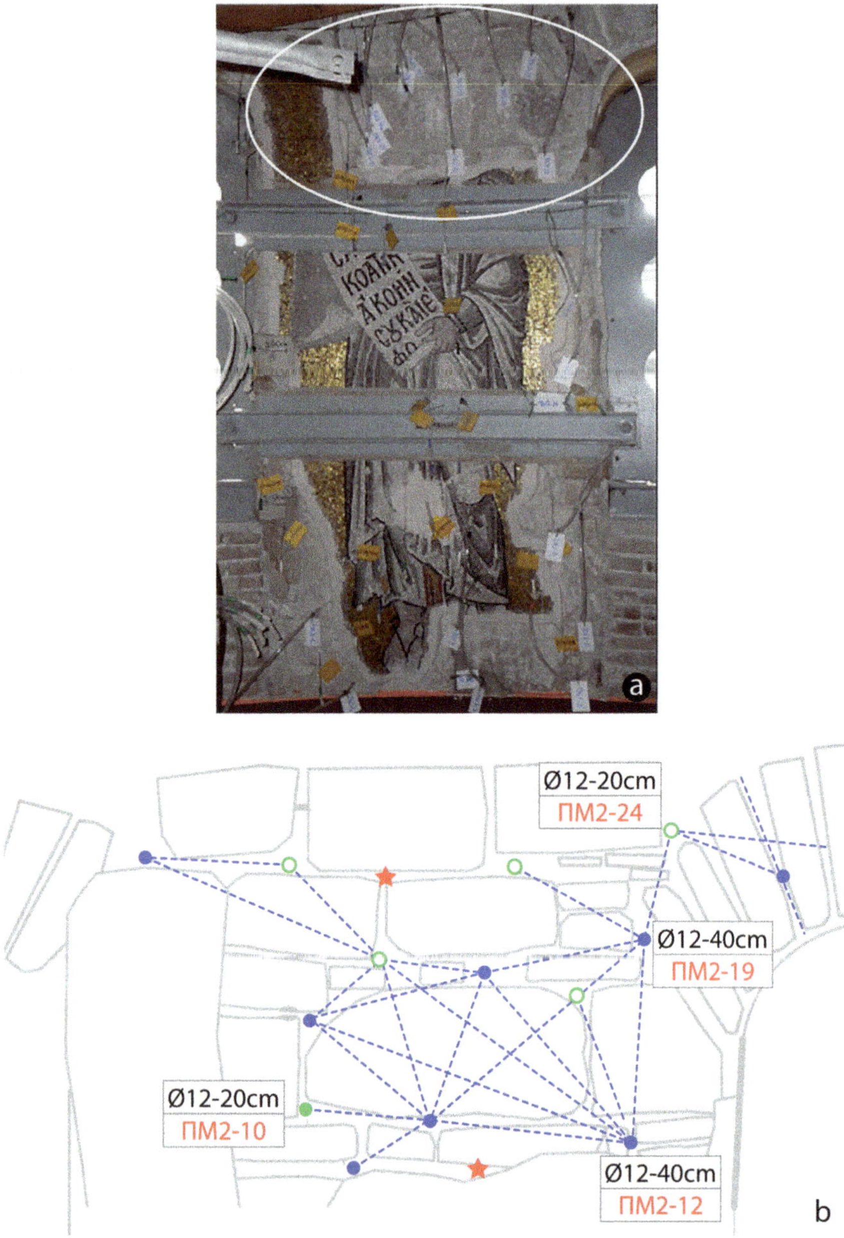


**Fig. 10.6** **a** Internal face of a pier of the drum of the cupola of the Katholikon of Daphni Monastery, Attica, Greece. **b** Drawing of the survey of the positions of holes of injection tubes on the upper part of the pier, and of their intercommunications. Full circles represent holes with a depth of 40 cm, and empty circles holes with a depth of 20 cm. Asterisks represent the holes which have no communication with other holes. Near to each hole, its name, its depth and its diameter is noted, as presented in selected holes on this drawing. Dashed lines between two holes show their intercommunication, detected with the use of compressed air. Reworked from Miltiadou et al. (2014). **a** Reproduced with the permission of the Ephorate of Antiquities of West Attica, Hellenic Ministry of Culture and Sports/Hellenic Organization of Cultural Resources Development

depth of 40 cm, and empty circles holes with a depth of 20 cm. Dashed lines between two holes show their intercommunication, detected with the use of compressed air. Asterisks represent the holes which have no communication with other holes. The name, the depth and the diameter of each hole are noted, as presented in selected holes on Fig. 10.6b. For the hole numbered ΠM2-12, an intercommunication with six adjacent holes was detected. As aforementioned, this type of information may be useful during injection procedure, as it is a first prediction of the possible paths of the grout inside the masonry during grouting.

In some cases of masonry of very good quality, without decorative plasters of other elements vulnerable to water infiltration, after the application of air under pressure, water may also be used to clean the holes and eventually check the intercommunication of discontinuities. Nevertheless, this may be a very delicate procedure, especially in presence of earth mortars; its application should be discussed with the responsible structural Engineer (see also Sect. 10.2.5).

### *10.2.5 Installation of Injection Tubes*

Subsequently, transparent plastic tubes are installed into the drilled holes and fixed in the wall, using the same mortar as the one for re-pointing.

The internal diameter of the plastic tubes is normally of the order of 10 mm to facilitate their insertion to the joints of masonry without extensive disturbance of the adjacent material.[6]

Half of the tubes should penetrate to a depth equal to 1/3 of the masonry thickness; the other half of them should go deeper into the 1/2 the thickness of masonry[7] (Fig. 10.7a).

In the exceptional case, where holes are drilled only on the one face of the masonry, tubes should penetrate to depths of 1/3, 1/2 and 2/3 of the thickness of masonry. Each category of these tubes, is characterized by a colour-mark on the protruding part of the tube, so that its depth is clearly recognized, during the injections (Fig. 10.5c).

In order to facilitate the flow of grout, the following measures are taken in shaping the in-masonry parts of each of these tubes (Fig. 10.7b):

- The end of the tube should be chamfered (45°), to allow the flow even in the case that the end of the tube would be in contact with some solid material
- The last inner part of the penetrated length of the tube should be perforated by two couples of holes, every approximately 5 cm.

---

[6] In some cases of thick loose rubble masonry, built with thick joints, when high consumption of grout is expected, larger diameters may be used, to facilitate the injection of a larger volume of grout (Van Rickstal 2000).

[7] These depths are needed in order to try to make grout available closer to vertical cracks opened at the interface of external leaf and internal infill of three leaf masonries, as well as to the cracks due to lateral dilatancy, near the middle of the thickness of the infill.

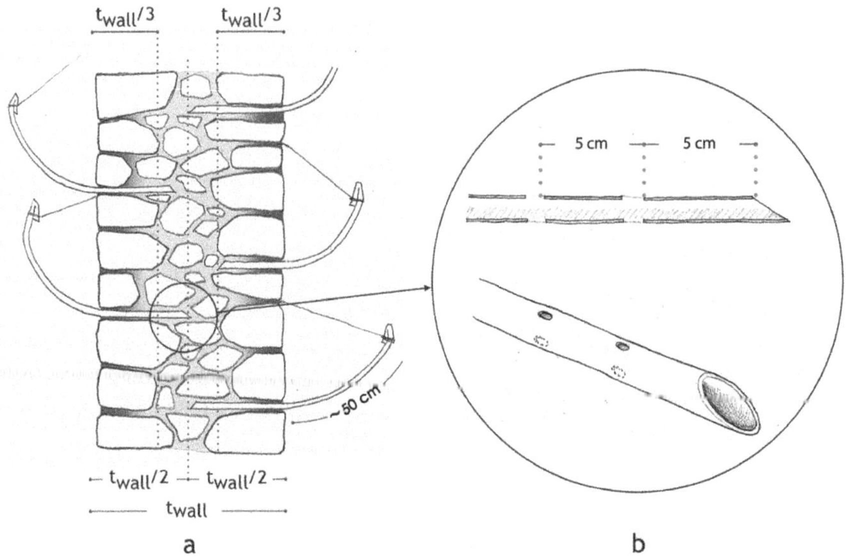


**Fig. 10.7** **a** Schematic presentation of the installation of tubes in a section of the wall. **b** Internal end of the tube chamfered (45°); two couples of perforated holes are presented

The protruding part of the tubes should be approximately 0.4–0.5 m long, in order to permit checking the consistency of the overflowing grout, as well as the continuity of the flow. These protruding parts of tubes should be smoothly curved upwards, and appropriately immobilized *after the injection* (Fig. 10.7a). In order to facilitate this immobilization after the injection, a practical solution may be to tie them on wires, purposely installed during the preparation of masonry and installation of tubes; this solution has been adopted in the case of the Katholikon of Daphni Monastery (Fig. 10.5b) and that of the Church of Panaghia Krina in Chios Island (Fig. 10.8). However, this is an optional solution[8]; any other solution is acceptable, provided that upwards curving and immobilization of tubes is achieved.

Plastic tubes reaching different depths in masonry, should be adequately marked as presented in Fig. 10.5c. They should be numbered consecutively from the bottom up to the top of the wall (Figs. 10.6 and 10.8). Their position and depth should be reported on appropriate drawings or sketches (Figs. 10.6b and 10.9). In case of large structures, a predefined way of tube numbering should be followed; a special code could be adopted comprising letters that relate to each face of the wall, accompanied by the number of the tube.

In Fig. 10.9 a drawing "in-progress" of the survey of the injection tubes of the North façade of the Katholikon of Daphni Monastery, prepared during the works is presented. At the upper part of the drawing only the survey of the position and number of the already installed tubes are shown, while at the lower part of the drawing some

[8] As shown in Fig. 10.5a, c, this solution was not adopted in these specific cases.

**Fig. 10.8** Injection tubes, adequately marked and numbered, installed on masonry elements of the Church of Panagia Krina in Chios Island, Greece, on: **a** an internal masonry pier, **b** the North façade. Reproduced with the permission of the Ephorate of Antiquities of Chios, Hellenic Ministry of Culture and Sports/Hellenic Organization of Cultural Resources Development

data regarding the grout consumption (see Sect. 10.7 of this chapter) have also been reported.

In order to reduce the dirtying of the surface of the wall because of the possible overflow of grout, it is suggested that small transparent plastic bags should be connected to the protruding ends of the tubes (Fig. 10.10); small ventilation holes should be provided on these bags.

It has to be noted that it is not suggested to proceed to an injection of water prior to the injection of grout. As studied by Miltiadou (1990), the excess of water leads to the formation of a thin layer of water around the in-situ materials, which does not permit the development of the bonding (mechanical and chemical) mechanisms; thus, bonding may be jeopardized. Moreover, as noted by Binda (1993), the danger of washing out of fine materials existing in the old mortars especially in case they contain clay components has not to be underestimated. Besides, wetting the mortars, especially the earth containing ones, results in undesirable loss of their strength and increase of their deformability, followed very often by expansion phenomena and creation of surface efflorescence due to soluble salt migration from the interior to the surface (Baronio and Binda 1983, as mentioned in Valluzzi 2003). Similarly, no water injection is used by Valluzzi (2003), Silva et al. (2014), Vintzileou and Miltiadou-Fezans (2008).

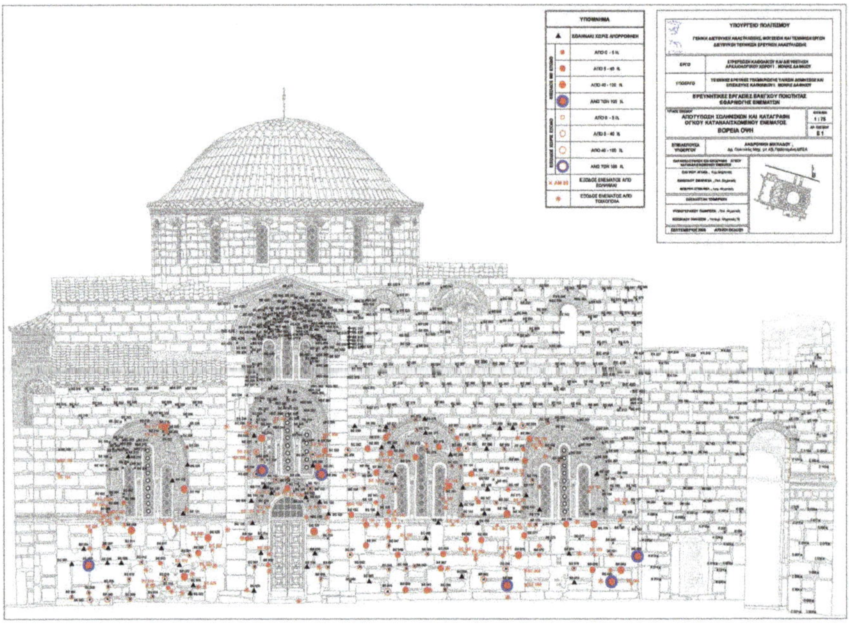

**Fig. 10.9** Drawing "in-progress" of the survey of the injection tubes of the North façade of the Katholikon of Daphni Monastery, prepared during the works

**Fig. 10.10** Transparent plastic bags are shown, connected to the protruding ends of the grout tubes in the masonry wall of the South façade of the Katholikon of Daphni Monastery, Attica, Greece. Reproduced with the permission of the Ephorate of Antiquities of West Attica, Hellenic Ministry of Culture and Sports/Hellenic Organization of Cultural Resources Development. Photo by N. Delinikolas

### 10.2.6 Installation of Fine Injection Tubes in Specific Cases

In case that a face of masonry is covered with frescoes or mosaics or decorative plasters, injection can be carried out without their removal. Additional holes should be drilled on the non-decorated face of the masonry, reaching depths equal to 2/3 of wall width approximately, and special care has to be taken by the competent Conservators for the appropriate preparation of the decorated surfaces for grouting injections.

However, it has to be noted that it is essential, even on the decorated face, to drill a minimum number of holes, mainly for the exit of the air and of the grout, but also for grout entrance, in case of wide cracks (Miltiadou-Fezans et al. 2008a). The presence of some holes and exit tubes in the decorated face is vital, in order to avoid possible damages due to grout pressure or uncontrollable leakage of the grout; thus, the areas reached by the grout may be checked, and the movement of the grout behind the decorated face be observed. These holes should be drilled on possibly existing cracks and on other areas to be selected by the competent Conservators (Anamaterou et al. 2017), in collaboration with the structural Engineer (Figs. 10.11 and 10.12). To this end, NDT's should be used to roughly determine the location of stones and joints,

**Fig. 10.11** Preparation for grouting of the mural mosaic depicting the Nativity of Virgin Mary of the Katholikon of Daphni Monastery, Attica, Greece. Thin grouting tubes of various diameters were used. Reproduced with the permission of the Ephorate of Antiquities of West Attica, Hellenic Ministry of Culture and Sports/Hellenic Organization of Cultural Resources Development. Photo by N. Delinikolas

**Fig. 10.12** Preparation for grouting of the fresco depicting Saint Andreas of the Katholikon of Daphni Monastery, Attica, Greece. Thin grouting tubes of various diameters were used. Reproduced with the permission of the Ephorate of Antiquities of West Attica, Hellenic Ministry of Culture and Sports/Hellenic Organization of Cultural Resources Development. Photo by N. Delinikolas

as well as of possible detachments underneath the surface (Côte et al. 2005, 2008); the use of local supports on the decorated face during injection may be necessary.

The diameter of the holes may be very small, depending on the width of cracks or other discontinuities; alternatively, other criteria imposed by the competent Conservators may be applied (Anamaterou et al. 2017). Special lime-pozzolan or hydraulic lime mortar compositions should be designed for the installation of the tubes on mosaics or frescoes, usually containing fine aggregates; their grading depends on the crack's width.

Moreover, it has to be noted that *small diameter* tubes (down to 0.8 mm) or even needles, may also be used for grouting cracked architectural members like the drums of the columns of Parthenon Opisthodomos in the Acropolis of Athens, presented in Fig. 10.13 (Miltiadou-Fezans et al. 2005).

Similarly, fine injection tubes may be selected to be installed on cracked masonry faces containing historic pointing mortars that have to be conserved in situ, or large building stones with fine cracks that should be strengthened (Fig. 10.14a, b).

**Fig. 10.13** Drums of the marble columns of Parthenon Opisthodomos in the Acropolis of Athens, Greece. **a** Fine diameter grout tubes and **b** needles were installed, in order to restore the drums with hydraulic grouts of high injectability. Reproduced from the Archives of Y.S.M.A./E.S.M.A. with the permission of the Committee for the Conservation of Acropolis Monuments and the Acropolis Restoration Service, Hellenic Ministry of Culture and Sports/Hellenic Organization of Cultural Resources Development

## 10.3 Main Characteristics of In Situ Grouting Equipment

The basic equipment used for grouting, consists of the following units: mixer, agitator, grouting pump, grout pipes, grout recorder (Fig. 10.15).

In what follows, a short and indicative description of this equipment is presented. Extended information is provided by the pertinent works of Bruce et al. (1997), Houlsby (1990), Weaver (1991), Jefferis (1982), Coulray and Carson (1982).

### *10.3.1 Mixer*

The mixing procedure of the grout is of a paramount importance to ensure adequate injectability. In fact, deflocculation of agglomerates, dispersion and homogeneous distribution of the grains of the solid phase in the water, as well as wetting of their entire surface are *sine qua non* conditions for high penetrability and adequate fluidity and stability of the mix (see Sect. 2.2.4b and 3.4 of this book).

**Fig. 10.14** Injection tubes of various internal thin diameters installed in the external façade of masonry of the Katholikon of Dafni Monastery, Attica, Greece: **a** in areas with old byzantine pointing mortars and **b** in cracked building stones that also had to be injected. Reproduced with the permission of the Ephorate of Antiquities of West Attica, Hellenic Ministry of Culture and Sports/Hellenic Organization of Cultural Resources Development

For the ultrafine materials used in modern masonry grouting, simple mixers by agitation (paddle mixers) are not satisfactory. Instead, the mixing process must create high shearing of the grout, in order to break flocculation of fine particles.

To this end, high turbulence colloidal mixers should be used for grout mixing (Fig. 10.15). They essentially operate as follows:

- First, water is fed in the mixing *tank*; then, solid materials are added;
- The slurry passing through a narrow opening, comes down to a "mixing and circulation pump", where, by means of an appropriate rapidly rotating device (>1500 rpm) producing a highly turbulent flow, is submitted to shearing.
- This slurry is then conveyed again to the mixing tank, several times, in order to be re-submitted to the same mixing procedure. Thus, uniform and deflocculated suspensions are produced.
- Mixing time depends on the grout materials used and on the characteristics of the mixer.

High turbulence mixing may be adequate when hydraulic lime, cement, hydrated lime or/and natural pozzolans are added to the mix.

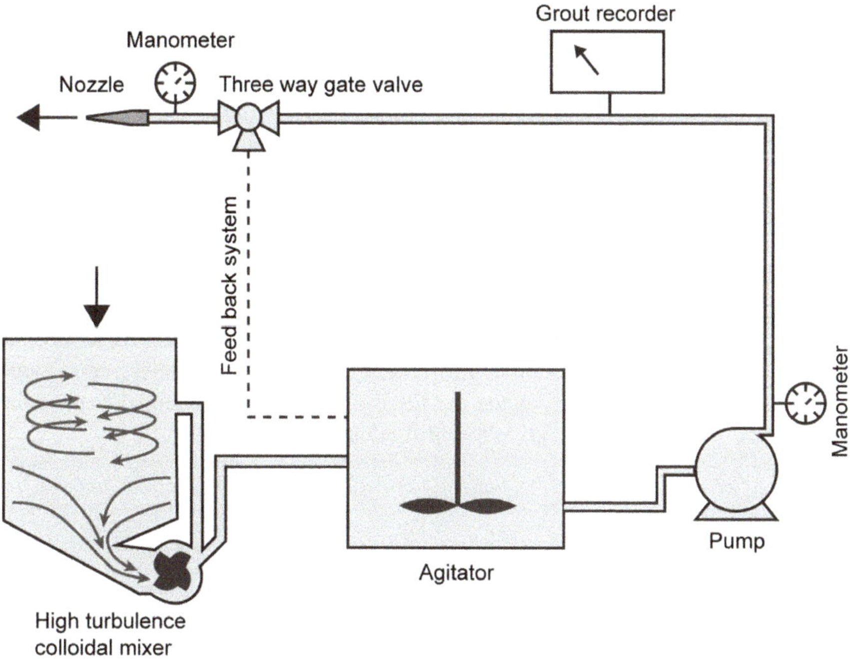


**Fig. 10.15** Schematic presentation of grouting equipment

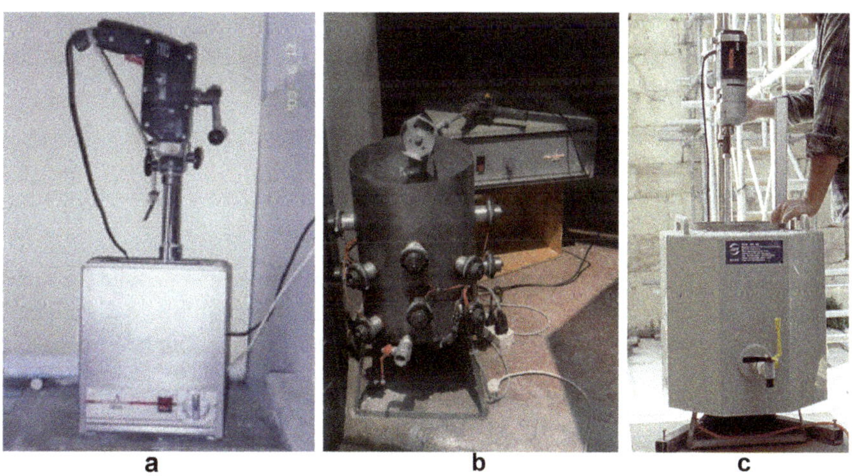


**Fig. 10.16** **a** Commercial ultrasound disperser (4.5 lt, 20 kHz frequency). **b** Prototype ultrasound apparatus having a capacity of 20 lt developed in LCPC. Reproduced from Miltiadou (1990). **c** Similar apparatus manufactured for the grouting interventions on cracked columns of the Parthenon of the Acropolis of Athens. Adapted from Miltiadou-Fezans et al. (2002)

An alternative and possibly better solution for adequate mixing of modern grouts is based on ultrasound dispersion.

The use of small volume commercial ultrasound dispersers of 20 kHz frequency (power of 250 W) are proved to be very efficient for relatively small quantities of grout up to 5lts, associated however with a simple paddle mixer. For larger volumes, a prototype apparatus was developed in LCPC[9] (Miltiadou 1990), having a capacity of 20 lt. A similar apparatus was also manufactured for the grouting interventions on cracked columns of the Parthenon of the Acropolis of Athens (Miltiadou-Fezans et al. 2005), as well as on the mosaic's substrata of the Katholikon of Daphni Monastery (Miltiadou-Fezans et al. 2007, 2008a).

However, ultrasound mixing equipment for larger quantities of grouts for normal masonry grouting may not yet be commercially available. Nevertheless, ultrasound mixing is absolutely necessary when densified silica fume is contained in the solid phase of the grout (Paillère et al. 1989; Miltiadou 1990).

### *10.3.2 Agitator*

After the appropriate mixing time, the grout is transferred to an agitator with slowly revolving multiple paddles, in order to keep the solid particles in suspension. The agitator serves as an intermediate vessel between the mixer and the grouting pump, in order to ensure continuous supply of grout. To this end, the agitator should be of sufficiently larger volume than the mixer. An appropriate filter is placed at the agitator's exit, in order to retain any impurities possibly derived during the production process of the grout.

### *10.3.3 Grouting Pump*

For the grout injection, pumps mechanically, electrically or compressed-air operating, are used. Only in small-scale applications, when electricity is not available, manually operated pumps can also be used, equipped however with control system and manometer.

Several types of pumps might be used, such as the following ones:

- Cavity (or helical screw) pumps, of high output, under a moderate but constant pressure.
- Piston pumps (mostly double-acting ones), of moderate flow rates and high-pressure capacity.
- Plunger pumps, of a large range of output and pressure characteristics.

[9] Laboratoire Central des Ponts et Chaussées; since 2011 LCPC is part of the «Institut français des sciences et technologies des transports, de l'aménagement et des réseaux» (IFSTTAR).

If the quantities of grout needed are relatively small, compressed air may be used, acting on grout contained in an appropriate pot; thus, by adjusting the input air, pressure is more easily controlled, and a relatively constant flow is achieved. Besides, such a "pressure pot" can be easily installed closer to the entrance tubes, so that the hydrostatic pressure is negligible.

In every case, the outflow pressure of the pump should balance the following pressures and frictions:

- The pressure needed at the nozzle of the entrance tube (see Fig. 10.15), in order to overcome the expected frictions in the body of masonry; this pressure, according to our experience, is roughly equal to 0.5–1.0 atm.[10] (See also Sect. 10.4.2a).
- The friction losses along the pipes from the production unit up to each specific nozzle.
- The hydrostatic pressure, when the grout goes up to a level higher than the position of the pump. For instance, considering that the apparent density of the grout is approximately 1.5 g/cm$^3$, a minimum pressure of 7.5 atm is needed for the grout to reach a height of 5.0 m from the level of the pump. It is therefore apparent that the pressure-control at each nozzle is *indispensable.*

To this end, the outflow pressure of the pump should be continuously checked by means of an intermediate manometer (Fig. 10.15). Furthermore, the pump should be equipped with a control system that does not permit the pressure to exceed an upper limit set in advance.

Moreover, it is apparent that, in order to avoid local masonry failures, the pressure control at each nozzle is absolutely indispensable. To this end the outflow pressure at each nozzle is also continuously checked by means of a nozzle manometer fixed at the end of the grout pipe line (Fig. 10.15).

Both the manometers of the pump and the one of the nozzles should be of sensitivity of 0.1 atm. Moreover, in order to avoid any undesirable side effect due to sudden overpressure, an adjustable "three-way-gate-valve" has to be installed before the manometer of the nozzle (Fig. 10.15); hence, if the pressure becomes too high, the three-way-gate-valve will direct the grout through a pipe into a vessel (or to the agitator if a feedback system is installed) and a pressure release will be accomplished.

### *10.3.4 Grout Pipe Lines*

Flexible pipes are used, connected with tight fittings. The adequate length of these pipes appropriately estimated should be made available to reach each grouted area. Easy flow of the grout requires sufficiently large diameter of these pipes. However,

[10] This range of pressures is suggested by other authors too (e.g. Valluzzi 2003, p. 34; Van Rickstal 2000, p. 169), as opposed to previous opinions favouring higher values. It is reminded that a considerable increase of grout consumption cannot be easily achieved by a mere increase of the injection pressure; instead, a finer and more fluid grout should preferably be envisaged, respecting all injectability criteria.

in order to minimise the risk of possible instability of the grout, very large pipe-diameters should be avoided.

### *10.3.5 Grout Recorder*

Pressure, flow rates and grout quantities should be continuously recorded; grouting is a very delicate operation—thus, an appropriate recorder is needed; this is an instrument of *fundamental* importance for a successful and rationally controlled grouting operation. Such recorders allow for the determination of the grout consumed at each masonry area. Thus, the Engineer may identify the areas inadequately reached by the grout, for possible further measures to be taken (such as possible modification of the tubes-grid or of the grout injectability characteristics, etc.). On the other hand, the knowledge of the volume of the grouted voids is necessary for the estimation of the strength of the masonry after grouting (see Sect. 6.3.2.1). It is important to note that these recorders should be of a high sensitivity, specifically needed in the case of masonry grouting (i.e., appropriate for low pressures and low flow rates).

## 10.4 Preparation of Grout and Execution of Injections

A grouting procedure comprises the preparation of the mix to be injected, the execution of the injections and, finally, the cleaning and finishing of the surface of the grouted masonry.

### *10.4.1 Mixing Procedure*

Mixing is recognized to be a stage of fundamental importance for a successful grouting, as explained in Sects. 2.2.4.2, 3.4 and 10.3.1.

Before mixing, all materials have to be separately weighed and should be ready to be introduced in the mixer, as the mixing time is rather short to allow for delays. It should be reminded that the mix proportions prescribed in the design are expressed in percentages or parts by weight. Thus, in order to facilitate the weighing at the worksite, the weight of each material is practically reported to the weight of one bag of the material participating in the composition in a higher percentage. A numerical example is given in Table 10.1.

Mixing should operate in the shade. Table 10.2 shows the addition-sequence and the suggested mixing times of materials for various grout compositions. In order to avoid overheating, the total mixing time should not exceed 8 min.

The mixed grout quantities are subsequently transferred to an agitator, as described in Sect. 10.3.2.

**Table 10.1** A numerical example of the weight of each material reported to the weight of one bag of hydraulic lime participating in the composition

| Grout composition | | Weight per one bag of NHL5 |
|---|---|---|
| NHL5 | 90% | 30 kg (one bag) |
| Pozzolan | 10% | 3.3 kg |
| Superplasticizer | 1% | 0.333 kg |
| Water | 80% | 26.7 kg |

Superplasticizer based on polycarboxylic ether. Water and superplastisizer percentage refer to the grout solids total mass

**Table 10.2** Sequence of adding, and suggested mixing times of materials for various grout compositions

| Examples of grout compositions | Sequence of adding materials in the mixer | Mixing times (min) | Total mixing time (min) |
|---|---|---|---|
| Cement, Lime and Pozzolan | (1) Water and Superplasticizer | $\sim \frac{1}{4}$ | 6 |
| | (2) Lime in powder | 2 | |
| | (3) Pozzolan | 2 | |
| | (4) Cement | $\sim 1\frac{3}{4}$ | |
| Natural Hydraulic Lime | (1) Water and superplasticizer | $\sim \frac{1}{4}$ | 4 |
| | (2) Natural Hydraulic Lime | $\sim 3\frac{3}{4}$ | |
| Natural Hydraulic Lime and Pozzolan | (1) Water and Superplasticizer | $\sim \frac{1}{4}$ | 4 |
| | (2) Pozzolan | 2 | |
| | (3) Natural Hydraulic Lime | $\sim 1\frac{3}{4}$ | |

A quality control of the mixing procedure is needed, as described in Sect. 10.5.1.

## 10.4.2 Injection Procedure

After the preparation of the grout (Sect. 10.5.1), a trial grouting application is carried out, as described in Sect. 10.5.2.

Before grouting, especially in summertime, the external façade of the areas to be grouted is thoroughly wetted with water, for two reasons: first in order to reinstate lost humidity of the surface material of masonry, and second to prevent the formation of strong grout spills on the façade of the building.

Subsequently, the injection operation starts from the bottom, proceeding *along the length* of the structural element to be grouted–up to its top. Care has to be taken to avoid injecting the masonry through injection holes in a vertical order, since, due to the hydrostatic pressure, there is a risk to develop inside the masonry pressures higher than its tensile strength.

The first "entrance-tube" to be injected is located around the middle of the wall-length,[11] preferably in the first series of tubes, starting from the bottom. The following rules are suggested.

1. During injection, pressure should be constantly checked at the "entrance-tube"; it should remain lower than the limit set by the relevant study. Usually, the pressure at the entrance tube is recommended to be kept lower than approximately 1 atm. In some cases of good quality masonry parts situated in low levels, the pressure may be allowed to reach 1.5 atm. On the contrary, in case of masonry bearing mosaics or frescoes or other valuable decorative plasters, the pressure is kept at lower levels, usually of 0.5–1.0 atm. The height of masonry to be injected the same day, has to be conservatively reduced to 2 m, taking into account the possible overpressure.
   In case of low-tension strength masonries, in order to avoid excessive internal pressure of the grout that could damage them, grouting should not be carried out to a height of masonry exceeding approximately 1 m per day. If any indication of local damage appears, immediate measures of reduction of grout pressure should be taken. In extreme cases of very low strength masonry, transversal support of masonry faces should be provided.
   Generally, after the application of grouting in one wall up to a certain level, grouting continuous first to all the other perimetric walls of the building until approximately the same level in height. Subsequently, grouting continuous to a higher level of the first wall.
2. When grouting is in process in one "entrance-tube", the grout supply should not be interrupted, except if pressure at the nozzle reaches the maximum accepted value, e.g., 1 atm. If a relatively long interruption happens, water contained in the already injected grout may be absorbed, producing possible clogging.
3. Normally, during the operation of injection at one "entrance-tube", the grout comes out from some of the tubes (this time called "outflow-tubes") in the vicinity of the "entrance-tube" (at the front or the back face of the wall). These outflow-tubes are not sealed, unless the following conditions are fulfilled:

   - The outflowing grout has approximately the same consistency as the injected one, without however containing air-bubbles.
   - The grout flows constantly throughout the end of the tube. If the grout merely appears in the protruding tube, but does not move further, this tube should not be sealed; it will remain under observation (see following point 5).

   If these conditions are fulfilled, the outflow-tube should be firmly tied, so that the grout is kept under pressure inside the masonry, until its hardening (Fig. 10.17). This sealing may be effectuated by means of a strong wire after full bending of the filled tube; alternatively, this is done by means of special rings.

[11] As also proposed by Van Rickstal (2000).

**Fig. 10.17** Grouting tubes after the injection, smoothly curved upwards and appropriately tied on purposely installed wires: **a** East façade of the Gate of the East entrance of Daphni Monastery, Attica, Greece. **b** West face of the N pier. **c** West face of the S pier. Reproduced with the permission of the Ephorate of Antiquities of West Attica, Hellenic Ministry of Culture and Sports/Hellenic Organization of Cultural Resources Development

In some cases, despite the good preparation of masonry, overflow of grout takes place at locations where no tubes had been installed. In such cases, if the leakage is moderate, a ready-made low water content paste (of pozzolan, of clay or of paper pulp) should be applied. If, however the leakage is more intensive, dry fine materials (pozzolan or clay) should be applied and firmly retained for some minutes on top of the leaking area. Occasionally, in the case of large leaking quantities, layers of fibres (e.g. cotton or wool swab) may be applied on the leakage area, covered by semi-dry paste of the aforementioned binders. Finally, an appropriate cleaning of masonry surface should then be carried out.

4. The injection which was started at an "entrance-tube" is considered as terminated, when there is no more absorption of grout, and the pressure is considerably increasing (practically above 1 atm). At this stage, the entrance-tube is also sealed.
5. If during the previously described operation, the injected grout appears in a neighbour tube but it does not outflow or even it moves back (see point 3 above), an injection should be carried out through this specific "quasi-outflow" tube, as soon as possible. Thus, it is hoped that the adverse conditions that have produced the previous unsatisfactory behaviour will be overcome.
6. After the completion of the grouting of an "entrance-tube", the following procedure is suggested:

   - A new "entrance-tube" is selected at the same level as the previous one, alternatively to the right or to the left side, among those which are not already sealed.
   - The location of the new "entrance-tube" will be selected on the basis of the location and the extent of grout-flow observed during the previous injection stage.
   - The same procedure will be applied at the same level of the other face of the wall, before the continuation of grouting at the other perimetric walls at a similar level, and then at higher levels.

7. Some points of basic importance should be made at this stage, assisting the Engineer in her /his everyday decision making:

   - Tubes where a sufficient outflow was observed, do not need to be grouted; thus, their protruding parts were sealed, as per § 3 above.
   - In an area where the intake was systematically very low, as compared to neighbour similar areas, or even when a refusal was observed, the Engineer may need to check again if the installed tubes are sufficiently clean, i.e., not containing obstructing debris. Subsequently, in such areas, additional holes should normally be opened, and a new series of grouting is to be carried out.
   - Special care should be taken in the case of the tubes located along discrete cracks. The movement of the grout along the crack and the filling of the cracked area has to be carefully monitored. Here again, in case that law intake and/or refusal is observed, checking of tubes obstruction has to be carried out, together with opening of additional holes and repetition of injections.

  - When a considerably high intake is observed, and there is not any outflow from the adjacent tubes, care has to be taken to find a possible well-hidden leakage (e.g., through possible internal gutters or towards an underground leak). In case such a leakage is not identified, the injection has to be stopped. As a possible remedy to this problem, a grout of lower penetrability (containing fine sand[12] ~25% of the solid phase) could be used; obviously, a new series of injection tubes has to be installed for this purpose.

8. It is recommended that grouting operations should not be carried out under extreme atmospheric temperatures. More specifically, temperatures higher than 32 °C are not recommended. The risk of freezing should also be avoided by taking into account recent meteorological forecasting of expected lowest temperatures during the night; pertinent recommendations regarding concreting conditions under low temperatures, may be helpful in this respect. A practical rule is that temperatures lower than 5 °C should be avoided.
9. Each day, during the injection, surface cleaning should be contacted to remove any spilled grout from the surface. Immediately after the completion of the works of every grouting day, an appropriate cleaning of the surfaces of the wall should be carried out to remove all remaining surface stains by means of water and a stiff brush.

### *10.4.3 Finishing of the Masonry Injected Face*

Soon after the completion of every day grouting and before hardening of the grout (one to three days), all protruding tube ends should be meticulously taken out of the wall; and after a final cleaning of the surface of masonry, the holes left in the joints should be repaired, using a mortar similar to the original one or to the re-pointing mortar. In case the tubes cannot be removed, they should be cut as deep as possible. If a general (new) re-pointing of the masonry is planned, it will also cover the areas of these cut ends. Otherwise, local replacement of the mortar of the joints is necessary in the areas of theses cut ends. Appropriate materials and techniques should be used to ensure in-time stability of these local repairs.

It is worth to remind here that after the removal of injection tubes, the appropriate cleaning of the faces and the repair of the holes, no traces are left of the application of this strengthening technique. Figures 10.18, 10.19 and 10.20 present three examples of this final stage.

[12] Having max grain diameter ~0.5 mm.

**Fig. 10.18** The North Façade of the Katholikon of Daphni Monastery, Attica, Greece: **a** Before grouting, after installation and numbering of the injection tubes. **b** After the completion of grouting, removal of injection tubes and local re-pointing of masonry joints. Reproduced with the permission of the Ephorate of Antiquities of West Attica, Hellenic Ministry of Culture and Sports/Hellenic Organization of Cultural Resources Development. Photos by N. Delinikolas

**Fig. 10.19** The mural mosaic depicting Archangel Gabriel of the Katholikon of Daphni Monastery, Attica, Greece: **a** Before grouting, after the installation of grout tubes. **b** After completion of grouting, removal of tubes and aesthetic restoration of mosaic surfaces by the competent Conservators. Reproduced with the permission of the Ephorate of Antiquities of West Attica, Hellenic Ministry of Culture and Sports/Hellenic Organization of Cultural Resources Development. Photos by N. Delinikolas

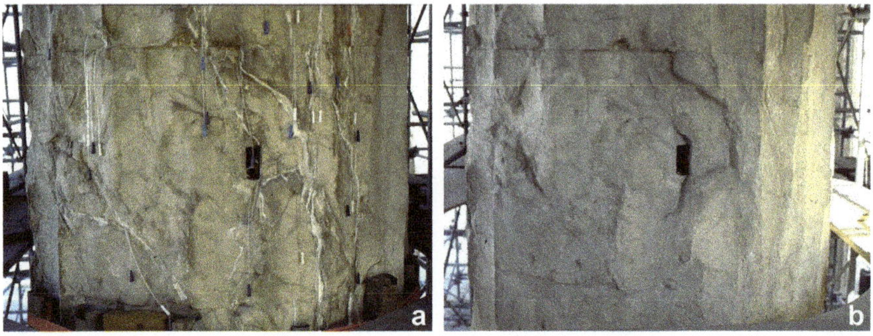


**Fig. 10.20** The column drum 2.4 of the Parthenon Opisthodomos of the Acropolis of Athens, Greece: **a** Prepared for grouting with the installation of very fine grout tubes. **b** After completion of grouting, removal of tubes and aesthetic restoration of drum surfaces by the competent Conservators. Reproduced from the Archives of Y.S.M.A./E.S.M.A. and with the permission of the Committee for the Conservation of Acropolis Monuments and the Acropolis Restoration Service, Hellenic Ministry of Culture and Sports/Hellenic Organization of Cultural Resources Development

## 10.5 On-Worksite Checking of the Prescribed Grout Design Data

Because of the sensitivity of the grout design due to possible differences of the real in situ conditions as compared with the in-laboratory ones, a checking of the suitability of the initial design is indispensable, before starting grouting operations.

### *10.5.1 Pilot Production of Grout in the Worksite*

To this end, a pilot production of grout using the in situ available materials and equipment is needed. Quite often, slight modifications of the mix proportions are necessary, related mainly to the percentage of water and admixtures. In fact, available mixing equipment, as well as even slight scatter in the physical-chemical properties of the raw materials (inherent even to industrial products) may significantly affect the percentage of water and admixtures that should be used. That is why the pilot production of the grout comprises the in-situ measurements of the injectability characteristics of the grout, as presented in Sect. 10.6.2.2, and the comparison of the values obtained with those prescribed in the job specifications. If the injectability characteristics satisfy the criteria set by the job specifications, the grout composition is considered adequate to be used in the in-situ pilot application (see Sect. 10.5.2). Otherwise, the appropriate adjustments should be made before the in-situ pilot application.

The pilot production of grout in situ should preferably be carried out at least one month before the scheduled date of starting the grouting operations. Thus, sufficient time will be available, not only regarding injectability checks, but also for mechanical and durability considerations (see Sect. 10.6.2).

Two examples of such checking may be of interest.

- The first example regards the following grout composition: 90% NHL5, Pozzolan 10%, SP 1% and water 80% (water and superplastisizer percentage refer to the grout solids total mass). Table 10.3 shows the values of various grout characteristics, as measured in the Laboratory and those preliminarily checked in the worksite. In this case, the suggested composition remained unchanged in the worksite (Miltiadou-Fezans et al. 2007, 2008a).
- The second example is a case that may frequently happen, due to (occasionally inevitable) variation of the properties of one of the basic binders. The suggested ternary grout contained 30% white cement, 20% hydrated Lime "A" in powder, 50% Pozzolan, 1% SP and 72.5% water (water and superplastisizer percentage

**Table 10.3** A numerical example of various characteristics of a grout to be used in a specific work, as measured in the Laboratory and preliminarily checked in the worksite

| Grout properties | Production of the grout | | Acceptable values | |
|---|---|---|---|---|
| | In lab (high turbulence laboratory mixer) | In worksite (high turbulence colloidal mixer) | | |
| Bleeding % | < 1% | 1% | < 3% | |
| Fluidity factor $F_l \times 10^3$ mm/s | 0.88 | 0.89 | > 0.7 | |
| Flow time: $t_{500\ ml,d=4.75}$ (in s), of 500 ml of grout (out of 1000 ml inserted) using a Marsh cone with 4.75 mm nozzle diameter, just after mixing | 21 | 22 | $20 < t_{500\ ml,d=4.75} < 25$ | |
| Apparent density (g/cm$^3$) Just after mixing | 1.51 | 1.50 | 1.5 + 1% | |
| Apparent density (g/cm$^3$) From outflow tubes | – | – | 1.5 ± 5% | |
| Flexural Strength 28 days (MPa) (3 specimens 40 × 40 × 160) | 1.30 | 1 | ~1 | Strength in worksite ~70% Strength in Lab |
| Compressive strength 28 days (MPa) (6 specimens from flexural test) | 2.50 | 2.0 [a] | ~2 | |

Reworked from Miltiadou-Fezans et al. (2007, 2008a)

[a] 1.5 MPa was found for the 28 days compressive strength, as measured in a second Lab

**Table 10.4** Another numerical example of various characteristics of a grout, as measured in the Laboratory and preliminarily checked in the worksite for two different hydrated limes

| Grout properties | Type of lime contained in the grout | | |
|---|---|---|---|
| | Lime A (finer) | Lime A (finer) | Lime B (coarser) |
| | Laboratory | Worksite | Worksite |
| Bleeding | 3% | 3% | 7% |
| Fluidity factor $F_l \times 10^3$ mm/s | 0.77 | 0.84 | 0.76 |
| Flow time: $t_{500\ \mathrm{ml},d=4.75}$ (in s), of 500 ml of grout (out of 1000 ml inserted) using a Marsh cone with 4.75 mm nozzle diameter (just after mixing) | 25 | 23.3 | 23 |

Reworked from Miltiadou-Fezans et al. (2006)

refer to the grout solids total mass). In the worksite, a different Lime "B" was purchased, which however produced unacceptably high "bleeding" values (see Table 10.4). Thus, the use of Lime "B" was rejected; Lime "A" was then brought in the worksite, with the satisfactory test results shown in Table 10.4 (Miltiadou-Fezans et al. 2006).

### 10.5.2 Pilot Masonry Application of Grouting

Finally, the verification of the grout-design in the worksite should also include a pilot injection operation on a representative wall, where all the preparatory works described in Sect. 10.2 were carried out, as also proposed by the "Guide de maîtrise d'ouvrage et de maîtrise d'œuvre: Ouvrages de maçonnerie (2006)". This pilot injection operation is carried out for the following reasons:

- to check (i) the adequacy of the preparatory works (sealing of cracks and other voids, density of injection tubes grid, cleaning of holes, installation of tubes, etc.), (ii) the level of injection pressure, and (iii) the capacity of the grout to flow through the masonry, while conserving its fluidity, stability and apparent density characteristics up to the most distant outflow-tube.
- To roughly estimate the volume of grout injected, normalized to the volume of the injected masonry, and to evaluate the effectiveness of grouting after a certain period of time, by means of one or more of the methods discussed in Sect. 10.8, as prescribed in the job specifications.

A masonry pier between two openings may be selected to this purpose, provided it has a considerable length.

In this pilot area, the works described in Sect. 10.4 will be executed (grout preparation and injection). The following matters have to be carefully examined:

- If the grout penetrates and outflows as prescribed in Sect. 10.4.2 and checked as in Sect. 10.6.2.3 (acceptable bleeding and density values of outflow-grout), the quality of grout is considered satisfactory. Otherwise, the grout composition and preparation should be re-checked, or, possibly, the initial mix-design should be reconsidered.
- If the volume consumed, normalized to the volume of the pilot wall, is of the order of the expected grout consumption prescribed by the relevant job specifications ($\pm 20\%$), the pilot operation is considered satisfactory.
- If the grout volume consumed substantially differs from that included in the job specifications, three cases may be distinguished.
  - When the grout outflows from the adjacent tubes, as prescribed in Sect. 10.4.2, and is checked as in Sect. 10.6.2.3 (acceptable bleeding, fluidity and apparent density values of outflow-grout), then a new pilot application should be organized in a different area of the masonry.
  - In case of higher intake without outflow from adjacent tubes, the previsions described in Sect. 10.4.2§7 should be followed.
  - If, the intake is substantially lower, and there is not a systematic outflow from adjacent tubes of grout or the grout quality checks (see Sect. 10.6.2.3) are not satisfactory, both the grout has to be re-examined, as well as the adequacy of preparatory works (density of grid of injection tubes, cleaning and installation of tubes, etc.).

In case there is not any estimation of the expected grout consumption in the job specifications, the following max indicative values could be considered: In low quality two-leaf masonries a possible max-value might be of the order of 8% of the volume of the wall. In three-leaf masonries such a max-value might be of the order of 15% of the volume of the wall.[13]

## 10.6 Quality Control of the Grout and of the Injection Procedure During the Execution of Works

### *10.6.1 Visual Inspection*

**Before** the injection, visual inspection of working place, working conditions, appropriate re-pointing and proper installation of tubes should be carried out.

**During** the injection of the grout, quality control according to the guidelines briefly presented in Sects. 10.2–10.5, should be carried out at regular time intervals,

[13] This upper level was found in the literature by many authors experimentally injecting three-leaf masonry walls. In this connection the “Fiche d’aide à la rédaction des documents de marché – Consolidation des maçonneries par injections de coulis minéral naturel” [FARCC No 03.0412,. 01. 01, Ministère de la Culture et de la Communication, Wallonie, Belgique (s.t)], note the following “the normal quantity of grout to be injected varies from 8 to 10% of the volume of the wall”.

in order to detect any possible errors. Thus, possible measures to remedy such errors, will be taken early enough.

**Immediately after** the completion of the daily works, a final inspection of the work-place may also identify possible omissions or failures.

### *10.6.2 Checking Grout Characteristics*

Fluidity, stability and apparent density of the grout should be frequently tested at the worksite, according to the instructions of the Engineer, and at least twice per day (Fig. 10.21). Apparent density and, if possible, bleeding and fluidity should also be tested twice a day on some grout-outflows, collected from distant exit tubes: Possible density changes of the grout (due to water absorption as it moves through porous material) will be detected. Similarly, measuring the bleeding of this outflow may be profitable in several ways, such as possible incidents resulting in increase of bleeding or checking of bleeding reduction as a hint about possible decrease of fluidity. Finally, the measurement of fluidity of the outflow- grout offers a direct information of the quality of grout in-situ.

Moreover, mechanical properties should also be tested (i) in the beginning of the project on samples taken during the pilot production of grout, (ii) at every new delivery of raw materials and (iii) on a regular basis during the works, as prescribed by the job specifications.

All measurements should be noted in details on the special job-calendar, accompanying the general project-calendar of the works. An example is presented in Fig. 10.22.

Hereafter, a description of these quality control methods is presented.

a b c

**Fig. 10.21** Testing of grout at the worksite: **a** Fluidity (Marsh cones of 3 mm nozzle diameter, to the right, and of 4.75 mm nozzle diameter, to the left). **b** Bleeding (NF P 18-359). **c** Apparent density measurement

| DAILY PROTOCOL FORM OF IN-SITU GROUT PROPERTIES | | | |
|---|---|---|---|
| Project:<br>Date:<br>Grout composition:<br>Number of batches: | Drawing or photo to note daily the position of grouting application | | |
| **1. Fluidity control** | | | |
| | Batch number | Batch number | Batch number |
| **1.1 Fluidity Factor Test (Marsh cone with $d_{nozzle\ orifice}$ = 3mm)** | | | |
| Flow time of 100ml of grout after mixing | | | |
| Calculated Fluidity Factor of grout after mixing | | | |
| Flow time of 100ml of outflow-grout | | | |
| Calculated Fluidity Factor of outflow-grout | | | |
| **1.2 Flow time (sec) from Marsh cone with $d_{nozzle\ orifice}$ = 4.75mm** | | | |
| Flow time of 500ml of grout after mixing | | | |
| Flow time of 500ml of outflow-grout | | | |
| **1.3 Alternatively...** | | | |
| | | | |
| **2. Stability control** | | | |
| **2.1 Bleeding measurement (3hs)** | | | |
| Grout after mixing | | | |
| Grout from outflow-tube | | | |
| **2.2 Density control (gr/cm$^3$)** | | | |
| Grout after mixing | | | |
| Grout 60 min after mixing | | | |
| Grout from outflow-tube | | | |
| **3. Mechanical Properties Tests (if any)** | | | |
| Prismatic or cylindrical specimens' number | | | |
| Specimens' code | | | |

**Fig. 10.22** Example of a Daily protocol form of in-situ grout properties

#### 10.6.2.1 Fluidity Related Tests

**Fluidity Factor Test**

Fluidity factor measurement using the Fluidity Factor Test (FFT).

As presented in Sect. 3.2.3, a 3 mm nozzle-diameter Marsh cone is used, filled with 1000 ml of grout; the flow time $t_f$ is measured for a flow of only $Q = 100\ \text{cm}^3$ of grout to pass through. The fluidity factor is calculated $F_l = \frac{Q}{A \times t_f}$, where "$A$" denotes the area of the cross section of the nozzle (Miltiadou-Fezans and Tassios 2012).

This value should be close enough to the fluidity factor value prescribed in the design (on the basis of the minimal value $W_{nom}$ of the voids of the masonry, see Sect. 3.3). In Sect. 3.6 of this book, an example of the in-situ measurements of the Fluidity Factor to check the quality of a ternary grout during the works is presented (Table 3.7).

In case that the job specifications prescribe the use of Flow time test, the following data may be useful.

**Flow Time Test**

Measurement of flow time using a Marsh Cone with the smallest possible nozzle diameter (indicatively NF P18-358, EN 445, EN 12 715, ASTM D6910).

The flow time of the prescribed quantity of grout through a cone having a prescribed nozzle-diameter is measured (Fig. 10.23). As analyzed in more details in Sect. 3.2.2, for masonry grouts a Marsh cone of 3mm nozzle diameter is preferable. In case this is not possible a 4.75mm nozzle diameter should be used instead.

As an example, Fig. 10.23, for a Marsh cone with 4.75 mm nozzle diameter, shows the average values of flow time of 500 ml of grout (out of 1000 ml inserted in the cone), in-situ measured after mixing (twice per day, during the grouting period, 20 weeks in total). It was the case of the Katholikon of Daphni Monastery (Miltiadou-Fezans et al. 2008a). These values were very close to the values measured in the laboratory and at the worksite during the pilot production tests, and within the limits set by the design ($20 < t_{500\ \text{ml},d=4.75} < 25$ s). The small variations were daily assessed, together with the bleeding and density results, and were discussed taking also into account the worksite conditions (temperature, new delivery of superplasticizer, etc.).

#### 10.6.2.2 Stability Related Tests

**Bleeding Measurements (Indicatively: Standard NF P18-359)**

As presented in Sect. 4.2.2.1, the grout is sealed into three transparent clean and dry cylinders of 100 ml, of a diameter of 25 mm and of height equal to 250 mm. The grout should remain at a shadowed place for three hours; subsequently, bleeding water volume is measured. Since the initial grout volume is equal to 100 ml, bleeding percentage can be measured by counting on the volumetric cylinder the volume of the bleed water in cubic centimeters. Each measurement should be repeated three times; an average value is taken. Normally, this should be lower than the limit set by the design, and in any case lower than 5% (see also Sect. 4.2.2.1).

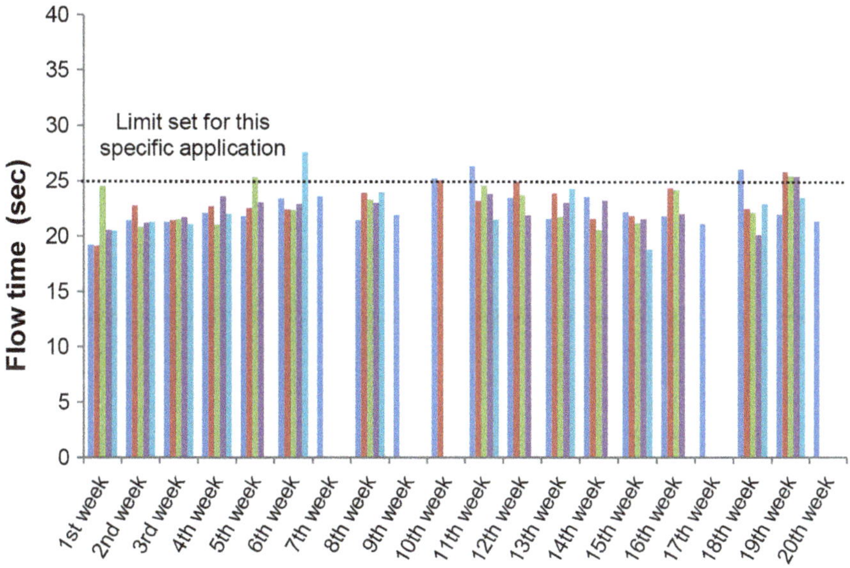


**Fig. 10.23** Average values of Flow time of 500 ml of grout (out of 1000 ml inserted in the cone) measured in situ after mixing, twice per day during the entire grouting period in the Katholikon of Daphni Monastery, using a cone with 4.75 mm nozzle diameter. Reworked from Miltiadou-Fezans et al. (2008a)

In Fig. 10.24 the results of in situ measurements in the case of the Katholikon of Daphni Monastery are presented. According to the limits set by the design for this specific case, the grout bleeding had to be lower than 3%. The measured bleeding variations were considered generally acceptable, since the grout was prepared in the worksite and, during the first six weeks, the ambient temperature was relatively high (summertime conditions).

**Apparent Density Test**

Grout is placed into transparent volumetric tubes (of 1000 or 2000 ml). Apparent density is determined by weighing out a specific grout volume (e.g., 50 ml) collected from the upper third of the height of each tube.

Apparent density measurements are carried out after mixing, as well as one hour after it. Each time, grout is collected from the same position at the upper third of the testing tube. This experimental procedure is essential for checking the quality of grout after mixing and possible tendencies to segregation after an interval of one hour. For this purpose, a precision scale (~5 kg) and some volumetric vessels are needed. Variations in apparent density should not exceed 1%.

In Fig. 10.25 the results of the in-situ measurements of apparent density variations after mixing of the grout used in the Katholikon of Daphni Monastery are presented. Apparent density was determined by weighing a specific grout volume (50 ml), collected from the same position of a 2000 cc volumetric tube (usually from the

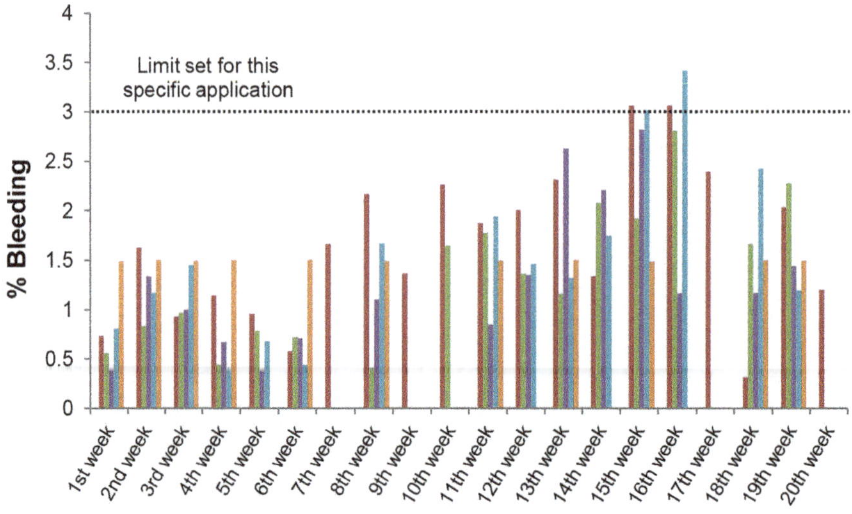


**Fig. 10.24** Bleeding values of grout, measured in situ after mixing, twice per day during the grouting period in the Katholikon of Daphni Monastery. Reworked from Miltiadou-Fezans et al. (2008a)

upper third of its height). The apparent density values measured in situ had to be 1.50 gr/cm$^3$, with an acceptable variation of $\pm 1\%$. As shown in the Fig. 10.25, all apparent density values measured in situ during the whole project satisfied this limit.

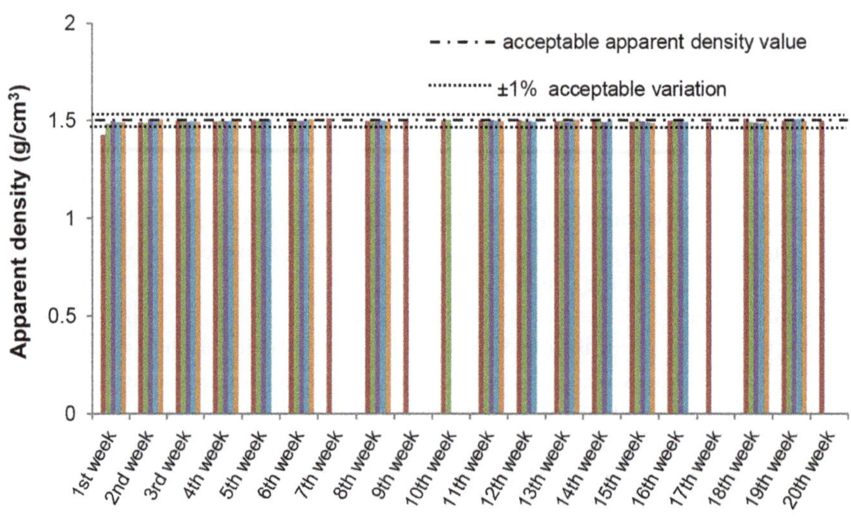


**Fig. 10.25** Apparent density values measured in situ after mixing, twice per day, during the entire grouting period in the Katholikon of Daphni Monastery. Reworked from Miltiadou-Fezans et al. (2008a)

#### 10.6.2.3 Checking the Quality of Grouting Operation

The quality control of grouting operation is carried out by means of the following measurements of the characteristics of the grout outflowing the protruding ends of grouting tubes. Measurement of apparent density and if possible, bleeding and fluidity are conducted in situ on samples collected from grout outflowing from distant exit tubes, at least twice per day. This testing provides information about the possible alteration of the quality of the grout during its penetration through the porous materials of the masonry. When the characteristics of the exit grout do not exhibit significant difference compared to the ones of the initial grout composition, it is ensued that the grout has retained its properties during its movement through masonry.

Table 10.5 presents indicative apparent density values of grout after mixing and of grout selected from corresponding outflow-tubes, checked in the case of the Katholikon of Daphni Monastery. The acceptable variation between these two values was set at ±5%.

It has to be noted that even in the case of outflow-tubes situated at a distance of two or three meters, the variation of the apparent density was retained at a limit lower than ±5%. This means that the grout has retained its properties during its movement through the porous masonry materials. This confirms both the quality of the grout (e.g., no segregation or important water loss due to absorption) and the soundness of its injection procedure.

**Table 10.5** Apparent density values of grout after mixing and of grout collected from distant outflow-tubes, measured during the masonry grouting works in the Katholikon of Daphni Monastery

| N° of entrance tube | N° of exit tube | Grout Density (gr/cm$^3$) | | |
|---|---|---|---|---|
| | | After mixing | Outflow | Variation (%) |
| AE74 | AE78 | 1.4885 | 1.5254 | 2.48 |
| AM18 | AE178 | 1.5056 | 1.5298 | 1.61 |
| AM288 | AM276 | 1.5013 | 1.5126 | 0.75 |
| BE68 | BE65 | 1.4936 | 1.5024 | 0.59 |
| BE86 | BE3 | 1.4858 | 1.5015 | 1.06 |
| BE517 | BE493 | 1.5022 | 1.5240 | 1.45 |
| NE36 | NE43 | 1.5023 | 1.5298 | 1.83 |
| KP405 | KP385 | 1.4989 | 1.5143 | 1.03 |
| KP404 | KP328 | 1.4983 | 1.5196 | 1.42 |
| Θ431A | Θ422 | 1.5048 | 1.5152 | 0.69 |
| Δ45/1155 | X260 | 1.5048 | 1.5068 | 0.13 |
| T358 | T361 | 1.4982 | 1.5045 | 0.42 |
| K251 | K260 | 1.5038 | 1.5036 | −0.01 |
| K116 | K109 | 1.4809 | 1.4953 | 0.97 |

Adapted from Miltiadou-Fezans et al. (2008a)

**Fig. 10.26** Specimens for the measurement of mechanical characteristics of grouts, taken (**a**) and cured in situ (**b, c**)

### 10.6.2.4 Mechanical Tests

As aforementioned, regarding the checking of mechanical characteristics of the grout, samples should be taken during the pilot application, in every new delivery of raw materials and on a regular basis in the worksite, as prescribed in the specifications of the job description (see also Sect. 10.6.2). Flexural and compressive strengths of the grouts will be tested at the age of 7, 28 days, 3 and 6 months on prismatic specimens (40 × 40 × 160 mm) (Fig. 10.26a); preferably cylindrical specimens (aprox. Ø = 20, L= 40 mm) should be used, if possible.

Curing will be effectuated in a plastic vessel, suitably covered by an air tight plastic cover (Fig. 10.26b). These vessels are large enough (~30 × 40 cm) as to allow for additional small pots full of water to be included in the plastic vessel (Fig. 10.26b), in order to ensure sufficient relative humidity; these vessels are kept in a protected area at the worksite. After 7 days, each specimen is taken off the moulds and put in closed plastic sacks, kept in a protected area at the worksite (Fig. 10.26c). After one month (28 days) the samples should be cured in Laboratory in 20 °C and 90 ± 5% RH, whenever possible.

Worksite values of the mechanical properties at least 70% of those prescribed may be considered as satisfactory.

As an example, Table 10.6 summarizes the results of mechanical properties of worksite specimens tested in the case of the Katholikon of Daphni Monastery.

**Table 10.6** Comparison of Strength of the grout prepared and cured in Lab (L) and in the worksite (W) of a specific job

| Age (days) | 28 | | 180 | |
|---|---|---|---|---|
| Place of production and curing | W | L | W | L |
| Flexural strength (MPa) | 1.00 | 1.30 | 2.40 | 2.00 |
| Compressive strength (MPa) | 2.00 | 2.50 | 6.04 | 5.34 |

Reworked from Miltiadou-Fezans et al. (2008a)
Grout composition: NHL5 90%, Pozzolan 10%, SP 1%, Water 80%. Water and superplastisizer percentage refer to the grout solids total mass

**Table 10.7** Compressive and flexural strength of grout composition, as prescribed in the mix-design and as they were modified in the worksite

| | | Compressive strength (MPa) | | | Flexural strength (MPa) | | |
|---|---|---|---|---|---|---|---|
| | | 28 | 180 | 360 | 28 | 180 | 360 |
| Design | Water: 80% | 3.00 | 8.20 | 10.54 | 1.25 | 3.10 | 3.40 |
| Modified in worksite | Water: 72.5%, SP 1% | 3.80 | 12.1 | 15.00 | 1.70 | 2.80 | 3.10 |

Reworked from Miltiadou-Fezans et al. (2006)
Water and superplastisizer percentage refer to the grout solids total mass

A comparison with respective values of the initial design (specimens cured in laboratory) is also made (Miltiadou-Fezans et al. 2008a), whenever available.

Sometimes, the in-situ application of the grout may lead to a reconsideration of the mix-design, suggesting some modifications to account for the real conditions of available materials, equipment and personnel. Similarly, new target values of strength might also be considered. In fact, since several of these materials in-situ may prove to be somehow different than those originally studied in Laboratory, slight modifications of the prescribed composition may be needed. Such was the case of the East and West ranges of cells of the South court-yard of Daphni Monastery, where a ternary grout was used: white cement 30%, hydrated lime 25%, Pozzolan 45% (see Table 10.7).

In this case, a reduction of water content, together with the use of a superplasticizer was imposed, in order to achieve similar injectability characteristics (Miltiadou-Fezans et al. 2006); expected strengths were modified, as shown in Table 10.7.

## 10.7 Final Report of the Execution of Injections

1. For each entrance tube, the following data should be recorded during the execution of the injection:
   - Which batches were used
   - Significant changes of pressure observed, if any
   - Quantity of grout passing through the specific entrance tube, based on the data derived from the relevant grout recorder [14]
   - Duration of grouting operation through this entrance tube
   - Code-numbers of the tubes where outflow of grout was observed, together with time indications
   - Location of the areas where uncontrolled grout-leakage was observed, together with time indications.

[14] Alternatively, the use of a flow meter may be selected or of a debit meter per tube, as proposed in the above mentioned FARCC No 03.0412,. 01. 01 of the Belgian Ministry of Culture (see footnote 13).

| DAILY PROTOCOL OF INJECTION AT ENTRANCE TUBES | | | | | |
|---|---|---|---|---|---|
| Project: | | | | Drawing or photo to note daily the position of grouting application | |
| Date: | | | | | |
| Code numbers of the batches: | | | | | |
| Number of page: | | | | | |
| N° of entrance tube | Code number of the outflow-tube or location of leakage area | Times | Pressure at entrance tube (Atm) | Grout quantity entered (lit) | Uncontrolled leakage areas |
| X14 | - | 10:00 | | | |
| | AE18 | 10:01 | | | |
| | | | | | |
| X29 | - | 10:20 | | | |
| | XE32 | 10:21 | | | |
| | | | | | |

**Fig. 10.27** Example of a Daily protocol of injection at each entrance tube

These data should be noted on an appropriate daily protocol, like the one shown in Fig. 10.27.

2. The data collected are reported on the drawings presenting the positions and numbers of the grouting tubes, so that a good estimation of the grout flow and local consumption can be made, for each region of the structure.

An example is shown in Fig. 10.28a, b. The drawings for the internal and external façade of the East wall of the Katholikon of Daphni Monastery are presented. One can easily recognize the areas with high grouting consumption, and those with a lower one. Tubes with no consumption are also noted. Thus, for this specific wall of the eleventh century, the volume of the grout consumption was estimated to be 6.5% of the total volume of the wall. In the case of the West wall of the monument, which was reconstructed during restoration interventions of the year 1895 (Delinikolas et al. 2003; Miltiadou-Fezans et al. 2004), the estimated grout consumption attained only 2.5% of its total volume (Miltiadou-Fezans et al. 2008a).

A useful representation of the results of local consumptions of grout at each entrance tube is also shown in Fig. 10.29, from the strengthening intervention of the walls of the Pisa Tower: dark colour circles depict the results of the primary round of grouting, and light colour circles those of the secondary round (Macchi and Ghelfi 2005, p. 188).

Such drawings, constitute the "as built-drawings" of such a non-visible intervention, and they offer an overall assessment of the grouting efficiency. Their meticulous

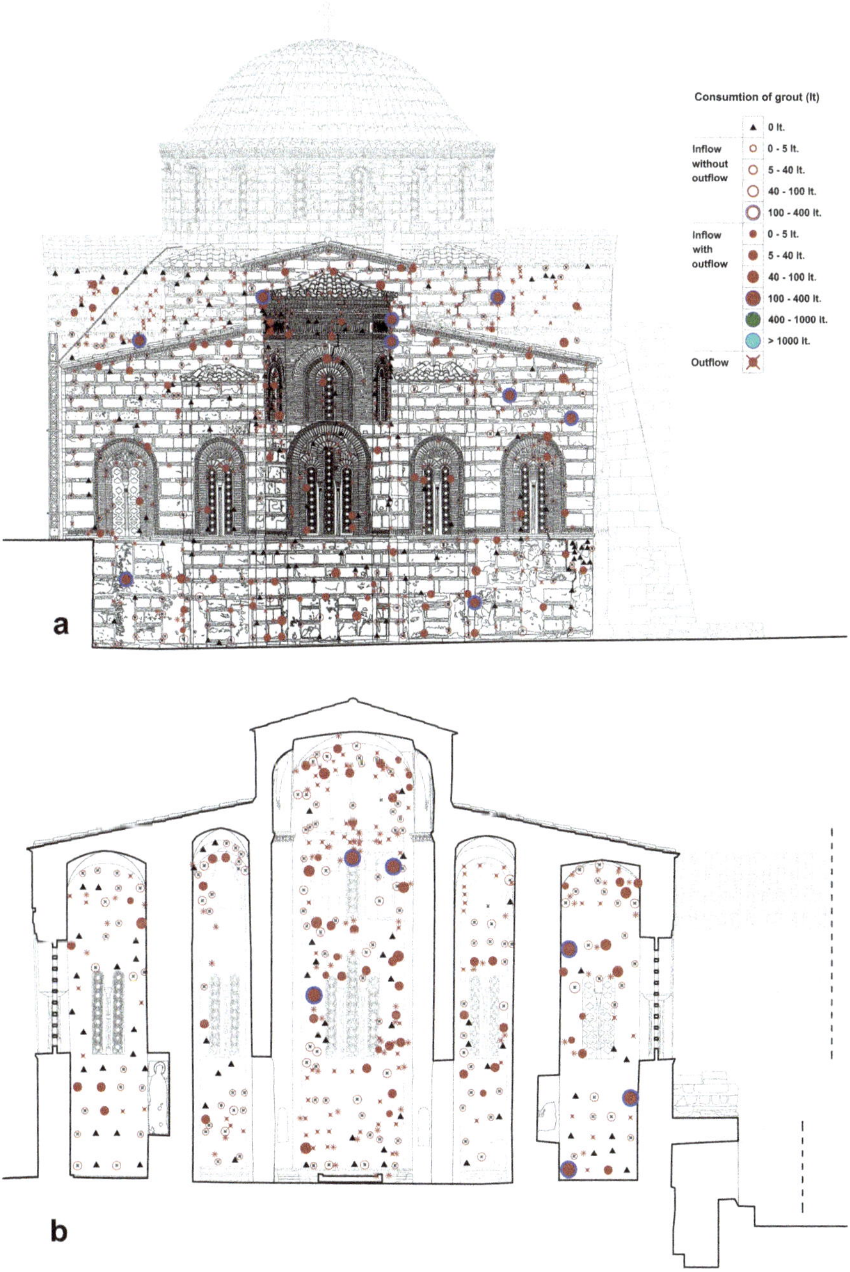


**Fig. 10.28** External and internal East façade of the Katholikon of Daphni Monastery—Schematic presentation of grouting consumption. Adapted from Miltiadou-Fezans and Delinikolas (2019)

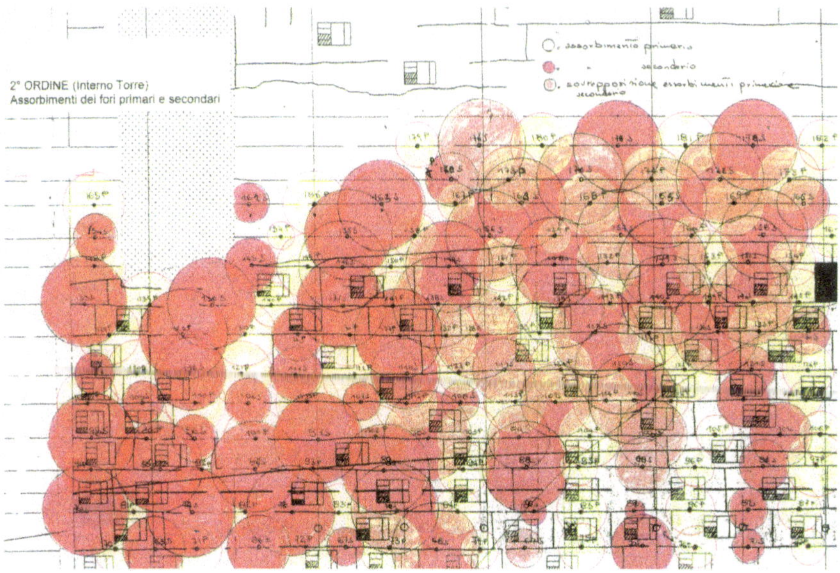


**Fig. 10.29** Local consumptions of grout at each entrance, from the strengthening intervention of the walls of the Pisa Tower. Reproduced from Machi and Ghelfi (2005), p. 188, with the permission of the Authors and of Bolletino d'Arte, Ministero per i Beni e le Attività Culturali, Italia

character is required in any *structural* grouting intervention, independently of the possible monumental value of the building.

## 10.8 Assessment of the Grouting Effectiveness After the Completion of the Works

The effectiveness of the intervention can be profitably evaluated by means of non-destructive techniques, frequently combined with minor-destructive techniques, before and after grout applications.

For this purpose, the design documents should possibly provide a description of the tests that have to be executed, as well as the criteria that may serve the purpose of assessing the effectiveness of the grouting intervention.

Minor-destructive tests (MDT) like core extraction may be used, together with endoscopy and flat jacks. Intensive research was conducted towards the use of non-destructive tests (NDT) for assessing grouting efficiency (Nappi 1996; Valle et al. 1999; Van Rickstal 2000; Maierhofer and Leipold 2001; Van Rickstal et al. 2002; Binda et al. 2000, 2001, 2003; Binda and Maierhofer 2006; Schuller 2006; Anzani et al. 2006; Côte et al. 2005, 2008; Miltiadou-Fezans et al. 2008b, c; Silva et al. 2014; Uranjek et al. 2012; Moropoulou and Lambropoulos 2015).

Techniques like sonic and ultrasonic methods, radar methods, electric resistivity methods applied by specialized personnel, can provide interesting results regarding the filling of voids of the masonry. Their effectiveness will be briefly discussed in the subsequent paragraphs.

Available techniques have to be applied to selected areas of the structure, before and after grouting. The comparison of the results allow for a qualitative assessment of the effectiveness of grouting intervention. In case of test results not conforming with specific design requirements (if any) the responsible team should evaluate the measurements and decide whether any re-testing or any remedial measures and/or any complementary interventions are needed.

### *10.8.1 Core Taking*

Core taking is strongly suggested, even in the case of buildings of historical value. Cores may offer valuable information regarding the inner masonry conditions. Suggested dimensions of cores are: $D_{min} = 10$ cm, with a length equal to two or three times the diameter, but not shorter than 2/3 the masonry thickness, if possible.

Visual inspection of cores offers the following information (see Fig. 10.30):

- Degree of filling of cracks or voids after grouting.
- Condition of the material of the grout contained in these discontinuities.
- Possible identification of non-filled discontinuities.

Moreover, compression or splitting tests on these cores can also provide information about the mechanical characteristics of the sampled masonry material after grouting, provided however that undisturbed cores were in fact taken.

It is however noted that these characteristics cannot be reliably correlated with the strengths of other sufficiently large parts of masonry.

The number of cores to be taken depends on the structural uniformity of the masonry and the possible limitations imposed in case of monuments. A minimum of two cores is however recommended to be taken out of walls under comparable conditions, whenever possible.

Finally, the most fruitful use of the results of core-taking is their consideration in connection with the data of the Final Report (see Sect. 10.7), regarding the same places of the injected masonry. Thus, a sort of calibration of the quantitative data included in the Final Report may be feasible.

### *10.8.2 Endoscopy*

Endoscopes (borescopes) are inserted through small diameter holes (drilled preferably into mortar joints if possible).

**Fig. 10.30** Cores taken after the application of grouting. The grout: **a**, **b** of white colour has filled the discontinuity in mortar-brick interface and all intercommunicating voids of the mortar itself, **c** of whitish colour has filled the cracks, the most of the interfaces mortar/aggregates and has been mixed with loose material of disintegrated mortar (right part of the figure), **d, e** of dark grey colour has filled all very fine discontinuities present in the mortar and in mortar/stones or bricks interfaces. **a, b** Adapted from Penelis and Stylianidis (1996). Courtesy of I. Papayianni

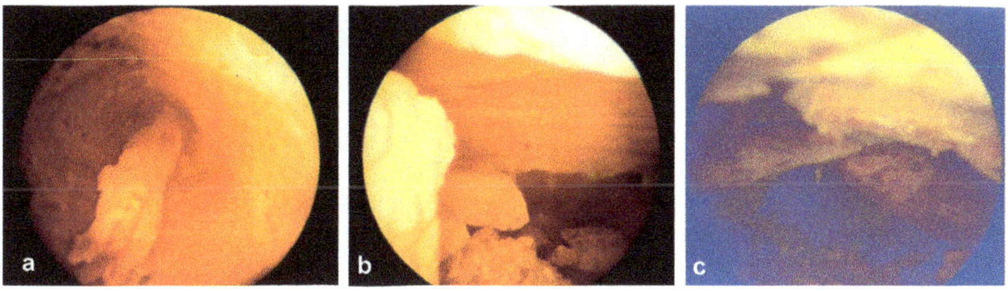


**Fig. 10.31** Indicative endoscopy images taken in the masonry of the Church of Panagia Krina in Chios Island, Greece, presenting: **a** A disintegrated timber lace inside the masonry, surrounded by void space, mortar and stones. **b** A timber lace in the upper part of the image, together with void space, disintegrated mortar and pieces of stones at the lower part of the image. **c** Two timber laces, the one perpendicular to the other, void space, mortar and stones. **a**, **b** Adapted from Vintzileou (2006). **c** Adapted from Palieraki et al. (2007). Courtesy of the Authors

Optical fibres and appropriate internal light permit the visual (possibly quantitative) inspection of the masonry (see i.e. Palieraki et al. 2007, 2008). An example is given in Fig. 10.31 (see also Figs. 5.7 and 5.8 of Chap. 5).

This method can also be combined with core extraction. For important monuments, a more sophisticated technique (Borehole Image Processing System) offers a colour picture of the entire internal surface of the hole (360°) developed on a plane. Thus, a detailed observation of the entire hole may be carried out (Macchi and Ghelfi 2005, p. 184), as reproduced in Fig. 10.32.

### *10.8.3 Sonic/Ultrasonic Methods*

**Sonic tests**. As it is known, sonic and ultrasonic methods are based on the transmission and reflection of pressure waves (P-waves) through a medium, at sonic and ultrasonic frequencies. The most commonly used sonic methods include the sonic transmission and sonic tomography. An instrumented "hammer" applies stress waves on the face of masonry, whereas a "captor" records the oscillations received at the opposite face of the wall. Thus, one basic information obtained is the velocity of the energy transmitted through the wall. As it is known, this velocity is connected to the elastic properties of the medium, and (indirectly though) to its compressive strength. However, it has to be admitted that in view of the remarkable inhomogeneity of historical masonries, such a relationship between compressive strength and sonic velocity is rather aleatoric. But the importance of the method of sonic testing in checking the effectiveness of grouting remains very high.

In fact, the method may first provide qualitative information about the homogeneity of the masonry. Sonic tests with low frequency (0.3–5.0 kHz) are more suitable to detect (before grouting) relatively large discontinuities within masonry of usual thickness; such discontinuities are compatible with the wavelength of low

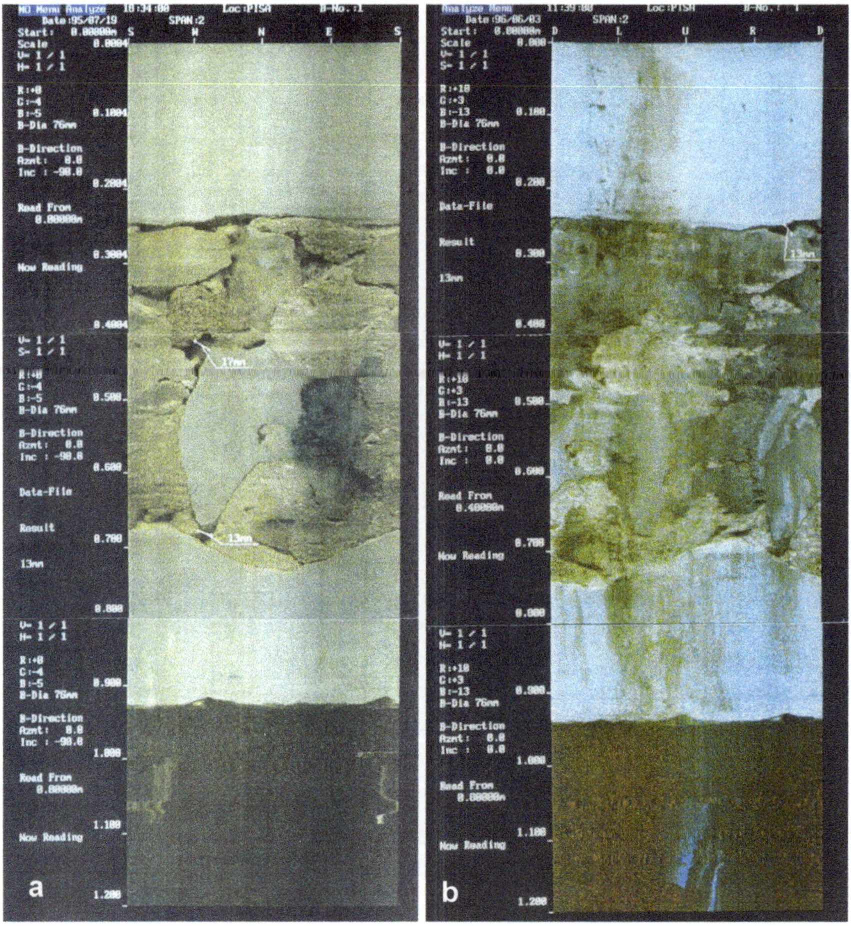


**Fig. 10.32** Picture of the entire internal surface of a borehole (360°) developed on a plane: **a** before and **b** after grouting in the Pisa Tower masonry. Reproduced from Machi and Ghelfi (2005), p. 184, with the permission of the Authors and of Bolletino d'Arte, Ministero per i Beni e le Attività Culturali, Italia. The high-definition scan was made available from the Università Iuav di Venezia, Archivio Progetti, Fondo Studio Tecnico prof. ing. Giorgio Macchi, dott. ing. Stefano Macchi

frequency tests ("hammering"). High frequency tests (20–100 kHz) are suitable only for more uniform and thinner walls.

The effectiveness of the grouting intervention may be evaluated by comparing wave *velocities* before and after grouting. As expected, an increase of velocity is an indication of the beneficial impact of the intervention. It has to be noted, however, that the method exhibits much higher attenuation characteristics in masonry, than in other more homogeneous materials (e.g., isolated stones). Thus, an appropriate grid is required on the wall surfaces, so that the scanning process to involve exactly the

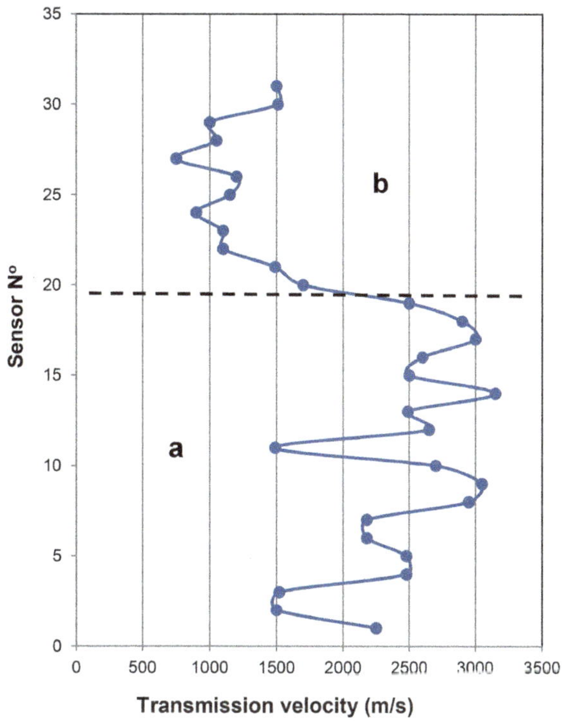


**Fig. 10.33** Sonic transmission velocities across a vertical line situated in the middle of a wall: **a** in a grouted lower area and **b** in a non-grouted upper area. Adapted from Côte et al. (2008)

same locations before and after grouting. An average pulse-velocity value (around a specific "place") may be calculated out of a sufficiently large number of measurements (e.g., 25) taken at distances depending on the mean value of the size of stones.

Figure 10.33 illustrates the results of an application of sonic test in a byzantine masonry wall (Côte et al. 2008); the lower area of the wall was grouted, whereas the upper area remained non-grouted. Sonic testing was carried out after approximately 6 months after grouting, along a *vertical* line situated in the middle of the wall. As shown in Fig. 10.33, average values of velocities equal to 2500 m/s and 1200 m/s were observed in the grouted and the non-grouted areas, respectively.

A similar test was carried out in another byzantine wall (Côte et al. 2008); this time however, velocity measurements were made along two distinct horizontal lines: One within the grouted area, and the other within the non-grouted area (approximately 2.0 m higher). Figure 10.34 illustrates the results. In this case, average velocity values were roughly equal to 2000 m/s and 750 m/s in the grouted and non-grouted areas, respectively.

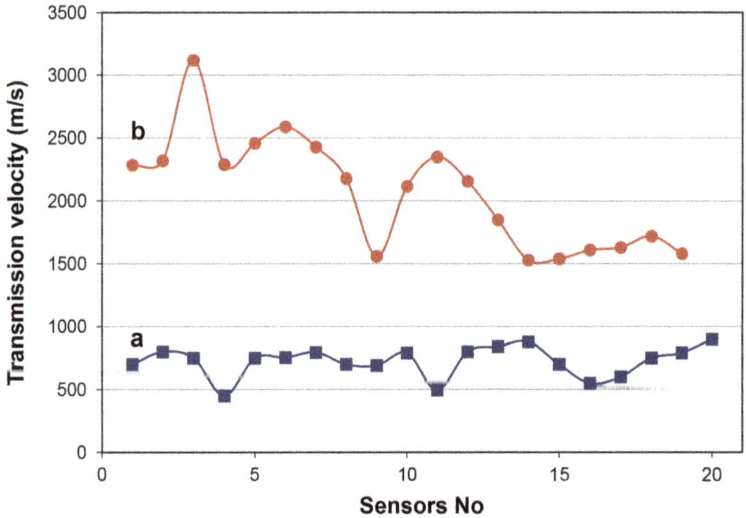


**Fig. 10.34** Sonic transmission velocities across **a** a non-grouted area of a wall and **b** a grouted area of the same wall. Adapted from Côte et al. (2008)

In both cases the usefulness of sonic testing is apparent. The sonic method is therefore applicable in order to check the effectiveness of grouting, before and after the operation, around a specific "place' of the injected wall.

Experienced researchers are in fact of the opinion that "sonic tests are a reliable technique to evaluate the effectiveness of grout injection" (Binda et al. 2001, p. 137).

One of the most emblematic applications of this checking method was the case of the Pisa Tower (Macchi and Ghelfi 2005), where the "cross hole" sonic method was applied: The transmitter and the receiver are introduced in two adjacent holes, horizontally through a very thick wall. Figure 10.35 shows the results of such sonic tests carried out on the (4.0 m thick) wall of the Tower, before and after grouting (Macchi and Ghelfi, op. cit., p. 196). The beneficial results of the grouting become evident, especially in the inner infilled spaces of the wall. It is also worth to note the more *uniform* distribution of velocities values after grouting.

Valuable information on the equipment and the conditions of the sonic tests in masonry are offered in the document RILEM TC 127-MS: Tests for masonry materials and structures, MS-D.1 Measurement of mechanical pulse velocity for masonry, 1996, pp. 459–475.

In conclusion, sonic testing proved to be a relatively simple and low-cost method for checking the effectiveness of grouting of masonry. If, however, a somehow broader and more detailed information is needed regarding the inner spaces of masonry, before and/or after grouting, "Sonic Tomography" method is more suitable, and may be applied as a complement of sonic tests.

**Sonic Tomography**. Sonic transmission velocities across a wall may contribute towards an overall image regarding the internal homogeneity of masonry; possible

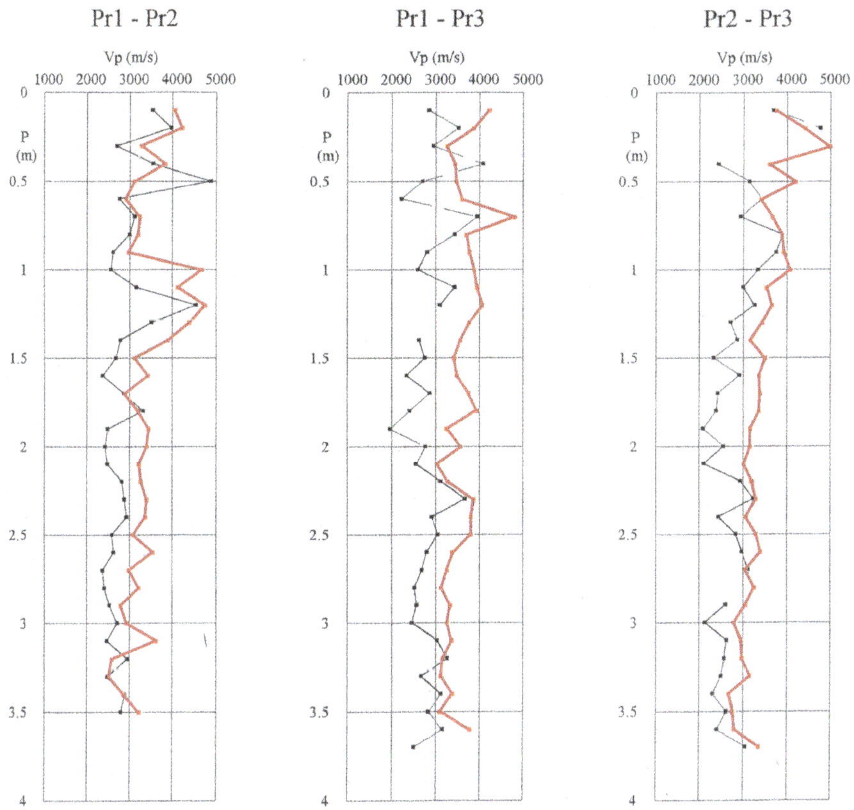


**Fig. 10.35** Results of "cross hole" sonic tests carried out within three pairs of holes on the (4.0 m thick) wall of the Tower of Pisa, before (thin line) and after (bold line) grouting. Reproduced from Machi and Ghelfi (2005), p. 196, with the permission of the Authors and of Bolletino d' Arte, Ministero per i Beni e le Attività Culturali, Italia

internal distinct masonry-layers, possible inclusion of other materials (wood or metal), as well as possible large voids, may be identified.

To this purpose, a series of receivers are installed along a straight line on one face of the wall. A transmitter (coupled with an accelerometer) is "hammering" the other side of the wall at appropriate distances, along a straight line within the same horizontal plane as the receivers, perpendicular to the wall. Thus, a large number of "ray paths" is realised between each hammered point and all receivers (Fig. 10.36a). Subsequently, by means of an appropriate algorithm (s. Côte et al. 2008, p. 1154), *local zones* of different velocities of P-waves may be identified (Fig. 10.36b). By means of such a tomography, local alterations of masonry may be observed or possible internal distinct layers of masonry can be identified.

More specifically now, assessment of the grouting may be achieved by comparing the pre-grouting and post-grouting tomography-images at the same cross section of

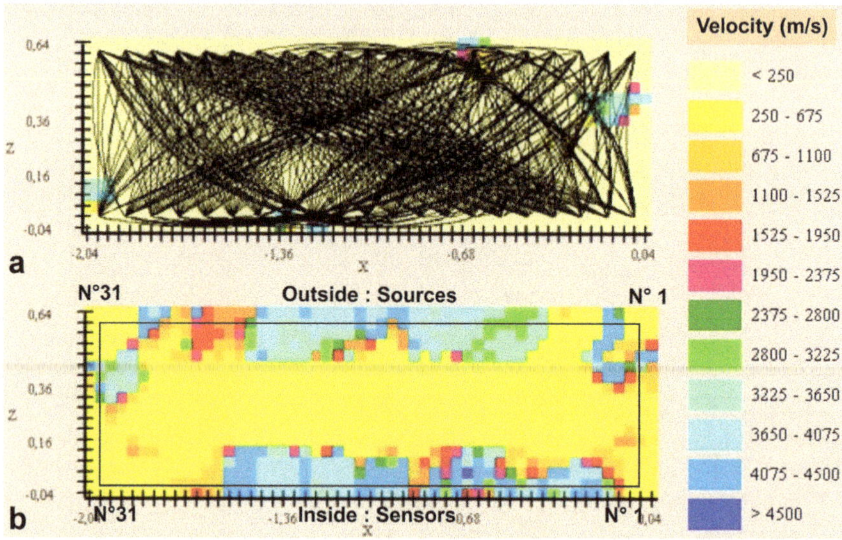


**Fig. 10.36** Tomographical "reconstruction" in a non-grouted area of a three-leaf masonry wall. **a** "Ray paths" between each hammered point and all receivers. **b** Local zones of different velocities of P-waves. Adapted from Côte et al. (2008)

the wall, as in the case of Fig. 10.37, where a considerable increase of velocity-values is observed (Macchi and Ghelfi 2005, p. 197).

Finally, it has to be noted that tomography of S-waves distribution may increase the resolution of results (see On site for masonry project report, 2006, page 51). Additional information may also be offered by frequency content and amplitude alterations observed in the received signal.

### *10.8.4 Radar Technique*

1. Using a radar system, consisting of a central unit and various antennas, electromagnetic signals are emitted through any kind of material. An antenna is moved along a straight line on the surface of the material under consideration. When the signal meets (e.g., inside a masonry) an "interface", part of the emitted radiation is reflected, and it is recorded by the central unit, whereas part of the same signal travels deeper into the material. Such "interfaces" may be cracks or any contacts of different materials with differing *dielectric* properties, such as stone and mortar, mortars of different qualities, air voids, steel, timber, moisture, etc. By appropriately processing the recorded reflected pulse, one may have a picture, more or less clear of the in-depth geometry of the investigated element. The Radar technique was first applied in the field of geophysical (GPR) investigation. Its application especially in masonry structures is still rather limited,

despite its importance for structures of high monumental value. It has to be admitted that the numerous interfaces that the electromagnetic pulse can meet when traveling through masonry, are of unknown nature and may cause multiple reflections of the pulse, thus giving a rather unclear picture of the internal state of masonry. Therefore, the results of the radar technique are expected to be less conclusive in case of a rubble three-leaf stone masonry than for other more

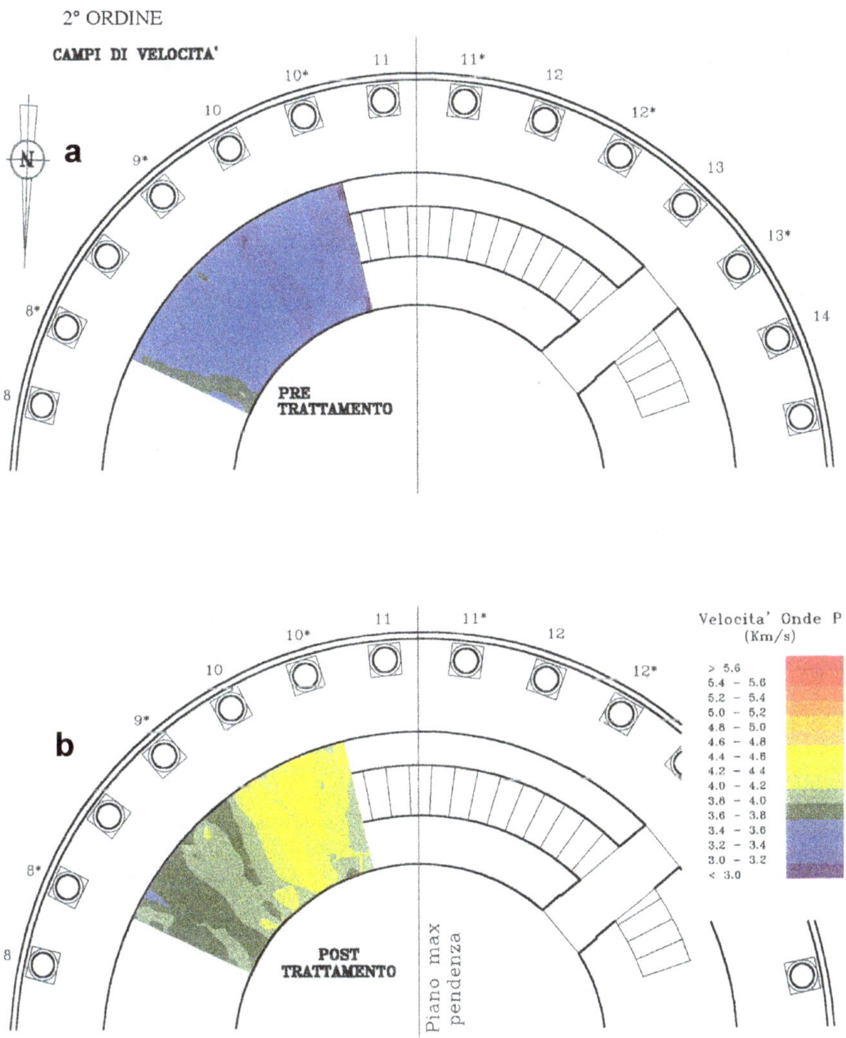


**Fig. 10.37** Sonic tomography images in a wall of the Pisa Tower, before **a** and after **b** grouting. Reproduced from Machi and Ghelfi (2005), p. 197, with the permission of the Authors and of Bolletino d' Arte, Ministero per i Beni e le Attività Culturali, Italia

regular types of masonry. Nevertheless, in recent years, several applications of this technique in monuments and historic centres are reported (see e.g. Binda et al. 2005).

The following provisions are included in the RILEM Recommendation "MS.D.3 -Radar investigation of masonry":

"The data is stored in the time domain and may be displayed either on a computer screen or printed out on a chart recorder. The centre frequency should normally be in the range between 0.1 GHz and 1.5 GHz, depending on the size and condition of the subject structure. [...]. Tests will be conducted under ambient conditions; however, the work should not be carried out in heavy rain or other conditions, as these will cause severely erroneous results. Severe water ingression will significantly affect the results of a radar survey; saturated walls may cause such high attenuation that the returned signal is lost in the noise."

2. The following steps in the application of the method may be distinguished.

   - An appropriate antenna is moved along a straight line (horizontal or vertical) on the surface of the masonry wall of a well-known thickness. The central unit of the equipment is recording the reflected waves corresponding to each position of the antenna.
   - A graphic representation of this recording is possible: To each position of the antenna, possible disturbances of the emitted signal are depicted along an "emittance and reflexion" time-axis. It is reminded that such disturbances are due to differences of dielectric properties of the materials in contact. The intensity of these signal disturbances may be illustrated by means of various tones of grey colour.
   - A similar representation may be made after a conversion of the "emittance-reflexion time" into geometrical units, if a rough average value of the velocity is found by means of an in-situ calibration. Thus, a two-dimensional picture of the position of the heterogeneities is finally produced

As an introductory example, the two PVC bars incorporated in concrete as in Fig. 10.38a, will produce a disturbance like the one shown in this Fig. 10.38c. Note that such disturbances do not reproduce the geometrical form of the encountered interface. Places of the examined material located "under" (or "close" to) a discontinuity,

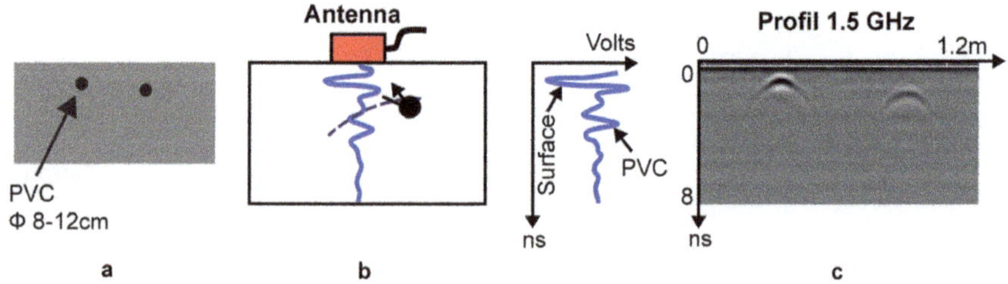


**Fig. 10.38** **a** Two PVC bars incorporated in concrete. **b** Schematic presentation of GPR application. **c** Disturbance produced in the radar profile by the PVC bars. Courtesy of X. Derobert

exhibit larger reflexion times; thus, in the graphic representation, these places may also be shown as "anomalies". That is why these representations need an appropriate numerical and graphical processing, before being used for predicting the conditions inside the examined structural element.

The following applications (Fig. 10.39) of the method were useful in examining the conditions inside a given masonry wall (Vintzileou 2003; Palieraki et al. 2008).

More importantly, the method may be very useful in assessing the results of grouting in masonry walls.

Figure 10.40 shows four GPR profiles through 40 cm thick masonry walls before **a** and after grouting **b**. It follows that the number of "discontinuities" and their importance have been considerably reduced after grouting, except for the middle area where a chimney was located.

### *10.8.5 Other Non-destructive Methods*

To the same end, to evaluate the effectiveness of grouting operations in masonry walls, several other N.D. methods have been proposed in the literature. Among these methods, the following ones should also be mentioned here:

**Electrical conductivity investigation method. Indicative references**:

- RILEM TC 127-MS: Non-destructive tests for masonry materials and structures. Recommendation: MS.D.8: Electrical conductivity investigation of masonry (2001). In Materials and Structures, Vol. 34, April 2001, pp 134–143.

**Electrical resistivity measurements. Indicative references**

- Venderickx K, Van Gemert D (2000) Geo-electrical survey of masonry for restoration projects. International Journal for Restoration of Buildings and Monuments, Aedificatio Verlag, 2000, Heft 2, pp. 151–172.
- Schueremans L, Van Rickstal F, Verderickx K, Van Gemert D (2003) Evaluation of Masonry Consolidation by Geo-Electrical Relative Difference Resistivity Mapping. Materials and Structures, Vol. 36, January–February 2003, pp 46–50.
- Keersmaekers R, Van Rickstal F, Van Gemert D (2005) Geo-electrical techniques as a non-destructive appliance for restoration purposes. In: Proc. 4th Inter. Sem. on Struct. Anal. of Historical Constructions, Padova, Italy, pp. 343–350.
- Van Rickstal F, Van Gemert D, Keersmaekers R, Posen D (2008) Enhancement of geo-electrical techniques for NDT of masonry. In: Proc. 6th Intern. Conf. on Structural Analysis of Historical Constructions, Bath, UK, pp. 1053–1060.

**Thermography. Indicative references**

- Stockton G, Allen L (1999) Using infrared thermography to determine the presence and correct placement of grouted cells in single-width concrete masonry unit (CMU) walls. Proc. of SPIE, 3700, Thermosense XXI, 1999.

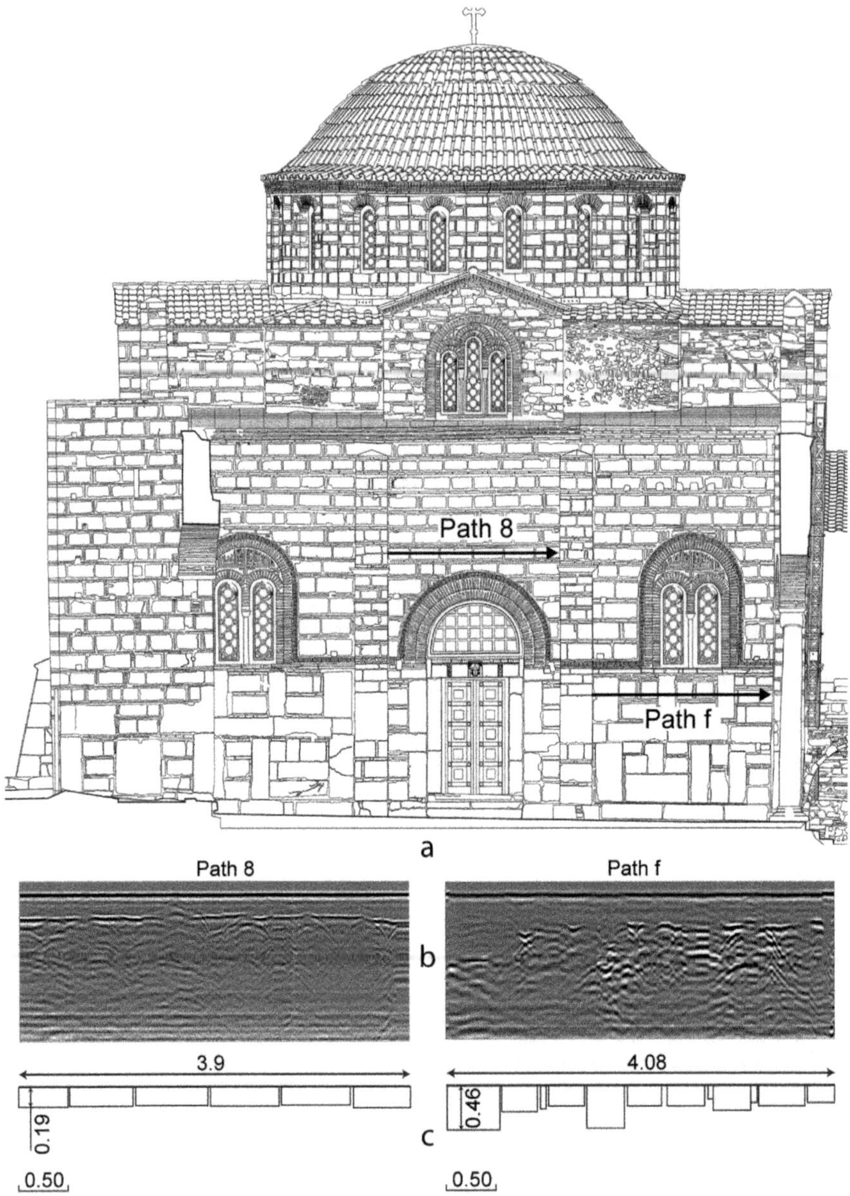


**Fig. 10.39** **a** Presentation of the position of two paths (path 8 and path f) on the drawing of the West Façade of the Katholikon of Daphni Monastery. Reworked from Delinikolas et al. (2003). **b** The two corresponding radar profiles after processing. **c** Presentation of the results in drawings (horizontal section of masonry, outer leaf of masonry). **b** and **c** Reworked from Vintzileou (2003), Courtesy of the Author and V Palieraki

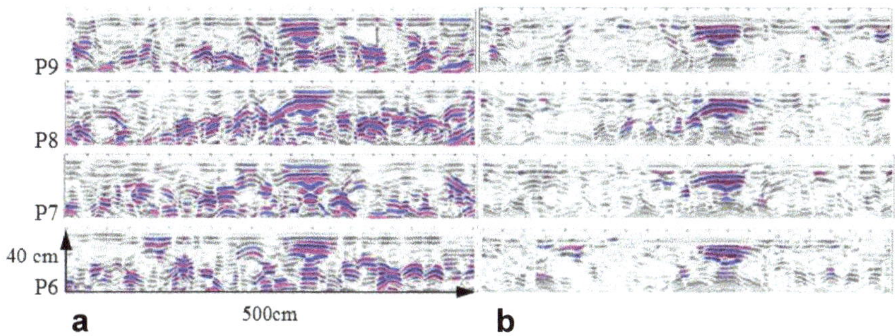


**Fig. 10.40** GPR profiles on walls at height 70–160 cm in intervals of 30 cm: **a** before and **b** after grout injection. Adapted from Uranjek et al. (2012). Courtesy of M Uranjek


- Maierhofer C, Brink a, Rolling M, Wiggenhauser H (2001) Transient thermography for non-destructive investigation of building structures in near surface region. In Proc, Workshop of RILEM TC &MDT, Mantova, Italy, 2001.
- Moropoulou A, Avdelidis NP, Koui M, Kanellopoulos NK (2000) Dual band infrared thermography as a NDT tool for the characterization of the building materials and conservation performance in historic structures. In: T. Matikas, N. Meyendorf, G. Baaklini, R. Gilmore (eds.), Nondestructive Methods for Materials Characterization, vol. 591 (pp. 169–174). Pittsburgh: Materials Research Society.
- Colantonio A (1997) Thermal performance patterns of solid masonry exterior walls of historic buildings. J Thermal Insul. and Bldg. Envs. 21:185–201.


### *10.8.6 Structural Dynamic Measurements*

The structural behaviour of masonry buildings is influenced by the stiffness of their elements. On the other hand, stiffness is the basic parameter shaping the frequency of transversal free vibrations of the structure, after an appropriate base excitation. In case of important masonry buildings, it is possible to measure their eigen-frequencies before and after grouting operations, in order to assess their effectiveness. As base excitations may be specifically selected micro-tremors or ambient vibrations.

It is worth to be noted that in the case of Monuments on which accelerometers were permanently installed as a monitoring system before grouting operations, the consequences of strong ambient vibrations, and or future earthquakes, may also be registered and offer information regarding frequencies before and after grouting.

This was the case with the Katholikon of Daphni Monastery grouting operations, with the results shown in Fig. 10.41: Approximately, frequency (in two different directions) seems to be almost doubled (Mouzakis et al. 2008), after the finalisation of grouting works.

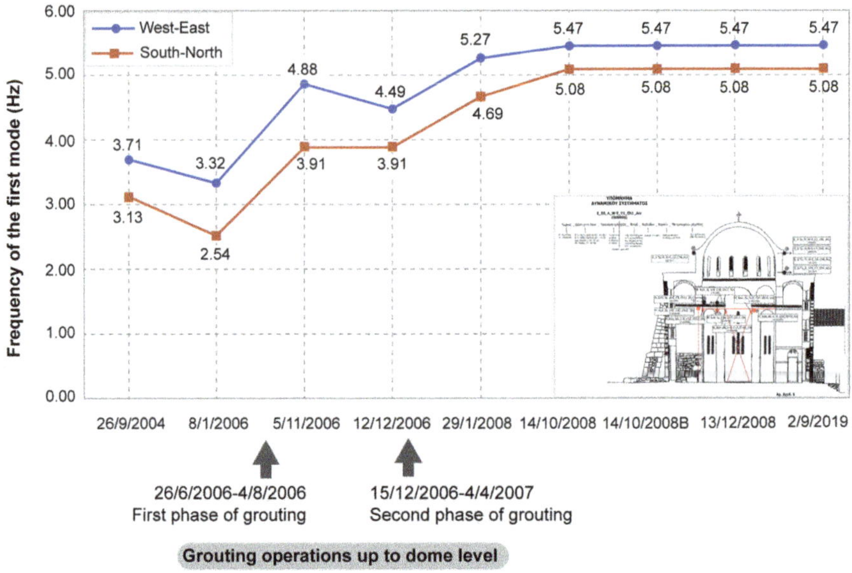


**Fig. 10.41** Frequency of the first mode for various seismic events occurred before, during and after the completion of grouting operations up to the dome level in the Katholikon of Daphni Monastery, according to the data registered by the accelerometers of a monitoring system installed before the application of any intervention. Reworked from Mouzakis et al. (2008), with the addition of unpublished data for 4 earthquakes occurred in 2008 and one occured in 2009, courtesy of Ch Mouzakis

# References

Anamaterou E, Deliprimi M, Getimoglou F, Krini M (2017) The Mosaics of Daphni Monastery: A Discussion of the Dilemmas and the Decisions Made throughout the Ten-Year Conservation Program. In Teutonico JM, Friedman L, Ben Abed A, Nardi R (Eds) Proceedings of the 12th Conference of the International Committee for the Conservation of Mosaics, Sardinia, October 27–31 2014, Published by the Getty Conservation Institute, Los Angeles 2017, pp 121–126

Anzani A, Binda L, Lualdi M, Tedeschi C, Zanzi L (2006) Use of sonic and GPR Tests to control the effectiveness of grout injections of stone masonry. In: Proceedings of ECNDT 2006, Th.2.4.3, p 7

Ashurst J (1989), Grouting masonry walls, Stone Masonry. Chapter 3, English Heritage Technical Handbook, Gower Technical Press

Ashurst J, Dimes FG (1990) methods of repairing and consolidating stone buildings. Conservation of buildings and decorative stone, chapter 1, vol 2. Butterworth Heinemann Ltd., London, pp 1–54

ASTM D6910-04 Standard test Method for Marsh Funnel Viscosity of Clay Construction Slurries

Baronio G, Binda l, Modena C (1992) Criteria and methods for the optional choise of grouts according to the characteristic of masonry. In: Proceedings of Inter. workshop CNR-GNDT on the Effectiveness of injection techniques for retrofitting of stone and brick masonry in seismic areas, Politecnico di Milano, 30–31 Mar, Milan, Italy

Binda L (1993) Strengthening of masonry by injection technique. 6th North American Masonry Conference, Philadelphia USA

Binda L, Maierhofer C (2006) Strategies for the assessment of historic masonry structures. Proceedings of NSF/RILEM Workshop In-situ Evaluation of Historic Wood and Masonry Structures, 10–14 July 2006. Czech Republic, Prague, pp 37–56
Binda L, Modena C, Baronio G, Abbaneo S (1997) Repair and investigation techniques for stone masonry walls. Constr Build Mater 11(3):133–142
Binda L, Saisi A, Tiraboschi C (2000) Investigation procedures for the diagnosis of historic masonries. Constr Build Mater 14(2000):199–233
Binda L, Saisi A, Tiraboschi C (2001) Application of sonic tests to the diagnosis of damaged and repaired structures. NDT&E Int 34(2001):123–138
Binda L, Baronio G, Tiraboschi C, Tedeschi C (2003) Experimental research for the choice of adequate materials for the reconstruction of the Cathedral of Noto. Constr Build Mater 17:629–639
Binda L, Zanzi L, Lualdi M, Condoleo P (2005) The use of georadar to assess damage to a masonry Bell Tower in Cremona, Italy. NDT&E Int 38(93):171–179
Bruce DA, Littlejohn GS, Naudts A (1997) Grouting materials for ground treatment: a Practitioner's guide. Grouting—compaction, remediation and testing. In: Vipulanandan C (ed) Proceedings of sessions sponsored by the grouting committee of the Geo-institute of the American Society of Civil Engineers, Logan UT, Geotechnical Special Publication No. 66, 16–18 July, pp. 306–334
Colantonio A (1997) Thermal performance patterns of solid masonry exterior walls of historic buildings. J Thermal Insul and Bldg. Envs. 21:185–201
Côte P, Dérobert X, Miltiadou-Fezans A, Delinikolas N (2005) Mosaic-grouting monitoring by ground-penetrating radar. In: Modena C, Lourenco P, Roca P (eds) Proceedings of IV international seminar on structural analysis of historical constructions, Padova, Italy, 10–13 Nov 2004, Taylor and Francis Group, London, ISBN 04 1536 379, vol 1, pp 401–406
Côte P, Dérobert X, Miltiadou-Fezans A, Delinikolas N, Durand O, Alexandre J, Kalagri A, Savvidou M, Chryssopoylos D, Anamaterou L, Georganis, F (2008) Application of non-destructive techniques at the Katholikon of Daphni Monastery for mapping the mosaics substrata and grouting monitoring. In Proceedings of the 6th International Conference on Structural Analysis of Historic Construction, 2–4 July, Bath, UK, edited by D' Ayala D. and Fodde, CRC Press/Balkema, Taylor and Francis Group, Vol. II, pp 1149–1156
Coulray AW, Carson CS (1982) Grouting plant and equipment. In: Proceedings of conference on grouting in geotechnical engineering ASCE, New Orleans, Louisiana, 10–12 Feb 1982
Delinikolas N, Miltiadou-Fezans A, Chorafa E, Zaroyianni E (2003) Restoration design of the Katholikon of Dafni Monastery, Phase A´- Architectural and Historical Survey, Design Report, Directorate for the Restoration of Byzantine and Post-Byzantine Monuments, Hellenic Ministry of Culture and Sports (in Greek)
EN 12 715—Execution of special geotechnical work—grouting
EN 445: 2007—Grout for prestressing tendons—Test methods
Fiche d'aide à la rédaction des documents de marché – Consolidation des maçonneries par injections de coulis minéral naturel", FARCC No 03.0412.01.01, Direction de l'Architecture et du Patrimoine, Ministère de la Culture et de la Communication, Wallonie, Belgium
Giaretton M, Valluzzi MR, Mazzon N, Modena C (2017) Out of plane shake-table tests of strengthened multi-leaf stone masonry walls. Bull Earthq Eng. https://doi.org/10.1007/s10518-017-0125-7
Houlsby AC (1990) Construction and design of cement grouting. Wiley, New York
Jefferis S (1982) Effect of mixing on bentonite slurries and grouts. Grouting in Geotechnical Engineering. In: Proceedings of the conference sponsored by the Geotechnical Engineering Division of the American Society of Civil Engineers, New Orleans, LA, 10–12 Feb, pp 62–76
Keersmaekers R, Van Rickstal F, Van Gemert D (2005) Geo-electrical techniques as a non-destructive appliance for restoration purposes. In: Modena C, Lourenco P, Roca P (eds) Proceedings of IV international seminar on structural analysis of historical constructions, Padova, Italy, 10–13 Nov 2004, Taylor and Francis Group, London, pp 343–350
Luso E, Lourenco P (2019) Mechanical behaviour of two leaf masonry wall-strengthening using different grouts. J Mater Civ Eng 31(7):04019096

Machi G, Ghelfi S (2005) Interventi definitivi di stabilizzazione strutturale. In: Fermo A, Francescone L, Lunetti D, Tursi L. Valente D (eds) La Torre restituita: Gli studi e gli interventi che hanno consentito la stabilizzazione della Torre di Pisa, Bollettino d' Arte, Ministero per i Beni e le Attivita Culturali, Italy, Volume Speciale 2005, Volume III, Roma, pp 173–215
Maierhofer C, Leipold S (2001) Radar investigation of masonry structures. NDT&E Int 34:139–147
Maierhofer C, Brink A, Rolling M, Wiggenhauser H (2001) Transient thermography for non-destructive investigation of building structures in near surface region. In: Proceedings of workshop of RILEM TC NDT-MDT, Mantova, Italy, 2001
Miltiadou AE (1990) Étude des coulis hydrauliques pour la réparation et le renforcement des structures et des monuments historiques en maçonnerie. Thèse de Doctorat de l'Ecole Nationale des Ponts et Chaussées. Pub. by LCPC in Collection Etudes et recherches des Laboratoires des Ponts et Chaussées, série Ouvrages d'art, OA8 ISSN 1161-028X, LCPC, Décembre 1991, Paris, France, p 278
Miltiadou Fezans A (2003) Structural Restoration Works of the North East range of Cells of Hosios Loukas Monastery, Boeotia, Greece. In: Proceedings of International Seminar Restoration of Historic Structures in Seismic Areas: The case of settlements in the Aegean, European Centre on Prevention and Forecasting of Earthquakes, Earthquake Planning and Protection Organization of Greece, Departement of Geography of the University of Aegean and School of Architecture of the National Technical University of Athens, Lesvos Island, Greece, 24–26 May 2001, Publ. by OASP, pp 155–165
Miltiadou-Fezans (2004) Structural restoration of the west part of the church of Aghia Eirini in Athens. In: Trakossopoulou K, Doussi M, Xatzitrifon N (eds) Proceedings of the 2nd National congress Appropriate interventions for the safeguarding of monuments and historical buildings, 14–16 Oct 2004, Thessaloniki, Greece, Hellenic Ministry of Culture and Technical Champer of Greece, vol 2, pp 173–183
Miltiadou-Fezans A, Delinikolas N (2019) The Restoration of the Katholikon of Daphni Monastery: a challenging project of holistic transdisciplinary approach, novel methodologies and techniques and digital modelling. In: Moropoulou A et al (eds) Transdisciplinary multispectral modelling and cooperation for the preservation of cultural heritage, Part 2. TMM_CH 2018, CCIS 962, Springer Nature-Communications on Computer and Information Science, vol 962, pp 59–77. https://doi.org/10.1007/978-3-30-12960-6_4
Miltiadou-Fezans A, Tassios TP (2009) Modern practical guidelines for the application of hydraulic grouts in historic constructions. In: Doussi M, Nikiforidis P (eds) Proceedings of 3rd National Congress Appropriate interventions for the safeguarding of monuments and historical buildings. New design tendencies, 9–11 Apr 2009, Thessaloniki Greece, IANOS Publications, pp 485–495
Miltiadou Fezans A, Tassios TP (2012) Fluidity of hydraulic grouts for masonry strengthening. Mater Struct. https://doi.org/10.1617/s11527-012-9872-8
Miltiadou-Fezans A, Papakonstantinou E, Zambas K, Panou A, Frantzikinaki K (2002) Structural restoration of the drums of the columns of Parthenon Opisthodomos with hydraulic grouts of high penetrability: research, design and application. Proceeding of the 5th International meeting for the Restoration of the Acropolis Monuments, Committee for the Conservation of the Acropolis Monuments, Hellenic Ministry of Culture, 2–6 Oct 2002, Athens, Greece, pp 161–178
Miltiadou-Fezans A, Delinikolas N, Chorafa E, Zaroyianni E (2003) Design of the structural restoration of the Katholikon of Dafni Monastery, Phase A´-Structural Survey, Analysis and remedial measures. Design Report, Directorate for the Restoration of Byzantine and Post-Byzantine monuments, Hellenic Ministry of Culture and Sports (in Greek)
Miltiadou-Fezans A, Vintzileou E, Delinikolas N, Zaroyianni E, Chorafa E (2004) Pathology of Dafni Monastery: survey, monitoring of cracks, interpretation and numerical verification. In: Modena C, Lourenco P, Roca P (eds) Proceedings of IV international seminar on structural analysis of historical constructions, Padova 10–13 Nov 2004, Taylor and Francis Group, London, pp 1285–1294
Miltiadou-Fezans A, Papakonstantinou E, Zambas K, Panou A, Frantzikinaki K (2005) Design and application of hydraulic grouts of high injectability for the structural restoration of the column

drums of the Parthenon Opisthodomos. In: Brebbia CA, Torpiano A (eds) Advances in Architecture 20, STREMHA IX Structural Studies, Repairs and Maintenance of Architectural Heritage, WIT Transactions on the Built Environment, vol 83, pp 461–471

Miltiadou-Fezans A, Anagnostopoulou S, Kalagri A (2006) Restoration mortars and grouts for the consolidation of the Cells of internal yard of Dafni Monastery. In: Proceedings of 1st conference on Restoration, Society for research and promotion of scientific restoration of monuments (ETEPAM), Thessaloniki (in Greek)

Miltiadou-Fezans A, Kalagri A, Delinikolas N (2007) Design of hydraulic grout and application methodology for stone masonry structures bearing mosaics and mural paintings: the case of the Katholikon of Dafni Monastery. In: Görün A (ed) Proceedings of the international symposium: SHH07-Studies on Historical Heritage, Antalya, Turkey, 17–21 Sept 2007, pp 649–656

Miltiadou-Fezans A, Kalagri A, Kakkinou S, Ziagou A, Delinikolas N, Zarogianni E, Chorafa E (2008a) Methodology for in situ application of hydraulic grouts on historic masonry structures. The case of the Katholikon of Dafni Monastery. In: D'Ayala D, Fodde E (eds) Proceedings of the 6th international conference on structural analysis of historic constructions, 2–4 July, Bath, CRC Press/Balkema, Taylor and Francis Group, vol I, pp 1025–1033

Miltiadou-Fezans A, Côte P, Dérobert X, Durand O, Alexandre J, Kalagri A, Savidou M, Delinikolas N, Chryssopoylos D (2008b) Methodology for the application of radar and seismic techniques in investigating Byzantine monuments bearing mural mosaics. The cases of the Katholikon of Dafni Monastery and the Katholikon of Hosios Loukas Monastery. In: Binda L, Di Prisco M, Felicetti R (eds) Proceedings of the international conference On site assessment of concrete, masonry and timber structures SACoMaTiS 2008, Varenna-Lake Como, Italy, 1–2 Sept 2008, RILEM Publ. S.A.R.L., RILEM PRO 59, vol II, pp 747–756

Miltiadou-Fezans A, Kalagri A, Koumantos A, Durand O, Savidou M, Toyntopoylos K, Chryssopoylos D, Georganis F, (2008c) Investigation of mosaic substrata in the Katholikon of Hosios Loukas Monastery Using ground penetrating radar and infrared thermography. In: Michaelidis D, Guimier-Sorbets AM, Nardi R (eds) Proceedings of 10th Conference of the International Committee for the conservation of mosaics, Palermo, Italy, 20–26 Oct 2008, Dipartimento dei Beni Culturali e dell' identita Siciliana, Regione Siciliana, pp 174–180

Miltiadou-Fezans A, Dourakopoulos J, Giannopoulos P, Manzo A (2014) The significance of a detailed survey for an adequate grouting application. In: Proceedings of structural faults and repair 2014, Imperial College, University of London, UK, e-book, paper 1248, pages 12

Mission ingénierie et références techniques de la sous-direction des monuments historiques et des espaces protégés et al. (2006) Guide de maîtrise d'ouvrage et de maîtrise d'œuvre : Ouvrages de maçonnerie. Ministère de la culture et de la communication, Direction de l'architecture et du patrimoine, France, Juin 2006, ISBN 978-2-11-129949-8, p 278

Moropoulou A, Lambropoulos K (2015) Non-destructive testing for assessing structural damage and interventions effectiveness for built cultural heritage protection. In: Asteris P, Plevris V (eds) Handbook of research on seismic assessment and Rehabilitation of historic structures, Advances in Civil and Industrial Engineering (ACIE) book Series, Engineering Science Reference (an imprint of IGI Global), USA, pp 448–498. https://doi.org/10.4018/978-1-4666-8286-3.ch015

Moropoulou A, Avdelidis NP, Koui M, Kanellopoulos NK (2000) Dual band infrared thermography as a NDT tool for the characterization of the building materials and conservation performance in historic structures. In: Matikas T, Meyendorf N, Baaklini G, Gilmore R (eds) Nondestructive Methods for Materials Characterization, vol 591. Materials Research Society, Pittsburgh, pp 169–174

Mouzakis C, Miltiadou-Fezans A, Touliatios P, Delinikolas N, Dourakopoulos J (2008) Earthquake based condition monitoring of the Katholikon of Dafni Monastery. In: D'Ayala D, Fodde E (eds) Proceedings of the 6th international conference on structural analysis of historic construction, 2–4 July, Bath, CRC Press/Balkema, Taylor and Francis Group, vol I, pp 621–628

Nappi A (1996) Non-destructive testing and system identification to evaluate diagnostics methods and reinforcement techniques applied to historical buildings. Final Report EC Contract N°: EV5V-CT92–0106

NF P18–358 (1985) Adjuvants pour bétons, mortiers et coulis - Coulis courant d' injection pour précontrainte - Mesure de la fluidité et de la réduction d' eau
NF P18-359 (juillet 1985) Coulis courants d'injection pour précontrainte – Mesure de l'exsudation (stabilité)
Onsiteformasonry project EUR21696 EN-2006. On-site investigation techniques for the structural evaluation of historic masonry buildings. In: Maierhofer C, Kopp C (coordinators) et al. (eds) European Commission, Directorate General for Research, European Communities 2006, p 141
Paillère A-M, Buil M, Miltiadou A, Guinez R, Serrano JJ (1989) Use of silica fume and superplasticizers in cement grouts for injection of fine cracks. In: Proceedings of the third international conference on use of fly ash, silica fume, slag and natural pozzolans in concrete, Trondheim, Norway, SP-ACI, vol 2, pp 1131–1157
Palieraki V, Vintzileou E, Miltiadou-Fezans A (2007) The use of radar technique and boroscopy in investigating historic masonry: application of the techniques in Byzantine monuments in Greece». In: Görün A (ed) Proceedings of the International Symposium: SHH07 Studies on Histori cal Heritage, Antalya, Turkey, 17–21 Sept 2007, pp 403–410
Palieraki V, Vintzileou E, Miltiadou-Fezans A, Delinikolas N (2008) The use of radar techniques and boroscopy in investigating old masonry: The case of Dafni Monastery. Int J Arch Heritage: Conserv Anal Restor 2(2):155–186
Penelis G, Stylianidis K (1996) Research Report no 10: Mortars and Grouts. In: Investigation of alternative solutions for the structural restoration of Panagia Acheiropoeitos, Thessaloniki, December 2006, Research report, Aristotle University of Thessaloniki, Laboratory of Reinforced Concrete (in Greek)
Recommendation: MS-D.1: Measurement of mechanical pulse velocity for masonry (1996) In RILEM TC 127-MS: Tests for masonry materials and structures. Materials and Structures, vol 29, Oct 1996, pp 459–475
Recommendation: MS.D.8: Electrical conductivity investigation of masonry (2001). RILEM TC 127-MS: Non-destructive tests for masonry materials and structures. In Materials and Structures, vol 34, Apr 2001, pp 134–143
Recommendation: MS.D.3: Radar investigation of masonry (2001). In RILEM TC 127-MS: Non-destructive tests for masonry materials and structures. Materials and Structures, vol 34, Apr 2001, pp 134–143
Schueremans L, Van Rickstal F, Verderickx K, Van Gemert D (2003) Evaluation of masonry consolidation by geo-electrical relative difference resistivity mapping. Mater Struct 36:46–50
Schuller MP (2006) Nondestructive testing and damage assessment of masonry structures. In: Proc. of NSF/RILEM Workshop In-situ Evaluation of Historic Wood and Masonry Structures, 10–14 July 2006, Prague, Czech Republic, pp 67–86
Schuller MP, Atkinson RH, Borgsmiller JT (1994) Injection grouting for repair and retrofit of unreinforced masonry. In: Proceedings of 10th international block masonry conference, Calgary, Canada, 5–7 July 1994
Silva B, Benetta MD, da Porto F, Valluzzi MR (2014) Compression and sonic tests to assess effectiveness of grout injection on three-leaf stone masonry walls. Int J Archit Heritage: Conserv, Anal Restor 8(3):408–435. https://doi.org/10.1080/15583058.2013.826300
Stockton G, Allen L (1999) Using infrared thermography to determine the presence and correct placement of grouted cells in single-width concrete masonry unit (CMU) walls. Proceedings of SPIE, 3700, Thermosense XXI, 1999
Toumbakari EE, Van Gemert D, Tassios TP, Vintzileou E (2005) Experimental investigation and analytical modeling of the effect of injection grouts on the structural behaviour of three-leaf masonry walls. In: Modena C, Lourenco P, Roca P (eds) Proceedings of 4th structural analysis of historical constructions, Padova, Taylor and Francis Group, London, pp 707–717
Uranjek M, Zarnic R, Bokan-Bosiljkov V, Bosilikov V (2012) Performance of NDT, MDT and DT techniques in assessing the effectiveness of grouting. In: Jasienko J (ed) Proceedings of international conference on structural analysis of historical constructions 2012, Wroclaw, Poland, pp 2544–2551

Valle S, Zanzi L, Rocca F (1999) Radar tomography for NDT: comparison of techniques. J Appl Geophys 41:259–269

Valluzzi MR (2003) Consolidamento e recupero delle murature. Collana Scientifica Recifere, a cora del Prof. Arch. Cesare Renzo Romeo, Ed. Gruppo Editoriale Faenza Editrice S.p.A., p 96

Valluzzi MR, da Porto F, Modena C (2004) Behaviour and modelling of strengthened three-leaf stone masonry walls. Mater Stuct 37:184–192

Van Rickstal F (2000) Grout injection of masonry, scientific approach and modeling. Katholieke Universiteit Leuven, Mei

Van Rickstal F, Schueremans L, Beeldens A, Toumbakari E, Ignoul S, Brosens K, Van Balen K, Van Gemert D (2002) New technologies for safeguarding cultural heritage—a state of the art of applied research and practice in Belgium. ITAM-ARCCHIP Workshop ARIADNE 12 Historic Structures, Prague, 1–7 June

Van Rickstal F, Van Gemert D, Keersmaekers R, Posen D (2008) Enhancement of geo-electrical techniques for NDT of masonry. In: D'Ayala D, Fodde E (eds) Proceedings of the 6th international conference on structural analysis of historic construction, 2–4 July, Bath, CRC Press/Balkema, Taylor and Francis Group, pp 1053–1060

Venderickx K, Van Gemert D (2000) Geo-electrical survey of masonry for restoration projects. Int J Restor Build Monuments, Aedificatio Verlag, 2000, Heft 2, pp 151–172

Vintzileou E (2003) The use of radar and endoscopy to investigate the masonry of the Katholikon of Daphni Monastery. In: Research report—Investigation of mechanical behaviour of brick enclosed byzantine masonry, bearing or not mural mosaics, and of the methods of its repair: the case of the Katholikon of Daphni Monastery. Laboratory of Reinforced Concrete, National Technical University of Athens, Directorate for Technical Research on Restoration, Hellenic Ministry of Culture and Sports (in Greek)

Vintzileou E (2006) Panagia Krina in Chios Island: Localization of timber reinforcements, investigation of their state of conservation and possible interventions. In: Research report, Laboratory of Reinforced Concrete, National Technical University of Athens, Directorate for Technical Research on Restoration, Hellenic Ministry of Culture and Sports (in Greek)

Vintzileou E, Miltiadou-Fezans A (2008) Mechanical properties of three-leaf stone masonry grouted with ternary or hydraulic lime-based grouts. Eng Struct 30(8):2265–2276

Vintzileou E, Tassios TP (1995) Three leaf stone masonry strengthened by injecting cement grouts. J Struct Eng ASCE 121(5):848–856 15

Weaver KD (1991) Dam foundation grouting. ASCE Publication, p 178

www.ingramcontent.com/pod-product-compliance
Ingram Content Group UK Ltd.
Pitfield, Milton Keynes, MK11 3LW, UK
UKHW022213051026
15624UKWH00001B/147
* 9 7 8 3 0 3 0 8 5 9 6 4 0 *